Technology-Assisted Problem Solving for Engineering Education: Interactive Multimedia Applications

Manjit Singh Sidhu
Universiti Tenaga Nasional, Malaysia

ENGINEERING SCIENCE REFERENCE
ENGINEERING SCIENCE REFERENCE
Hershey · New York

Director of Editorial Content: Kristin Klinger
Senior Managing Editor: Jamie Snavely
Assistant Managing Editor: Michael Brehm
Publishing Assistant: Sean Woznicki
Typesetter: Sean Woznicki
Cover Design: Lisa Tosheff
Printed at: Yurchak Printing Inc.

Published in the United States of America by
Engineering Science Reference (an imprint of IGI Global)
701 E. Chocolate Avenue
Hershey PA 17033
Tel: 717-533-8845
Fax: 717-533-8661
E-mail: cust@igi-global.com
Web site: http://www.igi-global.com/reference

Library of Congress Cataloging-in-Publication Data

Sidhu, Manjit Singh, 1971-
Technology-assisted problem solving for engineering education : interactive multimedia applications / by Manjit Singh Sidhu.
p. cm.
Includes bibliographical references and index.
Summary: "This book explores best practices in assisting students in understanding engineering concepts through interactive and virtual environments"--Provided by publisher.

ISBN 978-1-60566-764-5 (hardcover) -- ISBN 978-1-60566-765-2 (ebook) 1. Engineering--Study and teaching (Higher) 2. Interactive multimedia--Study and teaching (Higher) 3. Engineering--Computer-assisted instruction. I. Title.
T65.5.C65S53 2009
620.0071'1--dc22

2009018130

British Cataloguing in Publication Data
A Cataloguing in Publication record for this book is available from the British Library.

This book is dedicated to my wife Dr. Kirandeep Kaur Sidhu for her patience and forbearance and my daughter Tavleen Kaur Sidhu.

&

My late mother Bachan Kaur Sidhu

Table of Contents

Appendices

Preface

Many academicians are altering their teaching techniques today. Multimedia is captivating academicians because of its strengths to communicate difficult concepts in simple yet efficient ways. The key area lies in the learning environment. It is unstructured, and the learner is free to navigate the vast universe of information. Multimedia applications integrate animation, sound, graphics, and video to create an engaging, interactive and effective learning environment. Such software allows students to exercise more control over the pacing and sequencing of their own learning. With the availability of more sophisticated computers the potential to employ multimedia has grown tremendously.

In the case of engineering education which depends on static images, diagrams and complex mathematical computations in the traditional classroom, the employment of multimedia in teaching provides a better treatment in enriching the learning experiences by providing a multi-sensory perspective. In general, while it has been acknowledged that multimedia can be useful in the teaching of engineering subjects, non information technology academicians tend to believe that development of multimedia applications requires wide knowledge of high-end hardware, software and programming skills. This is true only partially. Today, with the availability of multimedia authoring tools such as Macromedia Flash®, Director®, Authorware®, click2learn ToolBook® and 3D modeling, animation and rendering tools such as Alias Maya®, Autodesk 3D Studio Max®, Maxon Cinema 4D®, NewTek Lightwave® etc, we can develop multimedia materials with a little practice of these tools.

However knowledge of the above mentioned authoring and modeling tools is not enough to develop a high quality teaching package that students could use in their learning. In order to be useful and transfer adequate knowledge skills to the learner, a multimedia programme design needs to have a sound pedagogical base. This book intends to help academicians with particular in the mechanical engineering field (although would be beneficial for other domains of engineering and sciences) in understanding the basic concepts of multimedia and various issues involved in the development of multimedia problem solving packages.

The main purpose of this book is to share knowledge of issues and trends in computer aided learning (CAL) in particular, for engineering, from the perspectives of problems faced by students and instructors, students preference of learning styles and a new approach to learning, visualizing and problem solving. In addition this book has presented some interesting challenges in the development of new problem solving packages that uses the principle of CAL. These integrated packages are termed as Technology Assisted Problem Solving or TAPS packages, which guide students step-by-step to complete various engineering mechanics problems.

As a case study the outcomes of the research presented in this book was focused on a higher learning institution in Malaysia. Since the use of CAL in higher learning institutions in Malaysia is still at its infancy, this study is mainly concerned with the development of effective TAPS packages in supplementing the teaching and learning of engineering mechanics.

The study adopted multi-design approach (i.e. 2-D, 3-D, coach-based, and desktop virtual reality) to simplify the underlying engineering principles and thereby accelerate the learning process of slow learners (i.e. learners experiencing difficulties with understanding Engineering Mechanics theories). Four TAPS packages were developed and tested by undergraduates to validate the design approach of the TAPS packages. These TAPS packages were developed using various 3-D modeling and multimedia authoring tools. The TAPS packages were structured according to the learners' needs based on the survey carried out using the Felder-Solomon's ILS questionnaires and the packages were evaluated using quantitative techniques for its effectiveness. Four groups of learners were identified i.e. sensory, visual, active and sequential. The results showed that different group of learners have different preferences of the features offered in the TAPS packages. Nevertheless, the study found that the step-by-step approach which was integrated in each of the four TAPS packages was beneficial in promoting learning and understanding of Engineering Mechanics concepts, particularly to slow learners. The outcome of this study indicates that the TAPS packages have great potential in aiding the learning of engineering and to enhance students' visualization in solving Engineering Mechanics problems.

ORGANIZATION OF THE BOOK

An introduction of this study and an outline of the problems and objectives are given in Chapter 1. Computer aided learning (CAL) is discussed in brief regarding its evolution and the role of new technologies. Problems associated with CAL, application of CAL in engineering education and problems encountered by engineering students are discussed in depth.

Theories and models of instructional design are discussed in brief in Chapter 2. Various cognitive styles have been described in the literature and many different measurement tools are available for these styles. Chapter 2 of this book presents a summary of these learning styles and measurement methods. The Chapter also discusses the appropriateness of Felder-Solomon's Learning Style Questionnaire (Felder and Silverman, 1988) in the context of computer-aided learning and the categorization of people according to their styles.

Chapter 3 of this book reviews different approaches in the user interface design and examines the role of interface design in various educational environments. The designing of the user interface has been discussed in details. This Chapter also reviews some problems in traditional user interface designs and lists some guidelines for better user interface design. Additionally, the role of the interface in educational environments and student's preferences towards interfaces in learning environments is briefly discussed.

Chapter 4 provides an overview and discusses issues related to multimedia and computer-aided learning where various aspects of CAL development are discussed. Several benefits and limitations of interactive multimedia and CAL are also reviewed. In addition, classifications of CAL in the context of learning content are listed and other trends of learning environments are discussed. The need for developing multimedia courseware for engineering has been explained in this Chapter.

Chapter 5 of this book describes the hardware and software required for multimedia development (i.e. multimedia courseware). These include the various input and output devices and configuration of a multimedia PC. Software such as painting and drawing tools, image, sound and video editing tools, 3D modeling and animation tools, desktop virtual reality tools and integrated design software have been discussed in details.

Chapter 6 of this book describes a new form of CAL that is termed as technology assisted problem-solving (TAPS) packages. The key concepts of a TAPS package, user interface, and contributing technologies that have been used to develop the packages are discussed in details. This new approach to learning, visualizing, and problem solving in engineering provides significant concepts to the development of the TAPS packages as described in Chapter 6. This Chapter gives an insight of multimedia effects on learning and how it can be used to support key aspects of learning and teaching.

A brief description of the Mechanical Engineering subjects and the usage of TAPS packages, are described in Chapter 7. In addition, all the TAPS packages developed for this study is discussed in details with its configurations and significance to the study. The TAPS packages are different from other CAL engineering packages in the sense that they provide multiple approaches to solve selected Engineering Mechanics Dynamics problems in 2-D and 3-D, static and dynamic illustrations, coach, desktop virtual reality (DVR), simple intelligence, and translation and rotational movement environment so that a student can visualize the engineering principles. Each package was developed as a separate component to solve different engineering problems. A summary of key features and differences of each TAPS package is provided.

Chapter 8 deals with the evaluation of interactive multimedia packages in general. Evaluation techniques used for CAL engineering packages have been described in this Chapter. The National Engineering Education Delivery System (NEEDS) which is an electronic database used for delivery and evaluating engineering education courseware is discussed in detail. Although most of the evaluation techniques were found to be suitable for the evaluation of educational software, there is no evidence that any single technique is suitable for all types of educational software. As such, further work is required to set a standard for the evaluation of educational software that could be globally accepted.

Chapter 9 discusses the research methodology employed for evaluation of students learning styles (using ILS questionnaires) and the TAPS packages. Statistical analysis results obtained from the evaluation of close-ended questionnaires based on the fourteen sections stated in section 9 are discussed in details. A summary of open-ended questionnaires which gives an insight to the strength and weakness of TAPS packages is stated in this Chapter. The analysis carried out based on the learning styles (four groups of learners i.e. sensory, visual, active and sequential), indicate that different groups prefer different features of the TAPS packages.

The effectiveness of the TAPS packages discussed in Chapter 10 is based on the questionnaires feedback and observational results. In addition, this Chapter also provides a brief account of the differences between the TAPS packages approach used in this study with that of commercial simulation packages accompanying the Engineering Mechanics Dynamics textbook.

Chapter 11 of this book discusses the challenges and trends of TAPS packages in enhancing engineering education. Additional hardware and software such as graphics tablet, interactive white boards and augmented reality are being used and tested for enhancing the existing TAPS packages.

The conclusions and further work of the study are reported in Chapter 12 of this book. Although the TAPS packages are only in its initial phase of development, this study provides sufficient evidence to continue its development especially to aid teaching and learning of mechanical engineering at UNITEN. The future work could include the development of more engineering problems that could be used by students in their learning e.g. within an entire semester. Future versions of TAPS packages should employ a standard user interface to enhance the problem-solving environment. The improvements should be based on the suggestions given in the evaluation of the TAPS packages carried out in this study.

Teaching conceptual and qualitative material effectively while leveraging the contents efficiently has been an elusive goal for many computer aided learning (CAL) packages in the past. With the advent of newer technologies such as multimedia and virtual reality these technologies are being researched and applied to various areas of educational settings, especially in science and technology. However the potential of these technologies has not been fully exploited, particularly in the teaching of engineering.

In this book, an innovative approach based on the principle of CAL is used to design and implement integrated packages known as Technology Assisted Problem Solving or TAPS packages, which guide students step-by-step to complete various Engineering Mechanics problems.

In summary, this book is concerned with the design, development, and evaluation of a new form of interactive multimedia CAL problem solving packages which replace traditional problem solving aspects of undergraduate level of teaching selected mechanical engineering subjects. These packages, subsequently named as TAPS packages, provide the instructors with an economical means of facilitating the teaching of engineering concepts in Dynamics to a large population of undergraduates. This study seeks to examine the overall effectiveness of these packages that provide an integration of traditional teaching (lectures) with technology assistance (software packages) to enhance the students learning in today's resource limited environment.

It is difficult to address every aspect of the technologies employed in the development of TAPS packages because of the rapid change, upgrades and evolvement of hardware and software. Nevertheless, hopefully, this book will provide useful and rich description through a reflective analysis of the technologies employed in this study. Alternative approaches are reviewed. This could provide a deeper insight into methods for educational research.

Majit Singh Sidhu
Author

Acknowledgment

I would like to thank IGP for granting me the opportunity to publish this book and share my knowledge with academicians, educationist, and individuals who wish to broaden their horizons about educational technologies pertaining to engineering education. Availability of the vast and untapped potentials of the latest technologies inspired me to integrate their benefits in educational software tools. In addition I would like to express my special gratitude to:

- Assoc. Prof. Ir. Dr. Ramesh Singh (Mechanical Engineering Department, UNITEN), for his constructive comments at all stages of the research presented in this book and especially during the development of the (TAPS) packages, advice on the course on performing research.
- Prof. Dr. Richard M. Felder and Barbara A. Solomon for the use of the Index of Learning Styles Questionnaire.
- My wife Dr. Kirandeep Kaur Sidhu for proof reading the book. This publication would not have been possible without her.
- Many thanks to the team at IGI Global, in particularly Joel Gamon, who have answered my queries, provided advice, and support.
- I also like to thank all the referees who provided constructive and comprehensive reviews of the contents of this book. They work anonymously in the background, but their tireless effort is much appreciated.

TRADEMARK NOTICE

Chapter 1
Introduction

BACKGROUND

Computer aided learning (CAL) is a terminology used for delivering educational experiences electronically. CAL materials may use any combination of teaching techniques including question and answer, simulation, multimedia, didacticism (tending to convey information), or problem solving. CAL environments increasingly are using a combination of interactive multimedia and virtual reality (VR) such as text, audio and video, graphics and images, two and three dimensional animations, and simulations in presenting learning materials. Interactive refers to the way the user engages in these environments to enhance his/her learning process.

EVOLUTION OF CAL

The first use of computers by educational institutions and the introduction of computers in classroom teaching and technical training began in the 1950's (Megarry, 1983). According to Robert (1994), the most pervasive tool to deliver education is the computer. This is probably due to the increasing popular-

DOI: 10.4018/978-1-60566-764-5.ch001

ity of personal computers (Law and Maguire, 1993). In the 1960's and 1970's, more teaching devices (computers) and teaching CAL applications were developed for teaching and job training. As a result, improvements in computer design during the 1960's enabled the integration of text and graphics; a move that encouraged the development of computer based training programs. Such use was, however, limited, as programming was costly, slow and tedious, and the computing relied on a mainframe, or central computer, to do all the work. The release of the first microcomputers during the late 1970's assisted computing in becoming more portable and affordable.

The paradigm shift from textbook learning to CAL started in the 1980's. Ease of use took a quantum leap with the introduction of the Apple Macintosh computer in 1984. Operation of the computer and its programs became more intuitive and user-friendly largely as a result of intense research in the area of human-computer interaction where graphical user interface was introduced (Kinshuk, 1996). The low cost of hardware equipment and educational software motivated the use of CAL in education and training. Today, as affordable personal computers become more powerful, the range of operations that can be reasonably performed has increased rapidly.

ROLE OF NEW TECHNOLOGIES IN CAL

The use of newer technologies in CAL packages i.e. multimedia and virtual reality, allows highly engaging activities such as interactivity and simulation. Interactive multimedia systems for learning came into existence in the early 90's (Robert, 1994). According to Negroponte (1995), multimedia is simply a mixture of data on digital basis. Multimedia systems support the physical and logical coexistence and interactive use of mixed media classes such as print, audio, and video in specific application environments. Cairncross and Mannion (1999) stated that interactive multimedia systems have the potential to create high quality learning environments that actively engage the learner. Additionally, Cairncross (2002) pointed out that the key elements of multiple media, such as user control over delivery of information and interactivity, could be used to enhance the learning process.

Another emerging technology that is being used in CAL is virtual reality (VR). VR systems were first introduced in the learning environment in mid 90's (Macpherson, 1998). The term 'virtual reality' is currently used to describe a range of computer-based systems in which a user can explore hardware or software generated 'micro world' (artificial environments) that allow close resemblance to reality. VR extends the interaction-oriented features of multimedia i.e. modeling objects and their behaviors in virtual environments, integrating position-tracked human-computer interaction devices, and performing numerically intensive computations for real-time navigation.

The prime feature of VR is 'visualization' followed by 'interactivity'. Special VR hardware and software are thus required to allow human-computer interaction to permit input of the user's actions and movement to the computer, and to provide corresponding simulated feedback to the user. An early application of such system was the flight simulator used to train pilots. However, it is in the area of hi-tech computer games that many of the application developments in this field have taken place.

Although VR has been used for educational purposes (Bell and Scott, 1995, Dede *et al*., 1996, and Kim *et al*., 2001), the potential of VR is just beginning to be exploited by a few science and engineering educators (Manseur, 2005 and Liarokapis *et al.* 2007). This will be discussed further in Chapter 2.

PROBLEMS ASSOCIATED WITH CAL

Despite many successes, CAL packages are bound by numerous limitations and drawbacks. Schank (1994) reported that early educational software offered tedious drill (exercise programs) and were primarily built on a set of pre-defined steps that did not engage the students in their learning process and thus failed to promote learning. According to Kinshuk (1996), the pre-defined steps incorporated into the computer software presented knowledge to learners through a specific order in a text form. The knowledge of a CAL package did not go beyond the information stored in its memory and has likely done little to improve learning. When using educational CAL packages, students typically rely on the package for feedback. Therefore, the prime challenges of a CAL package is to response to students' needs and provide appropriate feedback so that the students can understand the theories and be able to apply them in solving problems. However, according to Kinshuk (1996), the rigid and unresponsive nature made the instructional effectiveness less than satisfactory.

After the introduction of Artificial Intelligence (AI) to CAL, AI researchers in the 1970's started to transfer expertise from human professionals to a machine accessible form and then redirected this knowledge back to other human beings (Barr and Feigenbaum, 1982). AI researchers attempted to organize the problem-solving skill and expertise in a way that was suitable for teaching and consulting (Kinshuk, 1996). To develop such an intelligent teaching package, AI techniques such as knowledge representation, user modeling, and natural language processing have widely been incorporated into intelligent tutoring systems (ITS) (Kinshuk, 1996).

From the perspective of human computer interaction, the research on CAL is very focused on how to represent the learning content and tends to neglect the impact of the user-interface in the learning process (Guttormsen *et al.*, 2000). On the other hand, Cairncross and Mannion (2001) pointed out that there is a growing evidence that interactive multimedia is not being fulfilled. The author further argued that early designs were driven by technology (i.e. focusing mainly on physical interface) rather than pedagogy.

Additionally, Cairncross (2002) reported that it could take 40 hours to develop one hour's worth of quality interactive multimedia learning. This lengthy development time is extremely expensive because it is difficult and costly to find human experts to develop the interactive multimedia CAL packages. As an option with most other teaching media, instructors wanting to use interactive multimedia CAL package in their teaching can choose to develop simple unsophisticated materials locally or purchase more sophisticated, and thus more expensive, teaching materials from professional development units (commercial software house, etc).

Schank (1994) argued that most multimedia programs are not suitable for learning because they merely add video and graphics to page-turning programs. Kinshuk and Patel (2003) added that the collection of multimedia objects i.e. pictures, graphics, sounds, and video does not guarantee proper learning, especially when the complexity of the task, skill or learning increases. As such, further work is required to develop more effective approaches that could serve this purpose and task.

At the simplistic level, some multimedia programs and software have limited qualities and have been dubbed *'electronic page turners'* (Goldman and Torrisi-Steele, 2004). Interaction is limited to the student clicking a mouse to sequentially take them to the next page. To fully develop the enormous potential of multimedia as an interactive learning package, multimedia elements and activities must possess a multitude of interactivity option made available to the student.

Subsequently, a range of problems limit the use of hi-end hardware and software in the educational environments. These problems are mainly associated to VR technology. The main problem of VR technology in the educational environments is associated with high cost. Although prices are rapidly decreasing, immersive VR systems still cost hundreds of thousands of dollars (Rizzo *et al.*, 1997, Klett, 2002). Alternatively, cheaper options for VR technology are also available such as desktop virtual reality which only requires special browsers to interact and visualize the virtual objects (these are further explained in Chapter 5). Other problems that are restricting the pursuit of the VR vision in the educational environments include the limited availability of applications development expertise and software licenses for multiple users (learners).

APPLICATION OF CAL IN ENGINEERING EDUCATION

In general, education, in higher learning institutions in countries with developing economics still focuses on older educational models of linear progression or surface learning, whereas counterparts from developed nations provide predominantly high-impact audio-visual perception.

The western countries, particularly the UK and USA have used computers and CAL packages to motivate students of higher learning institutions since the 1960s (Ismail, 2001). Although most higher learning instutitions would employ and use new technology in teaching, several academicians commented that they still lack widespread practical experience of developing multimedia-learning materials (Julia *et al.*, 2002).

However since the emergence of newer hardware and software technologies for multimedia and VR, educational practitioners began to research further on the pedagogical effectiveness of these technologies. In a developing country such as Malaysia, multimedia technology was first briefly introduced in the late 1990s and became popular with the launch of Multimedia Super Corridor (MSC) (Norhayati *et al.*, 2001). To achieve the substance of Vision 2020, the government has set up a blueprint for the MSC. The MSC is a massive 750-square-kilometer high-tech information zone encompassing the Kuala Lumpur City Center (KLCC), Putrajaya (administrative center) and Kuala Lumpur International Airport (KLIA). To spearhead the development of the MSC and give shape to its environment, seven initiatives for multimedia applications have been identified. These initiatives are borderless marketing, smart schools, electronic government, multi-purpose card, telemedicine, research and development world wide manufacturing webs. Of these, the smart school initiative is regarded by the previous Prime Minister as a specific response to Malaysia's need to make the critical transition from an industrial economy to a knowledge-based society (Mahathir Mohammed, 1998). Subsequently VR hardware and software are being used in various research fields such as medicine, manufacturing and for scientific visualization.

The nation is devoting the abovementioned massive MSC to create the perfect environment for companies and education sector wanting to develop, distribute, and employ multimedia products and services. One of MSC's primary areas of multimedia applications includes "*smart schools*" where educational software packages are being customized to facilitate teaching and student learning purposes in primary and secondary schools. In general, although the educational sector is aware of the presence of MSC, these new technologies (multimedia and VR) are hardly being researched further for its utility and potential in teaching engineering. Additionally, multimedia and VR systems are not available on a large scale to support learning environments. Since these technologies are in its infancy state especially in the higher learning institutions further research is needed to address its usefulness and benefits.

Already locally commercialized CAL packages (custom made packages developed in Malaysia) are available for learning purposes, but these CAL packages vary in subjects and are more suitable for primary and secondary school levels. There are still many areas that need improvement because these CAL packages are simply computerized textbooks with page turning architecture that do not promote learning (lack of capability to motivate user to participate actively in the learning process). Additionally, most commercial CAL packages are based on outdated theories of learning. For example, long series of short answer questions are provided to students in the CAL packages but the CAL packages are not intelligent enough to provide appropriate reasons and solutions to the incorrect answers given by the students (most commonly alert messages i.e. '*correct*' or '*wrong*' are used).

Thus to improve the learning process, particularly in higher learning institutions, suitable CAL packages tailored to the needs of students to enhance learning should be employed and be made an integral part of the current curriculum of the educational system. However, in order for CAL to become an effective learning platform, certain criteria must be met. For example, the instructors should be prepared to accept and encourage students to use the technology as an additional learning aid in an effective manner. The full potential of CAL cannot be realized if computers are merely used by students for preparing laboratory reports and assignments.

In general the development of CAL packages is a tedious process. Even simple graphics and animations require significant effort to develop the contents of CAL packages that employs multimedia and VR technologies.

For engineering and technology education, CAL multimedia and VR applications can include computer simulation, numerical analysis, computer aided design (CAD), computer aided manufacture (CAM), and electronic communications (Palmer, 2000).

Although many CAL learning environments have emerged in general, in this book the term technology assisted problem solving (TAPS) packages will be used to refer to packages that were developed in this study and used by students to assist them in their learning. The main difference between conventional CAL and TAPS packages is that TAPS packages do not include text of theories learnt by students during normal classroom lectures but rather constituted selected engineering problems that are difficult for students to understand from text books. The TAPS packages aimed at coaching students, particularly who need additional support in applying principles presented in lectures to problems, on the best approach to solve a particular engineering problem in step-by-step or logical approach. Although a TAPS package may be considered as a form of CAL, more specifically, TAPS packages were designed to allow independent problem solving, develop logical thinking, and promote learning of the subject matter. At present, the developed TAPS packages are not a complete set of engineering problems that could be used throughout an entire semester but it is envisaged that more TAPS packages could be developed if these are proven to be effective. TAPS packages are discussed further in Chapters 6 and 7 of this book.

PROBLEMS ENCOUNTERED BY ENGINEERING STUDENTS

This study focuses on the problems faced by students in the field of Mechanical Engineering, in particularly the Engineering Mechanics Dynamics course. This subject is chosen because a number of academicians in the field as reported in the literatures found that the main problem faced by students is visualization of dynamic motion of particles or rigid bodies.

For example Katarzyna (2002), reported that the problems many undergraduate students face while studying the Engineering Mechanics Dynamics course is the difference in understanding with regard to what is being taught in the classroom. Undergraduate students often expect a variety of teaching methods to be used in their learning. Although, the lecture method is a common way of delivering knowledge to students, it treats all students on the same level of the basic acquired knowledge. However, most of the students do not bring the same academic preparation (do not have the same foundation, motivation, interest, ability to learn/grasp) and come from different disciplines with varying learning styles. This results in different starting points, progress rates, and ultimately different levels of satisfaction, academic progress, and performance.

Some entry-level undergraduate students find certain engineering subjects difficult to understand and this discourages learning from taking place. As a result of this problem, if the lectures are too fast, this group of students may not be able to keep pace with the rest of the class thus the gap in their knowledge will only get wider as compared to the more brilliant students. In this situation, the instructors are forced to find alternative methods (for example conducting extra classes) to help these students in understanding the subject matter. Since some students may take more time to understand the problem solving techniques and may require the lesson to be repeated several times before they understand, it is proposed to employ and use multimedia and VR technologies to help them understand the engineering problems.

A major challenge facing instructors in teaching the Engineering Mechanics Dynamics course at University Tenaga Nasional, Malaysia is helping students relate the theory to the physical world. Past experiences in teaching the first and second year students indicated that there were students who found it difficult to visualize some of the concepts in engineering and to apprehend theory and practical. In general, engineering textbooks cannot represent mechanical actions such as movement of linkages, pistons, and crankshaft in the form of dynamic illustrations (animated forms). Since these mechanical actions are very important in visualization experiments, the motions and actions of these objects maybe worthwhile to be shown in animated forms. Therefore, it can be clearly seen that educating students in engineering related field is made difficult by the complex ideas and phenomena demonstrated by conventional methods.

However one of the enormous challenges that any institution looks forward to employ new technologies is the training of instructors to integrate these new technologies in their teaching of engineering. In general, instructors must be computer literate to develop and utilize multimedia and VR technologies. Therefore, properly trained instructors are required if multimedia and VR technologies are to be integrated and implemented within the classroom. In this book the term slow learners has been used for those students who have difficulties in understanding Engineering Mechanics Dynamics subject only and does not reflect on their general academic performance.

This book is concerned with the design, development and evaluation of low cost computer based problem solving aids to deliver knowledge in engineering courses. The discussion and case studies of the book is based solely on the development of the TAPS packages that focuses on the study in Mechanical Engineering at University Tenaga Nasional, Malaysia.

The aim of the work reported in this book was to enhance the understanding and problem solving skills of students (slow learners) taking the Engineering Mechanics Dynamics course at University Tenaga Nasional (UNITEN), Malaysia. This was done by analyzing the learning styles, student needs, development of computer based problem solving packages, and exposure to modern technologies. The main development path chosen was to introduce and employ new educational computer technologies.

Initially, the perceived need was to use computer based learning aids to show selected engineering problems involving motion of bodies and to provide a convenient platform for students to visualize and study them.

More specifically, the objectives of the study explained in this book can be summarized as follows:

1. To evaluate whether slow learners could better visualize and solve engineering problems with the aid of technology assisted problem-solving (TAPS) packages.
2. To explore how best to assist students to understand engineering concepts via interactive and virtual environment.
3. To determine the best approach in integrating the various elements in the TAPS packages with the aim of promoting learning and understanding of engineering concepts suitable for slow learners.

Whilst the aim was to design and develop TAPS packages for the projection of selected engineering problem solving tasks of the subject matter in a virtual environment, the study:

1. Limits its scope to the topics in the selected Engineering Mechanics Dynamics problems.
2. Employs a multi approach learning and problem solving environment in engineering i.e. 2-D, 3-D, coach based and desktop virtual reality for student learning (particularly slow learners).
3. Customizes the TAPS packages to facilitate learning of Dynamics course at UNITEN, Malaysia.

This study began with an initial hypothesis that the design of interactive multimedia problem-solving applications for engineering could be improved through the development of productive computer packages and employment of some principles of good user interface design.

Initially multiple user interface designs were planned with the aim to determine how multimedia attributes could be exploited to enhance problem solving. One of the important issues that the initial user interface design raised was to integrate multimedia attributes into the engineering problem solving steps leading from problem statement to solution which is fundamental. The study further explored this issue.

The book describes a new form of CAL that is termed as technology assisted problem-solving (TAPS) packages. The key concepts of TAPS package, user interface, and contributing technologies that have been used to develop the packages are discussed in details in Chapter 6. The TAPS packages are different from other CAL engineering packages in the sense that they provide multiple design approaches to solve selected Engineering Mechanics Dynamics problems in interactive 2-D and 3-D environment, static and dynamic illustrations, coach, desktop virtual reality (DVR) and simple intelligence, so that a student can visualize the engineering principles. Each package was developed as a separate component to solve different engineering problems. This new design approach to learning, visualizing, and problem solving in engineering provides significant concepts to the development of the TAPS packages.

OUTLINE OF THE BOOK

An introduction of this study and an outline of the problems and objectives are given in Chapter 1. Computer aided learning (CAL) is discussed in brief regarding its evolution and the role of new technologies. Problems associated with CAL, application of CAL in engineering education and problems encountered by engineering students are discussed in depth.

Theories and models of instructional design are discussed in brief in Chapter 2. Various cognitive styles have been described in the literature and many different measurement tools are available for these styles. Chapter 2 of this book presents a summary of these learning styles and measurement methods. The Chapter also discusses the appropriateness of Felder-Solomon's Learning Style Questionnaire (Felder and Silverman, 1988) in the context of computer-aided learning and the categorization of people according to their styles.

Chapter 3 of this book reviews different approaches in the user interface design and examines the role of interface design in various educational environments. The designing of the user interface has been discussed in details. This Chapter also reviews some problems in traditional user interface designs and lists some guidelines for better user interface design. Additionally, the role of the interface in educational environments and student's preferences towards interfaces in learning environments is briefly discussed.

Chapter 4 provides an overview and discusses issues related to multimedia and computer-aided learning where various aspects of CAL development are discussed. Several benefits and limitations of interactive multimedia and CAL are also reviewed. In addition, classifications of CAL in the context of learning content are listed and other trends of learning environments are discussed. The need for developing multimedia courseware for engineering has been explained in this Chapter.

Chapter 5 of this book describes the hardware and software required for multimedia development (i.e. multimedia courseware). These include the various input and output devices and configuration of a multimedia PC. Software such as painting and drawing tools, image, sound and video editing tools, 3D modeling and animation tools, desktop virtual reality tools and integrated design software have been discussed in details.

Chapter 6 of this book describes a new form of CAL that is termed as technology assisted problem-solving (TAPS) packages. The key concepts of a TAPS package, user interface, and contributing technologies that have been used to develop the packages are discussed in details. This new approach to learning, visualizing, and problem solving in engineering provides significant concepts to the development of the TAPS packages as described in Chapter 6. This Chapter gives an insight of multimedia effects on learning and how it can be used to support key aspects of learning and teaching.

A brief description of the Mechanical Engineering subjects and the usage of TAPS packages, are described in Chapter 7. In addition, all the TAPS packages developed for this study is discussed in details with its configurations and significance to the study. The TAPS packages are different from other CAL engineering packages in the sense that they provide multiple approaches to solve selected Engineering Mechanics Dynamics problems in 2-D and 3-D, static and dynamic illustrations, coach, desktop virtual reality (DVR), simple intelligence, and translation and rotational movement environment so that a student can visualize the engineering principles. Each package was developed as a separate component to solve different engineering problems. A summary of key features and differences of each TAPS package is provided.

Chapter 8 deals with the evaluation of interactive multimedia packages in general. Evaluation techniques used for CAL engineering packages have been described in this Chapter. The National Engineering Education Delivery System (NEEDS) which is an electronic database used for delivery and evaluating engineering education courseware is discussed in detail. Although most of the evaluation techniques were found to be suitable for the evaluation of educational software, there is no evidence that any single technique is suitable for all types of educational software. As such, further work is required to set a standard for the evaluation of educational software that could be globally accepted.

Chapter 9 discusses the research methodology employed for evaluation of students learning styles (using ILS questionnaires) and the TAPS packages. Statistical analysis results obtained from the evaluation of close-ended questionnaires based on the fourteen sections stated in section 9 are discussed in details. A summary of open-ended questionnaires which gives an insight to the strength and weakness of TAPS packages is stated in this Chapter. The analysis carried out based on the learning styles (four groups of learners i.e. sensory, visual, active and sequential), indicate that different groups prefer different features of the TAPS packages.

The effectiveness of the TAPS packages discussed in Chapter 10 is based on the questionnaires feedback and observational results. In addition, this Chapter also provides a brief account of the differences between the TAPS packages approach used in this study with that of commercial simulation packages accompanying the Engineering Mechanics Dynamics textbook.

Chapter 11 of this book discusses the challenges and trends of TAPS packages in enhancing engineering education. Additional hardware and software such as graphics tablet, interactive white boards and augmented reality are being used and tested for enhancing the existing TAPS packages.

The conclusions and further work of the study are reported in Chapter 12 of this book. Although the TAPS packages are only in its initial phase of development, this study provides sufficient evidence to continue its development especially to aid teaching and learning of mechanical engineering at UNITEN. The future work could include the development of more engineering problems that could be used by students in their learning e.g. within an entire semester. Future versions of TAPS packages should employ a standard user interface to enhance the problem-solving environment. The improvements should be based on the suggestions given in the evaluation of the TAPS packages carried out in this study.

In summary, this book is concerned with the design, development, and evaluation of a new form of interactive multimedia CAL problem solving packages which replace traditional problem solving aspects of undergraduate level of teaching selected mechanical engineering subjects. These packages, subsequently named as TAPS packages, provide the instructors with an economical means of facilitating the teaching of engineering concepts in Dynamics to a large population of undergraduates. This study seeks to examine the overall effectiveness of these packages that provide an integration of traditional teaching (lectures) with technology assistance (software packages) to enhance the students learning in today's resource limited environment.

REFERENCES

Barr, A., & Feigenbaum, E. A. (1982). *The Handbook of Artificial Intelligence*, 2. Los Altos, CA: Kaufmann.

Bell, J. T., & Scott, H. F. (1995). The Investigation and Application of Virtual Reality as an Educational Tool. *Proceedings of the American Society for Engineering Education Annual Conference* (pp. 1718 – 1728).

Cairncross, S. (2002). *Interactive Multimedia and Learning: Realizing the Benefits*. PhD. Thesis, Napier University, Scotland, UK.

Cairncross, S., & Mannion, M. (1999). How multimedia functions in engineering education. *Engineering Science and Education Journal*, *8*(3), 100–107. doi:10.1049/esej:19990302

Cairncross, S., & Mannion, M. (2001). Interactive Multimedia and Learning: Realizing the Benefits. *Innovations in Education and Teaching International*, 156–164.

Dede, C. Salzman, M. & Loftin, B. (1996). The Development of a Virtual World for Learning Newtonian Mechanics. In P. Brusilovsky, P. Kommers, & N Streitz, (Eds), *Multimedia, Hypermedia, and Virtual Reality*. Berlin: Springer.

Felder, R. M., & Silverman, L. K. (1988). Learning and Teaching Styles in Engineering Education. *Journal of Engineering Education, 78*(7), 674–681.

Goldman, J. D. & Torrisi-Steele, G. (2004). Constructive Pedagogies of Interactivity on a CD-ROM to Enhance Academic Learning at a Tertiary Institution. *International Journal of Educational Technology*.

Guttormsen, S., & Krueger, H. (2000). Using New Learning Technologies with Multimedia. *IEEE MultiMedia*, 40–51.

Ismail, J. (2001). The design of an E-learning system: beyond the hype . *The Internet and Higher Education Journal*, *4*(3-4), 329–345. doi:10.1016/S1096-7516(01)00069-0

Julia, M. Niall, S. & Su, W. (2002). *Report on the British Council Seminar and Asian Association of Open Universities Exhibition in Malaysia*. Retrieved from http://cvu.strath.ac.uk/niall/malaysia/

Katarzyna (2002). *Knowledge Representation, Content Indexing and Effective Teaching of Fluid Mechanics Using Web-Based Content.* Masters Thesis, Massachusetts Institute of Technology, Cambridge, MA.

Kim, J. Heon, P. S. Tae, Lee, H. & Yuk, K. C. (2001). Virtual Reality Simulations in Physics Education. *Interactive Multimedia Electronic Journal of Computer-Enhanced Learning, 3*(2).

Kinshuk, R. (1996). *Computer Aided Learning for Entry accountancy students.* PhD. thesis, De Montfort University, UK.

Kinshuk, R., & Patel, A. (2003). Optimizing Domain Knowledge Representation with Multimedia Objects. *Learning and Teaching with Technology: Principles and practices*. London: Kogan.

Klett, K. (2002). Improving Engineering Student Learning in a Web-Based Learning Space Due to Virtual Reality Techniques and Advanced Interactivity. *Proceedings of the 2002 eTEE Conference Switzerland* (pp. 195 – 101).

Law, R. C. H., & Maguire, R. B. (1993). A critical survey of intelligent tutoring systems. *Congressus Numerantium, 92*, 55–64.

Liarokapis, F. Mourkoussis, N., White, M., Darcy, J., Sifniotis, M., Petridis, P., Basu, A., Lister, P.F. (2007). Augmented Reality and Web 3D to Support Engineering Education. *Proceedings of the Engineering and Technology Education Conference, Greece,* (pp. 24 -28).

Macpherson, C. (1998). Virtual Reality: What is the state of play in education? *Australian Journal of Educational Technology*, *14*(1), 60–74.

Mahathir, M. (1998). *Multimedia Super Corridor.* Subang Jaya, Malaysia: Pelanduk Publications (M) Sdn. Bhd.

Manseur, R. (2005). Virtual Reality in Science and Engineering Education. *35th ASEE/IEEE Frontiers in Education Conference*, (pp. 8 – 13).

Megarry, J. (1983). Thinking, Learning and Educating: The role of the computer. [New York: Nichols Pub. Co.]. *Computers & Education*, 15–28.

Negroponte, N. (1995). *Being digital.* New York: Alfred A. Knopf.

Norhayati, A. M. Halimah, B. Z. & Siew, P. H. (2001). Courseware Development to Motivate Life-Long Reading Habits. *Multimedia at Work,* (pp. 76 – 81). Washington DC: IEEE.

Palmer, S. (2000). On-and off-campus computer usage in engineering education. *Computers and education Journal, Elsevier Science 33*(3), 141 – 154.

Rizzo, A. A., Buckwalter, J. G., & Neumann, U. (1997). Virtual reality and cognitive rehabilitation: A brief review of the future. *The Journal of Head Trauma Rehabilitation, 12*(6), 1–15. doi:10.1097/00001199-199712000-00002

Robert, D. (1994). *The Development, Formative and Summative Evaluation of a Computer Multimedia Tutorial: A Case Study*. PhD. thesis, Walden University.

Schank, R. C. (1994). *Active Learning through Multimedia* (pp. 69 – 78). Washington, DC: IEEE Multimedia.

Chapter 2
Instructional Design for Multimedia and Theory of Learning Styles

INTRODUCTION

The effectiveness of any instructional programme or instructional material depends upon an appropriate planning or designing, what is called in professional parlance, "Instructional Design". In general, instructional design is relatively a young discipline (Usha, 2003). In its literal meaning, instruction means a set of events that facilitate learning. On the other hand the word design is a generic term, which means "a creative model". Instructional design includes several processes such as the use of knowledge, observation, and creativity to plan and create situations that enhance learning opportunities of the individuals. However, to accomplish the aforementioned processes, the instruction has to be planned to be effective and designed in some systematic approach. Learning theories have significant bearing on instructional design, as there is a logical development from learning to instruction (Usha, 2003). Instructional design optimizes learning outcomes while learning theories are the backbone of any instructional design. Instructional design is the articulation or the manifestation of the learning theories, and its main aim is to optimize learning by using the known theories of learning (Usha, 2003).

According to Strain (1994), a wide divergence of views exist among the researchers in instructional design concerning the relative contribution of various schools of psychology and claims that instructional

DOI: 10.4018/978-1-60566-764-5.ch002

Table 1. Description of various learning theories

Category	Descriptions	Psychologists
Behaviorism	• Research has been conducted on animals but is related to human behavior. • Based on observable changes in behavior which can be measured. • Learning results from the classical conditioning of simple reflexes. • Learning is the formation of a connection between stimulus and response.	John B. Watson Ivan Pavlov E.L. Thorndike B.F. Skinner
Cognitivism	• Research related to human behavior. • Theory is based on the thought process behind the behavior. • Learning involves associations established through contiguity and repetition. • Stressed on the role of reinforcement which provides feed back about the correctness of responses. • Learning involves subsuming new materials to existing cognitive structure	Jean Piaget Lev Vygotsky Bruer Jerome David Ausubel
Constructivism	• Learners construct their own perspective of the world, through individual experiences and scheme. • Learners construct their own knowledge. Learners are encouraged to search for other related relevant information. • Prepare the learner to problem solving ambiguous situations.	George Herbert Mead D. H. Jonassen D. N. Perkins

design has grown out of the systems approach with its roots firmly in behaviorists psychology that has dominated instructional design since the 1960s. However, Hannafin and Reiber (1989) pointed out that instructional design developed in the 1980s by Gagne, Merrill, Reigeluth and Scandura is largely due to the influence of cognitive theories of learning. Usha (2003) pointed out that the emphasis of instructional design has been on how information is retrieved, selected, processed and perceived. More recent developments are due to constructivist learning theories. Instructional designers no longer rely on any one theory. They draw upon and incorporate from various learning theories, mix those with other information and apply the results to meet human needs (van Patten, 1989).

This Chapter first describes some meanings, theories, models and learning styles followed by the students learning characteristics under study, instructional design and multimedia and then reports the results of a study designed to measure and more clearly define the value of these characteristics with relation to the selected undergraduate engineering subjects. Instructional design for multimedia is also discussed. The main purpose of this Chapter is to determine the major characteristics affecting student learning in order to incorporate these in the design and delivery of the TAPS packages. Students in today's undergraduate level classrooms often display widely varying characteristics that extremely affect learning outcome. Although student learning characteristics have been widely studied in the more traditional teaching and learning environments, educators have just begun exploring the applications in interactive multimedia and its associated technological techniques.

GENERAL THEORIES OF LEARNING

This section examines three general categories of learning namely, behaviorism, cognitivism and constructivism. These three categories of learning have implications for instructional design. A brief introduction to the three categories of learning is given in Table 1.

Referring to Table 1, behaviorists believe that learning results in changing the learning behavior whereas cognitivists believe that learning occurs when learners add new concepts and ideas to their cognitive structure. Constructivists believe that the learners construct knowledge for themselves i.e. each learner individually. All the three learning categories have implications for instructional design.

THEORY AND MODELS OF INSTRUCTIONAL DESIGN

This section examines a few instructional design theories and models. A theory provides a general explanation of observations and explains the behavior whereas a model is a mental picture that helps us to understand something that we cannot see or experience directly (Dorin, Demmin and Gabel, 1990). There are various instructional design theories and models developed by various authors (Usha, 2003). Reigeluth (1999) defines an instructional design theory as the one that offers explicit guidance on how to help people learn and develop. The kinds of learning may include cognitive, emotional, social, physical and spiritual learning. Reigeluth (1999) states four major characteristics that all instruction design theories have in common. These are:

- Design orientation,
- Identification of methods of instruction and situations,
- Methods of instruction that can be broken into more detail components methods, and
- Choice of Probalistic Methods.

The design theories have become important as they help the stakeholders to develop a vision of the instruction early in the design process (Usha, 2003). This vision is in terms of ends (how learners will be different as a result of it) and the means (how those changes in the learners will be fostered). Banathy (1991) states that instructional design theories should allow much greater use of the notion of user designer. This means that the users play a major role in designing their own instruction.

These theories are also important as they provide guidance at three levels (Reigeluth, 1999). These are:

- Methods that best facilitate learning under different situations,
- Learning tool features that best allow an array of alternative methods to be made available to learners,
- System features that best allow an instructional design team to design quality learning tools.

Although there are over hundred different instructional design models (ISD), almost all of them are based on the concept of ADDIE (Analysis, Design, Development, Implementation, and Evaluation) model which is a generic, systematic approach to the instructional design process. It provides instructional designers with a framework in order to make sure that their instructional products are effective and that their creative processes are as efficient as they can possibly be. This model provides a systematic approach to course development efforts and it is a basic model that is suitable for any type of learning, including web-based. In Table 2 different models of instructional design are summarized with their features. All these models can be used to design instruction of course units (in print, multimedia and online) and have the following components in common:

Table 2. Models of instructional design

Models of Instructional Design	Description
Gagne-Briggs Model	To design instruction: • Categorize learning outcomes • Organize instructional events for each kind of learning outcome • There are nine instructional events • Events are tailored to the kind of outcome to be achieved • Model is adapted to Web Based Instruction
David Merrill	The model by David Merrill (Component Display Theory) is based on the following assumptions: • Different classes of learning outcomes require different procedures for teaching and assessment • Teaches individual concepts • Classifies objectives on two dimensions • Formats instruction to provide student directed teaching
Dick and Carey	This model: • Uses a systems approach for designing instruction • Identifies instructional goals in the beginning and ends up with summative evaluation • Is applicable for K-12 to business to government
Hannafin and Peck	The model has three phrases: • Need assessment is performed in the first phase • Second is the design phase • Instruction is developed and implemented in the last phase All the phases involve a process of evaluation and revision
Gerlach and Ely	The model: • Includes strategies for selecting and including media within instruction • Is suited to higher education
ADDIE Model	ADDIE stands for the steps of the model as following: • *Analyze*: define the needs and constraints • *Design*: specify learning activities, assessments and choose methods & media • *Develop*: begin production, formative evaluation, and revise • *Implement*: put the plan into action • *Evaluate*: evaluate the plan from all levels for next implementation Each step has an outcome that feeds the subsequent step. Evaluation is essential after each step.
Source: http://lts.ncsu.edu/guides/instructional_design/selecting_models2.html	

- Identify and analyze the instructional objectives,
- Plan and design solutions to the instructional objectives,
- Implement the solutions and evaluate & revise objectives, strategies, etc.

In this research the ADDIE model was adopted since it has the systemic approach to development and many advantages when it comes to the creation of technology-based training.

PEDAGOGICAL CHARACTERISTICS AFFECTING STUDENT LEARNING

In recent years, approaches to teaching have changed significantly and have led to a greater differentiation between teaching and learning. While studies on improving teaching have been ongoing for many years, on the other hand studies on improving learning have received great deal of attention only recently.

More researchers today are looking into what characteristics affect a student's learning curve given that the teaching techniques are close to optimal.

A variety of student characteristics impact a student's performance and ultimately individual achievements in the classroom. Huitt (2002) listed six main characteristics as follows:

- Intelligence, achievement, and prior knowledge
- Learning style
- Cognitive development
- Gender
- Race
- Moral and character development

STUDENT ATTRIBUTES AFFECTING LEARNING

The design approaches (of TAPS packages) under study have focused on the first year undergraduate level classroom that teaches Engineering Mechanics subjects. Considering that many first year undergraduates have different level of knowledge in science and mathematical subjects, the student characteristics list of learning as stated above can be extended as follows:

- *Basic knowledge background*: This characteristic represents the basic science and mathematics knowledge of the student. On a given scale, it shows whether, and how much, basic science and mathematics knowledge the student has. The scale is however multidimensional, showing not only the background knowledge in science and mathematics, but also knowledge of other categories required for a better understanding of the selected Engineering Mechanics subject. Engineering Mechanics subject is better understood if the student has an intermediate knowledge of topics such as calculus, science, mathematics, and physics.
- *Academic performance*: A student's prior academic performance is often a factor that is overlooked in a student's current academic achievements. A good or bad performance often affects a student positively or negatively, particularly during test, or quizzes.
- *Exposure to modern educational technologies*: This represents the experience that students already have in using modern technological learning aids such as computer and learning packages. The use of computer packages is more easily understood if students already have some elementary computing skills.
- *Learning style*: Student learning styles are probably one of the most researched factors affecting student cognition and learning rate. Many studies have been performed on student learning styles with many different categorizations made.

Learning styles are most often targeted in elementary education. A number of researchers have tried to categorize learning styles in different manners. Some of these are discussed in the subsequent Sections.

THEORY OF LEARNING STYLES

Literatures on learning styles have emphasized much awareness that students do have different learning styles, characteristics, strengths, and preferences in the ways they absorb and process information (Felder, 1996). Dunn (1999) described learning style as "*...the way each learner begins to concentrate, process and retain new and difficult information*, p.224". Although the theory of learning styles has been used in other educational domains, several practitioners within the field of science and engineering education have also noted the importance of teaching with learning styles (Jensen and Wood, 2000; Hein and Budny, 2001; Felder, 2002; Felder and Spurlin, 2005; Jessica and Tara, 2005; Knudsen, 2006, Milgram, 2007; Aidan, 2008). The learning style models approach can be used to measure an individual's preferences in way of thinking, learning and the degree to which a certain learning style is ineffective.

HISTORY OF LEARNING STYLES

There have been several learning style questionnaires (instruments) and models developed to categorize the way learners imbibe and process information. Some most quoted and popular ones found in the literatures include, the Myers-Briggs Type Indicator (MBTI), Kolb's Learning Style Model, Herman Brain Dominance Instrument (HBDI), McCarthy's 4MAT model, Dunn and Dunn Learning Style Model of Instruction, Felder-Silverman Learning Style Model and Honey and Mumford Learning Styles Evaluation. Research conducted with engineering students using any of these learning styles mentioned is reported to provide a positive involvement. However three models that have been applied extensively to engineering namely Kolb, Honey and Mumford, and Felder and Silverman are discussed in some detail in this book.

The issue of how to help people to learn effectively has been an active research area over the last decade (Mumford, 1982). Most individuals have different learning styles, which indicate preference for particular learning experiences. Witkin's (1976) work on field dependent and field independent cognitive styles concentrates on the differences in the way individual structures and analyses information. Pask and Scott (1972) identified holist and serialist strategies in problem solving. Pask argued that the holist and serialist strategies are the manifestations of the underlying differences in the way people approach learning and problem solving. Miller and Parlett (1974) described cue-consciousness and identified two distinct groups of students. The first group is respective to, and actively seeks out, clues and hints from their tutors regarding forthcoming examinations, these they termed as clues-seeking, whereas the second, who have less sophisticated strategies and do not pick up on available hints, are termed clue-deaf.

Dunn (1979) pointed out a person's learning orientation is perhaps the most important determinant of his/her educational attainment. Logically, the greater its congruence with the teaching method used, the greater the chance of success (Allinson and Hayes, 1990). Consequently, some instruments are available which seek to measure learning styles. In past years, a number of researchers have examined the concept of learning styles (Delahaye and Thompson, 1991). Marton and Saljo (1976a) believed that students' approaches to learning tasks could be categorized into two broad areas that they labeled as 'deep approaches' or 'surface approaches'. Deep approaches involved an active search for meaning underlying principles, structures that linked together different concepts or ideas and widely applicable techniques. Surface approaches, on the other hand, rely primarily on attempts to memorize course work, treating the material as if different facts and topics were unrelated.

Figure 1. The Kolb Learning Model

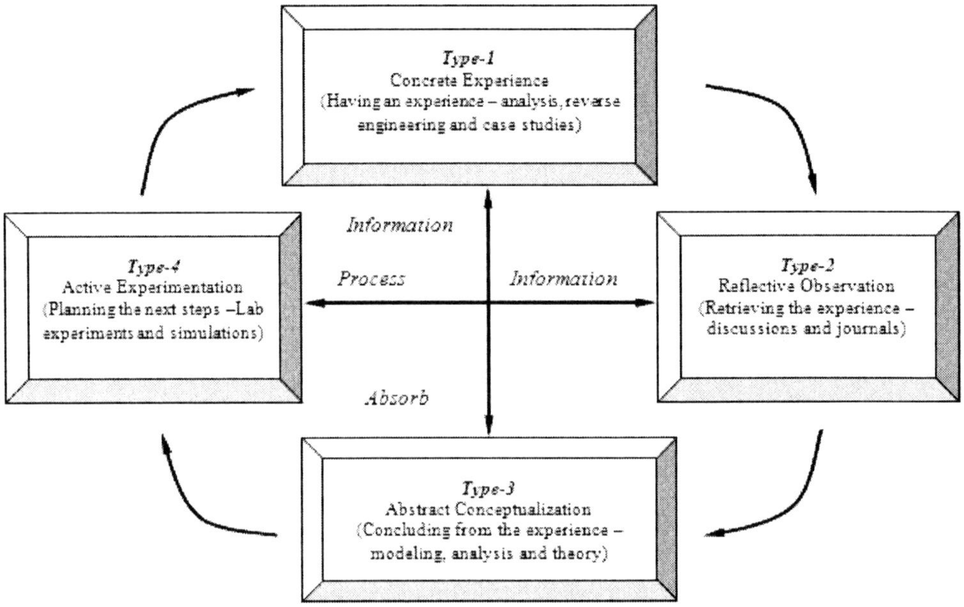

Follow-up studies by Marton and Saljo (1976b), and Svensson (1977), demonstrated that most students were somewhat versatile in their choice of learning approach. The students' choice depended on factors such as their interest in the topic, the nature of their academic motivations; the pressure of other demands on their time and energy; the total amount of content in the course; the way in which a task is introduced, and their perceptions of what will be demanded of them in subsequent evaluations or applications of the material (Kinshuk, 1996). Recent work in the field is more expansive (in those issues are assessment, instruction, personality and evaluation) as they relate to learning styles and strategies are comprehensively addressed (Weinstein *et al.*, 1988; Ginter *et al.*, 1989; Green *et al.*, 1990). However, the Kolb (1976) learning style model has motivated most researchers and used widely to measure student-learning style.

KOLB'S LEARNING STYLE MODEL

Kolb developed the learning style inventory (LSI) in 1976 and revised in 1985 (Tendy and Geiser, 1998). The LSI was a nine–item self-report questionnaire in which four words describing one's style by which respondents attempt to categorize their learning style. One word in each item was used to correspond to one of four learning modes as shown in Figure 1. The four stages cycle of the learning modes in the Figure are identified as Type-1: concrete experience (CE), Type-2: reflective observation (RO), Type-3: abstract conceptualization (AC) and Type-4: active experimentation (AE).

According to Figure 1 the concrete experience mode describes people who feel more than they think. Individuals in this mode tend to be very good at relating to others and they tend to be spontaneous decision-makers. Their characteristic question is "*Why*". To be effective with Type-1 students, the instructor should function as a motivator.

Figure 2. The Experiential Learning Model (after Kolb, 1984, p21) with the linked Honey and Mumford Learning Styles in bold (Honey and Mumford, 1986)

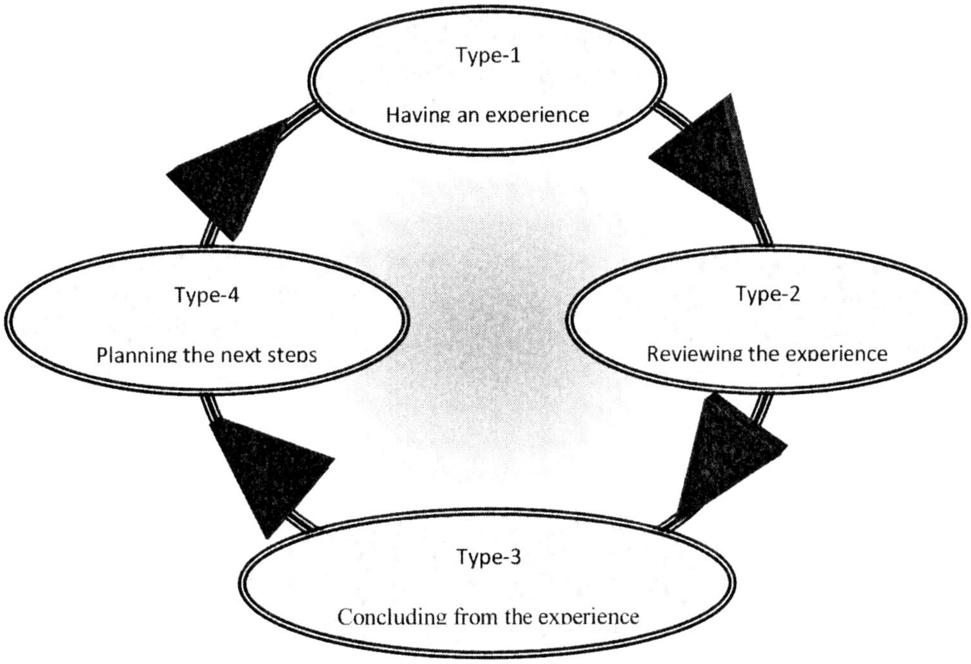

The reflective observation mode describes people who would rather watch and observe others than to be an active participant. Individuals in this mode tend to appreciate exposure to differing points of view. Their characteristic question is "*What*". To be effective with Type-2 students, the instructor should function as an expert.

The abstract conceptualization mode describes people who think more than they feel. Such people tend to have a scientific approach to problem solving as opposed to a more artistic approach. Their characteristic question is "*How*". To be effective with Type-3 students, the instructor should function as a coach, providing guided practice and feedback in the methods being taught.

Lastly, the active experimentation mode describes individuals who take an active role in influencing others as well as situations. These individuals welcome practical applications rather than reflective understanding as well as actively participating rather than observing. Their characteristic question is "*What If*". To be effective with Type-4 students, the instructor should pose open-ended questions and then get out of the way, maximizing opportunities for the students to discover things for themselves.

Most studies of engineering fields on the Kolb model find that the majority of the subjects are Types-2 and -3. For example, Bernold *et al.* (2000) found that of the 350 students in their study, 55% were Type-3, 22% were Type-2, 13% Type-4, and 10% Type-1. Sharp (2001) reported that 64 out of 1,013 engineering students that were tested, 40% were Type-3, 39% was Type-2, 13% was Type-4, and 8% were Type 1. Spurlin *et al.* (2003) reported on an ongoing study comparing freshman-engineering students using the four Kolb model. Their preliminary results showed that Types-2 and -3 students did better, and they are conducting further studies intended to pinpoint reasons for the relatively poor performance and high risk of attrition of the Types 1 and 4 students.

However critics of the application of Kolb's LSI suggest that its application for education research purposes was premature in the sense that the instrument's psychometric properties (reliability, validity, consistency) had not been sufficiently assessed.

As such Honey and Mumford (1992) proposed Learning Styles Questionnaire (LSQ) as an alternative to Kolb's LSI. Although LSQ has been criticized by some researchers for failing to construct validity and has correlations among its four learning styles (Goldstein and Bokoras, 1992; Tepper *et al*., 1993), this has been the most favored learning style instrument in the literature for evaluation of CAL modules (Hayes and Allinson, 1988; Furnham, 1992; Allinson and Hayes, 1990). Generally the LSQ has also been used for students in engineering courses.

HONEY AND MUMFORD'S LEARNING STYLES QUESTIONNAIRE

The Kolb model is the theoretical background to Honey and Mumford (1986) Learning Style Questionnaire, which has four styles i.e. Theorists, Activist, Reflector and Pragmatist. The questionnaire directly assesses the four basic types of style in Kolb's model as shown in Figure 2. This analysis has been used widely through business and education and most recently as a basis for selecting undergraduate engineering course (Halstead and Martin, 2002).

According to the theory, the hypothesized learning cycle can be entered at any stage but must be followed in sequence. A person could start, for example, at (Type-2) by acquiring some information and analyze it before reaching some conclusions, (Type-3), and deciding how to apply it, (Type-4).

The four stages, experiencing, reviewing, concluding and planning are mutually supportive. None is fully effective as a learning procedure on its own. Each stage plays an equally important part in the total process, though the time spent on each may vary considerably. Most people, however, develop preferences, which give them a liking for certain stages over others. The preferences led to a distortion of the learning process consequently greater emphasis were placed on some stages in comparison of others.

Here are some typical examples: -

- Preferences for concluding such that people have a compulsion to reach an answer quickly. This results in a tendency to jump to conclusions by circumventing (by-passing) the review stage, where uncertainty and ambiguity things to have (Kinshuk, 1996).
- Preferences for seizing on an expedient course of action and implementing it with inadequate analysis. This results in a tendency to go for 'quick fixes' by overemphasizing the planning and experiencing stages to the detriment of reviewing and concluding (Kinshuk, 1996).

The LSQ is designed to assess the relative strengths of four different learning styles: Activist, Reflector, Theorist, and Pragmatist. These four styles correspond approximately to those suggested by Kolb's (1976) LSI. According to Honey and Mumford (1986, 1992):

- *Activists*

Activists involve themselves fully and without bias in new experiences. They enjoy the present situation, and are happy to be dominated by immediate experiences. They are open-minded, not specialized, and this tends to make them enthusiastic about anything new. Their philosophy is: "*I'll try anything once*".

They tend to act first and consider the consequences afterwards. Their days are filled with activity. They tackle problems by brainstorming. As soon as the excitement from one activity has died down they are busy looking for the next. They tend to thrive on the challenge of new experiences but are bored with implementation and longer-term consolidation. These people constantly involve themselves with others but in doing so; they seek to center all activities around themselves.

- *Reflectors*

Reflectors like to stand back to ponder on experiences and observe them from many different perspectives. They collect data, both first hand and from others, and prefer to think about it thoroughly before coming to any conclusion. The thorough collection and analysis of data about experiences and events is what counts so they tend to postpone reaching definitive conclusions for as long as possible. Their philosophy is to be cautious. They are thoughtful people who like to consider all possible angles and implications before making a move. They prefer to take a back seat in meetings and discussions. They enjoy observing other people in action. They listen to others and get the drift of the discussion before making their own points. They tend to adopt a low profile and have a slightly distant, tolerant, unruffled air (proud of themselves) about them. When they act it is part of a wide picture which includes the past as well as the present and others' observations as well as their own.

- *Theorists*

Theorists adapt and integrate into complex but logically sound theories. They think problems through a vertical, step-by-step, logical way. They assimilate disparate facts into coherent theories. They tend to be perfectionists who will not rest easy until things are tidy and fit into a rational scheme. They like to analyze and synthesize. They are keen on basic assumptions, principles, theories, models and systems thinking. Their philosophy prizes rationally and logic. "*If it's logical it's good*". Questions they frequently ask are; "*Does it make sense*?" "*How does this fit and that*?" "*What are basic assumptions*?" They tend to be detached, analytical and dedicated to rational objectivity rather than anything subjective or ambiguous. Their approach to problems is consistently logical. This is their 'mental set' and they rigidly reject anything that does not fit with it. They prefer to maximize certainty and feel uncomfortable with subjective judgments, lateral (side, surface) thinking and anything flippant (playful).

- *Pragmatists*

Pragmatists are keen on trying out ideas, theories and techniques to see if they work in practice. They positively search new ideas and take the first opportunity to experiment with applications. They are the sorts who return from management courses complete with new ideas that they want to try out in practice. They like to get on things and act quickly and confidently on ideas that attract them. They tend to be impatient with ruminating and open-ended discussions. They are essentially practical, down to earth people who like making practical decisions and solving problems. They respond to problems and opportunities 'as a challenge'. Their philosophy is: "*There is always a better way*" and "*If it works its good*".

Each style is associated with a stage on the continuous learning cycle. People with Activists preferences, are well equipped for experiencing. People with Reflector approach, with their preferences for deliberating over data, are well equipped for reviewing. People with Theorist preferences, with their need to tidy up and have 'answers' are well equipped for conducting. Finally, people with Pragmatist preferences, with their liking for things practical, are well equipped for planning (Honey and Mumford, 1986). Whilst Honey and Mumford (1986) found Kolb's four-stage learning cycle acceptable, they were less satisfied with the LSI, questioning the use of one-word descriptors as a basis for attributing style, and expressing concern over the face validity of the styles themselves.

The LSQ is a self-administered inventory consisting of 80 items, with which respondents are asked to tick indicating agrees or cross indicating disagree respectively. The 80 items comprise four subsets of 20 randomly ordered items, each subset measuring a particular learning style. The vast majority of these items are behavioral and the aim is to discover general behavioral trends. The LSQ is scored by awarding one point for each ticked item and no item carries more weight than another. The items describe an action that someone might or might not take. Occasionally, an item probes a preference or belief rather than a clear behavior. Examples of items include:

1. "On balance I talk more than I can listen" (*Activist*).
2. "I tend to discuss specific things with people rather than engaging in social discussion" (*Reflector*).
3. "I am keen on exploring the basic assumptions, principles and theories underpinning things and events" (*Theorist*).
4. "I can often see better, more practical ways to get things done" (*Pragmatist*).

The score key 'decodes' the items and lists those that probe Activists tendencies, Reflector tendencies and so on. The LSQ is scored by awarding one point for each ticked item and no points for crossed items. Thus the maximum possible score for each learning style is twenty. Raw scores are meaningful only when viewed in the context of normative data. The norms are calculated by analyzing the actual scores of people who have completed the questionnaire. Honey and Mumford (1986) analyzed the scores of well over one thousand people by 1982 and divided them into five groups.

- *A* – the point at which 10% of the scores are above and 90% are below.
- *B* – the point at which 30% of the scores are above and 70% are below.
- *C* – the middle 40% scores with 20% above and 20% below the mean.
- *D* – the point at which 70% of the scores are above and 30% are below.
- *E* – the point at which 90% of the scores are above and 10% are below.

Each of the five groups arrived at in this way is indicative of a person's learning style preferences:

- Any scores in the *A* group indicate a very strong preference since statistically only 10% of the scores fall into this group.
- Scores in the *B* group indicate strong preferences.
- Scores in the *C* group indicate moderate preferences.
- Scores in the *D* group indicate low preferences.
- Scores in the *E* group indicate very low preferences since statistically only 90% of the scores are above this group.

Since the LSQ consists of 80 questionnaires, it was not found to be a suitable instrument for measuring the students learning style in this research.

FELDER-SILVERMAN LEARNING STYLE MODEL

The Felder-Silverman Learning Style Model classifies students along four dimensions: sensing/intuitive, visual/verbal, active/reflective and sequential/global as shown in Table 3. According to these dimensions, a student's learning style may be defined by the answers to four questions:

1. *What type of information does the student preferentially perceive?*

Sensing learners retain information obtained through their senses, while intuitive individuals are more likely to retain information obtained through their own memory.

2. *What type of sensory information is most effectively perceived?*

Visual learners prefer pictures, while verbal learners prefer the written and spoken word.

3. *How does the student prefer to process information?*

Active learners learn by experimenting, while reflective learners learn by thinking about a concept.

4. *How does the student characteristically progress toward understanding?*

Sequential learners learn in small incremental steps, while global learners need a strong understanding of the big picture (Felder *et al.*, 1996). A study of over 800 students at the University of Western Ontario (UWO), London, Canada, found that engineering students have strong sensing, visual, active and sequential preferences (Rosati, 1998).

In this research, the subjective views of students are examined for the evaluation of the developed TAPS packages and are used to analyze their learning styles. Respondents may be classified as adopting

Table 3. The four dimensions of Felder and Silverman's learning Styles

Sensory / Intuitive	Sensors prefer facts, data, experimentation, sights and sounds, physical sensations are careful and patient with detail, but may be slow. Intuitions prefer concepts, principles and theories, memories, thoughts, insights and may be quick but careless.
Visual / Verbal	Visual learners prefer pictures, diagrams, charts, movies, demonstrations and exhibitions. Verbal learners prefer words, discussions, explanations, discussions, written and spoken explanations formulas and equations.
Active / Reflective	Active learners learn by doing and participating through engagement in physical activity or discussion. Reflective learners learn by thinking or pondering through introspection.
Sequential / Global	Sequential learners take things logically step by step and will be partially effective with understanding. Global learners must see the whole picture for any of it to make sense and are completely ineffective until they suddenly understand the entire subject.

a particular learning style preference, based on the score obtained on individual scale using the Index of Learning Styles (ILS).

Structure of ILS

The index of learning styles is a forty-four-question instrument developed in 1991 by Richard Felder and Barbara Solomon (Felder *et al.*, 1996), to assess preferences on the four dimensions of the Felder-Silverman model. The index of learning styles is a self-scoring instrument that assesses preferences on the Sensing/Intuition, Visual/Verbal, Active/Reflective and Sequential/Global dimensions. The ILS is available at no cost to individuals who wish to assess their own preferences and instructors or students who wish to use it for classroom instruction or research. (A copy of the ILS questionnaire is enclosed in Appendix A for reference).

Scoring and Interpreting the ILS Questionnaire

Scoring the questionnaire is quite straightforward. When an individual submits a completed ILS questionnaire on-line, a profile is instantly returned with scores on all four dimensions, brief explanation of their meaning and links to references that provide more detail about how the scores should and should not be interpreted.

Each learning style dimension has associated with it 11 forced-choice questionnaires, with each option (*a* or *b*) corresponding to one or the other category of the dimension (e.g., visual or verbal). For statistical analyses, it is convenient to use a scoring method that counts '*a*' responses, so that a score on a dimension would be an integer ranging from 0 to 11. Using the visual-verbal as an example, 0 or 1 '*a*' responses would represent a strong preference for verbal learning, 2 or 3 a moderate preference for verbal, 4 or 5 a mild preference for verbal, 6 or 7 a mild preference for visual learning, 8 or 9 a moderate preference for visual, and 10 or 11 a strong preference for visual. This method was used in the statistical analyses reported in this book. The method used to score the on-line version of the instrument subtracts the '*b*' responses from the '*a*' responses to obtain a score that is an odd number between – 11 to + 11.

Studies Utilizing the ILS

A number of studies have collected the response data for the Index of Learning Styles. Some investigators simply measured and reported response profiles and drew conclusions from them regarding appropriate teaching methods for their classes, and others used the profiles to examine various aspects of student performance and attitudes. A summary of learning styles profiles reported in various studies can be found in Felder and Spurlin (2005).

MULTIMEDIA IN LEARNING

According to Usha (2003) media is a Latin word and is used to describe ways to convey messages and information. When we talk about media we think of newspapers, magazines, radio, TV, audio- video programmes, computers, etc. Many prefixes are used with the word Media like, etc. The most common buzzword used in education is Multimedia, which is the integration of text, audio, video, graphics and animation into a single medium. Instructional multimedia is the integration of various forms of media in the instructional process. It is the technology that combines print, radio, television, animation, photographs, and other forms of illustration. Integration of different media multiplies the impact of a message (Usha, 2003). The focus is on instruction and learning. According to the research reports by Mayer and McCarthy (1995) and Walton (1993) 'multimedia has gained acceptance with many benefits derived from its use. Learning gains are 56% greater, consistency of learning is 50-60% better and content retention is 25-50% higher'. Instructional multimedia focuses on what the learner is expected to do upon the completion of the instruction.

On one hand, research on multimedia has established learning gains of significant order over the conventional instructional strategies, and on the other, has shown how instructional design is a tested, well-researched mechanism of enhancing human learning. By logical extrapolation, we can say that instructional multimedia can be more effective, if it is backed up by scientific instructional design (Usha, 2003).

INSTRUCTIONAL DESIGN FOR MULTIMEDIA

The major challenge in designing instruction through multimedia is, therefore, the choice of media and their application for optimizing human learning with reference to the stated instructional objectives. According to Usha (2003), we must, hence, consider the various components that constitute the instructional design for multimedia learning system such as objectives, content, media options, and evaluation options.

Objectives: The first challenge is to specify the objectives of the multimedia learning. The objectives must be stated in behavioral and measurable terms. They can range from simple to complex, from lower to higher order learning. The objectives may belong to the domains of cognition, psychomotor and affection.

Content: The content of any instructional design is necessarily informed by stated objectives of learning. Depending upon the objectives the content will also range from simple to high level of complexity. The choice of content must also ensure that there is adequate and correct provision for the achievement of objectives.

Media Options: As mentioned above multimedia essentially incorporates several media like text (as in printed text), audio, video, graphics, animation etc. It is important to match the learning objectives and decide the media to synchronize the design and learning from it. Each media can offer either the whole or part of the content with or without referring to one another. For example, dissection of a frog can be shown through animation and also through a video programme. But as multimedia offers interactivity, learners can actually feel the dissection if it is animated and the multimedia programme runs like an actual dissection. Similarly, for language learning through multimedia, audio is very important.

Evaluation Options: evaluation is part of instructional design. Without evaluation, one would rarely understand the achievement of objectives, which is the primary goal of instructional design. Evaluation options must include both summative and formative evaluation. However, in both the cases of formative and summative evaluation, we can choose from online, offline, paper and pencil versus performance tests, etc.

In summary, this section dealt initially with fundamental issues of learning theories, and concept, theory and models of instructional design. Secondly, instructional design for multimedia learning system must be a document indicating the stated goals, choice of content with specifications of levels of difficulties, the choice of instructional methods and media, and strategies of evaluation. The documented design must incorporate instructional design of the micro components of the multimedia learning system as well.

This Chapter has reviewed some extensively used learning style measurement instruments and their efficacy in the evaluation of CAL. The vast majority of research conducted with engineering students using any of the learning styles questionnaires and models mentioned in this Chapter is reported to provide a positive intervention whereas students require greater flexibility in assessing variety of learning preferences (Halstead, 2003). In this study, the Felder-Solomon's Index of Learning Style Questionnaire was adopted because it is simple, can be easily implemented using web-based quiz and has been used to classify the learning styles of engineering students (Rosati, 1998, Allen and Mourtos, 2000). It is proposed that there is a scope to use such an instrument to assist engineering students in developing flexibility of thought as well as self-awareness. However, in reflecting on the evaluation results in Chapter 7, the book will also examine the instrument's underlying assumption that it is possible to categorize students according to their learning styles and hence the appropriateness of Felder-Solomon's Index of Learning Style Questionnaire for the present study.

The main reason for utilizing the Felder-Solomon's Index of Learning Style Questionnaire as an instrument for this study was to determine engineering students' (students who need additional support in applying principles presented in lectures to problems) most productive learning style and to incorporate the same in developing effective TAPS packages. For example, do these students prefer to see animated objects rather than static images, do they prefer text rather than sound or a mix of both media, and do they prefer to see the TAPS packages solving a problem or prefer to solve the problem themselves? The feedback from these questionnaires provided useful information to improve the contents of the TAPS packages so that the TAPS packages can be effective in helping the students in solving the engineering problems.

The aim of this research is to evaluate the interactive multimedia TAPS packages with the help of students' opinions about them. The effect of learning styles of students is an important factor in deciding the type of students that are likely to benefit from these packages.

The next Chapter focuses on the user interface design of learning packages where the principles and various approaches of the interface design are discussed with special focus on interfaces for learning environments.

REFERENCES

Aidan, D. (2008). Learning Styles of First Year Level 7 Electrical and Mechanical Engineering Students at DIT. In *International Symposium for Engineering Education,* Dublin City University, Ireland, (pp. 1 – 5).

Allen, E., & Mourtos, N. (2000). Using learning style preferences data to inform classroom teaching and assessment activities. In *Proceedings Frontiers in Education*, Kansas City, KS. (p. S2B/6).

Allinson, C. W., & Hayes, J. (1990). The learning style questionnaire: an alternative to Kolb's inventory. *Journal of Management Studies*, *25*(3), 269–281. doi:10.1111/j.1467-6486.1988.tb00036.x

Allinson, C. W., & Hayes, J. (1990). The learning style questionnaire: an alternative to Kolb's inventory. *Journal of Management Studies*, *25*(3), 269–281. doi:10.1111/j.1467-6486.1988.tb00036.x

Banathy, B. H. (1991) *Systems design of education: A journey to create the future.* Englewood Cliffs, NJ: Educational Technology Publications.

Bernold, L. E., Bingham, W. L., McDonald, P. H., & Attia, T. M. (2000). Impact of Holistic and Learning-Oriented Teaching on Academic Success. *Journal of Engineering Education*, *89*(2), 191–199.

Delahaye, B., & Thompson, B. (1991). Learning styles – What do they measure? In *Asia Pacific Human Resource Management* (pp. 60 – 68).

Dorin, H., Demmin, P. E., & Gabel, D. (1990). *Chemistry: The study of matter,* (3rd Ed.). Englewood Cliffs, NJ: Prentice Hall, Inc.

Dunn, R. (1979). Learning-A matter of style. *Educational Leadership*, *36*(6), 430–432.

Dunn, R. (1999). Understanding the Dunn and Dunn learning styles model and need for individual diagnosis and prescription. *Reading . Writing and Learning Disabilities*, *6*, 233–247.

Felder, R. (1996). Matters of Style. In *ASEE Prism*, (pp. 18 – 23).

Felder, R. M. (2002). The Effects of Personality Type on Engineering Student Performance and Attitudes. *Journal of Engineering Education*, *91*(1), 3–17.

Felder, R. M., & Spurlin, J. (2005). Applications, Reliability and Validity of the Index of Learning Styles. *Journal of Engineering Education*, *21*(1), 103–112.

Felder, R. M., & Spurlin, J. (2005). Applications, Reliability and Validity of the Index of Learning Styles. *Journal of Engineering Education*, *21*(1), 103–112.

Furnham, A. (1992). Personality and learning style: A study of three instruments. *Personality and Individual Differences*, *13*(4), 429–438. doi:10.1016/0191-8869(92)90071-V

Ginter, E. Brown, S. Scalise, J. & Ripley, W. (1989). *Computers in Teaching Initiative Report.* Bath, UK: University of Bath.

Goldstein, M. B., & Bokoras, M. A. (1992). Tilting at windmills: Comparing the learning style inventory and the learning style questionnaire. *Educational and Psychological Measurement*, *52*(3), 701–708. doi:10.1177/0013164492052003020

Green, D., Snell, J., & Parimanath, A. (1990). Learning Styles in assessment of students. *Perceptual and Motor Skills*, *70*, 363–369. doi:10.2466/PMS.70.2.363-369

Halstead, A. (2003). Engaging Students with Learning and Personal Development by Using a Thinking Styles Questionnaire. *Progress 3 Conference, Strategies for Student Achievement in Engineering* – University of Hull, Paper No. 5.

Halstead, A., & Martin, L. M. (2002). Learning styles: A tool for selecting students for group work. *International Journal for Electrical Education*, *39*(3).

Hannafin, M.J. & Rieber, L. (1989). Psychological Foundations of Instructional Design for Emerging Computer-based Instructional Technologies: Part I. *ETR&D, 37*(2).

Hayes, J., & Allinson, C. (1988). Cultural differences in the learning styles of managers. *Management International Review*, *28*, 75–80.

Hein, T. L., & Budny, D. D. (2001). Research on Learning Style: Applications in the Physics and Engineering Classrooms. *IEEE Transactions on Education*, *44*(3).

Honey, P., & Mumford, A. (1992 & 1986). *The Manual of Learning Styles*. Maidenhead, UK: Peter Honey.

Huitt, W. G. (2002). Moral and Character Development. Retrieved Jan 2003 from http://chiron.valdosta.edu/whuitt/col/student/studchar.html

Jensen, D., & Wood, K. (2000). Incorporating Learning Styles to Enhance Mechanical Engineering Curricula by Restructuring Courses, Increasing Hands-On Activities and Improving Team Dynamics. *ASME Curriculum Innovation*, (pp. 1 – 8).

Jessica, M., & Tara, M. (2005). Educational Applets for Active Learning in Properties of Electronic Materials. *IEEE Transactions on Education*, *48*(1).

Kinshuk, R. (1996). *Computer Aided Learning for Entry accountancy students*. PhD Thesis, De Montfort University, UK.

Knudsen, H. (2006). Experimental Modeling of Dynamic Systems: An Educational Approach. *IEEE Transactions on Education*, *49*(1), 29–38. doi:10.1109/TE.2005.853071

Kolb, D. A. (1976). *Learning Style Inventory: Technical Manual*. Boston: McBer.

Kolb, D. A. (1984). *Experiential Learning*. Upper Saddle River, NJ: Prentice Hall.

Marton, F., & Saljo, R. (1976a). On qualitative differences in learning, 1: outcome as a function of the learner's conception of the task. *The British Journal of Educational Psychology*, *46*, 4–11.

Marton, F., & Saljo, R. (1976b). On qualitative differences in learning, 2: outcome as a function of the learner's conception of the task. *The British Journal of Educational Psychology*, *46*, 115–127.

Mayer, S. & McCarthy (1995). *Multimedia Learning Day-by-Day, Point and Click*.

Milgram, D. (2007). Gender Differences In Learning Style Specific To Science, Technology, Engineering And Math - Stem. Retrieved August 15 2007 http://ezinearticles.com/?Gender-Differences-In-Learning-Style-Specific-To-Science,-Technology,-Engineering-And-Math---Stem&id=658953

Miller, C. M., & Parlett, M. R. (1974). *Up to the mark: a study of the examination game*. London: SRHE.

Mumford, A. C. (1982). Learning Style and learning skills. *Journal of Management Development*, *1*(2), 55–65. doi:10.1108/eb051522

Pask, G., & Scott, B. C. E. (1972). Learning strategies and individual competence. *International Journal of Man-Machine Studies*, *4*, 217–253. doi:10.1016/S0020-7373(72)80004-X

Reigeluth, C. M. (Ed.). (1999). *Instructional- Design Theories and Models, Volume II- A New Paradigm of Instructional Theory.* New Jersey: LEA.

Rosati, P. (1998). The learning preferences of engineering students from two perspectives. *In Proceedings Frontiers in Education*, Tempe, AZ, (pp. 29 – 32).

Sharp, J. E. (2001). Teaching Teamwork Communication with Kolb Learning Style Theory. *Proceedings, 2001 Frontiers in Education Conference*, Washington DC: ASEE/IEEE.

Spurlin, J. E. Bernold, L. E. Crossland, C. L. & Anson, C. M. (2003). Understanding How Freshman Engineering Students Think They Learn. In *Proceedings 2003, ASEE Conference and Exposition*. Washington, DC: American Society for Engineering Education.

Strain, J. (1994). Pathways to a Profession Designing Instruction for Instructional Designers. In M.S. Parer, (Ed.) *Unlocking Open Learning.* Victoria, Australia: Monash University.

Svensson, L. (1977). On qualitative differences in learning, 3: study skills and learning. *The British Journal of Educational Psychology*, *47*, 233–243.

Tendy, S. M., & Geiser, W. F. (1998). The search for style: It all depends on where you look. Nat. *FORUM Teacher Education Journal*, *9*(1), 3–15.

Tepper, B. J., Tetrault, L. A., Braun, C. K., & Romero, J. E. (1993). Discriminant and convergent validity of the problem solving style questionnaire. *Psychological Measurement*, *52*(2), 437–444. doi:10.1177/0013164493053002013

Usha, V. R. (2003). Multimedia as an Educational Tool. *Educational Multimedia Book* (pp. 4 – 9).

van Patten, J. (1989) What is Instructional Design. In K.A. Johnson, and L.J. Foa, (Eds.), *Instructional Design: New Alternatives for Effective Education and Training.* NewYork, Macmillan.

Walton, J. H. (1993). *Multimedia Learning: Today and Tomorrow*. Keynote address of the Orlando Multimedia, SALT conference.

Weinstein, C. E. Goetz, E. T. & Alexander, P. A. (1988). *Learning and Study Strategies*. San Diego, CA: Academic Press.

Witkin, H. A. (1976). Cognitive style in academic performance and in teacher-student relations. *Individuality in Learning* (pp. 38 – 72). San Francisco, CA: Jossey-Bass.

Chapter 3
User Interface Design Approaches in Learning Environments

INTRODUCTION

Interface design provides the practical information for the multimedia author to develop well-designed and usable interfaces. However, the design of the user interface for any learning package involves many interacting concerns. The developer of such a package needs to consider the tasks to be achieved using the particular learning package. Although most learning package developers may use their own choices to develop the package, various approaches to user interface design has been proposed by standardization bodies (e.g. ISO, CEC/CENELEC, BSI) to provide the basic mechanisms for developing, promoting and imposing standards in the user interface in designing learning packages (Hutchins 1987; Ianella, 1992, Pangalos, 1993; Deborah, 1997; Plass, 1998; and Carter 2002). Unfortunately, many learning package developers do not use a common standard user interface design when designing a learning package. This Chapter reviews some approaches that are available to learning package developers and suggestions for user interface design.

DOI: 10.4018/978-1-60566-764-5.ch003

Figure 1. Definition of user interface (Plass, 1998)

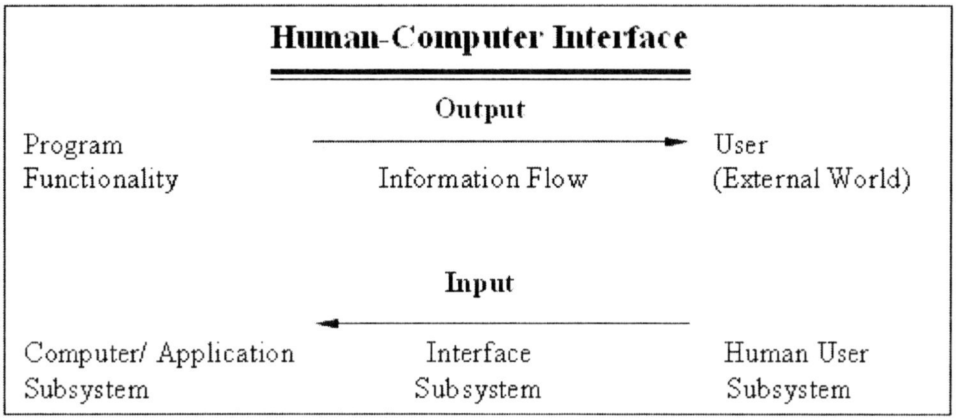

USER INTERFACE DESIGN APPROACHES

A discipline that is concerned with the design, evaluation and implementation of interactive computing systems for human is termed as human computer interaction (HCI). The user interface is an interface developed to enable HCI. Using the definition of the user interface as a communication channel between the user and the functional elements of the computer (Furnes and Barfield, 1995; Marchionini, 1991 and Waterworth, 1992, Plass, 1998), human-computer interaction can be seen as a system with three components: a computer/application, an interface, and a human user subsystem as shown in Figure 1.

According to this model (Figure 1) the interface subsystem serves two functions, firstly allocates user input to internal representations of the application and secondly internal representations of the application to output that is understandable to the user. The kind of input and output modes employed by the interface subsystem determines the type of the interface. For example a text based system uses only written verbal communication mode, whereas a direct manipulation system allows the user to manipulate objects and use visual, verbal and auditory representations of the systems state (Plass *et al.*, 1998).

In general, although most components of technology assisted problem-solving (TAPS) packages can be considered part of the user-interface in one way or another, it is useful to distinguish those aspects that are clearly concerned with communication between the package and the user. This includes the actual presentation of the information, as well as the acceptance of user-input. Because the general standard communication with the student is primarily done through the natural/high level computer language i.e. English, the package must have the ability to understand the student's response and generate text of its own. There is a strong argument for user interfaces that include not only text, but also visually exciting graphics and the ability for the student to interact with this graphical environment (Swaine, 1992).

Earlier software packages were command-driven or known as text-based command-line interface. For example if the user wanted to copy a file, the command "copy" had to be typed by the user followed by pressing the enter key on the keyboard to process the task. On a Unix platform the command "dw" has to be typed to delete a word. In some cases, this structure can influence the design of the user interface because it contains design features that are influenced by the internal structure of the computer. The user has to learn certain syntax of commands, parameters, and options that are closer to the machine code of a microprocessor than to a high computer language. Such commands had to be remembered by the users of the software. In addition, since communication between user and computer is purely textual, applications are programmed to include the use of function keys, single characters, short abbreviations, whole words, or combination of multiple commands. An issue with command interface is the number of keystrokes required to complete the command. Since such commands must be remembered, special care must be taken in choosing the commands for the application (Kinshuk, 1996).

Interactive learning packages can be difficult to be implemented because the order in which learning is done is often inflexible, it is difficult to recover from mistakes, and each learning package has its own interaction conventions. Although many factors contribute to these difficulties, it can be assumed that the essence of the problem is not the user interface as viewed over a single interaction, but rather the lack of support for the user's problem solving process especially over extended periods of time.

Plesis *et al.* (1995) pointed out that the interface for a CAL program should be sensitive to the age of the students, cultural backgrounds, computer literacy level and other related factors. In general, good interface design can be categorized by four principles i.e. clarity, consistency, common sense and comfort. These principles are to be applied in four key aspects of human computer interface: environment, appearance, support and interaction. These four aspects make it possible for the system developer and users to work together towards making systems easy to use (Plesis *et al.,* 1995, Kinshuk, 1996).

Although, there is no clear-cut classification of interfaces available in the literature (Kinshuk, 1996), a four-way categorization of styles or modes of interaction was initiated by Hutchins (1987) i.e. conversational interfaces, declaration interfaces, model-world interfaces and collaborative manipulation interfaces.

Conversational interfaces are those characterized most closely by command-language styles of interaction. They are based on a 'conversational metaphor' whereby interaction with the system is via some intermediary language. Input in such interfaces need not necessarily be via typed input. Menu driven systems can also be conversational in style. Traditional approaches to CAL packages can embody conversational style of interaction in the computer-as-coach paradigm (this includes directive teaching as well as coaching method). Metaphors are fundamental concepts, terms and images through which information is easily recognized, understood, and remembered. Metaphors include the essential means by which choices for command/control are communicated and the status of all data and functions are depicted (Kinshuk, 1996).

According to Kinshuk (1996) declaration type of interfaces are more direct than conversational interfaces and employ certain aspect of command languages or menu designs. Examples include the use of abbreviations for certain Disk Operating System (DOS) commands, such as 'cd' for 'change directory'.

Model-world interfaces are graphical environments that allow users to engage directly with a visualized world of objects, properties and relationships that model a given task domain. Model-world interfaces do not use any kind of language that, by virtue of flexibility of expression, entails the possibility of making errors. They contain a model world which constraints the set of possible actions so as to make an error almost impossible (Hutchins 1987, Kinshuk 1996).

The best features of declaration and model-world interactions are captured by collaborative manipulation interfaces. This involves giving the system access to the virtual world in which the user operates, to enable higher-level discussion of the domain. The 'display controller' display components, which the operator may then reject, indicating that they are not relevant in the current context. The display controller learns about the operator's preferences in the same way it has learned about the system's behavior, by observation. The display is thus a shared world of action, which can be collaboratively manipulated by the user and the display controller. The collaborative manipulation style out of the four styles of interaction described by Hutchins (1987) is probably most suited to the goals of guided learning systems as suggested by O'Mally (1987).

Another mode of interaction known is direct manipulation that is enabled by the provision of dragging and picking operations for all graphical objects (Plass, 1998). In a direct manipulation interface the user can manipulate physical actions of icons representing the object of the task domain directly by using a pointing device such as the mouse whose results appear immediately on the screen. Pangalos (1993) found that direct manipulation is attractive to the users because it is comprehensible, natural and fast.

PROBLEMS IN TRADITIONAL USER INTERFACE DESIGN

User interface design principles have been present in the computer science literature and industry for almost three decades. Some user interface design practitioners have provided reasons why there have been obstacles in traditional interface design practices in developing good user interfaces (Landseadel, 1995, Aldrich *et al.*, 1998). Kinshuk (1996) cited some of the reasons as follows:

- *Development strategies*

The tendency to first develop the functionality and only then map a user interface on it was the traditional development approach. The main focus was targeted not on better quality of interface but on the budget and schedule performance. No usability testing was done until later, when it was already too late for any changes and expensive to fix. A couple of problems with this design approach are related to usability. Firstly, a user interface, even a graphical user interface (GUI), does not ensure usability if the design is not suited to the user or to the natural flow of the task, using windows, icons, touch screens, or any other similar devices. Secondly, the developer may not foresee all of the usability issues, as the developer himself is not a user.

- *Flaws of software reuse*

Using an old code into a new system may give rise to potential usability issues. For example, the error handling software may have been implemented differently, resulting in a mismatch in the functionality. Benefits of proper software reuse maybe many, but they could be carefully weighted against any drawbacks associated with inconsistency.

- *Organizational barriers*

The user interface is often designed using the institutions' programmers and human computer specialists are not included by many organizations. Even in organizations that include human factor specialists, their analysis is often applied to the design after implementation is well in progress and too late for strategic changes. Therefore, the design must be based on the need and interest of the users in order to create a successful design.

DESIGNING THE USER INTERFACE

According to Kinshuk (1996), computing systems are also becoming increasingly interactive. As they do so, the amounts of code written for input and output (i.e. interface) have risen. A variety of technologies and tools are being developed to create and improve user interfaces. The arrival of interactive systems for developing user interfaces seems to be one of the most promising current developments in the area of user interface development of software (Kinshuk, 1996). Various methodologies and design principles are also being established to develop user interfaces. In addition choices about user interface design options can be assisted by reference to standards and guidelines. The following are a few design guidelines and methods for user interfaces as outlined by Pangalos (1993):

- *User interface consistency*

The three different aspects of consistency such as semantic consistency, syntactic consistency and physical consistency are important to the user across all applications in the user interface area. Semantic consistency refers to the meaning of the elements that make up the interface, such as the result of invoking a particular command. Syntactic consistency refers to the words used for the commands and the sequence of the appearance of the elements comprising the interface, such as the particular word used to invoke a function. Lastly, physical consistency refers to the hardware and how it is used, e.g. which key is pressed for a particular command.

A consistency has to be maintained on all aspects of the objects that are visible to the user and the actions that can be performed on these objects in order to achieve semantic consistency. The same terms should be used to express the same meaning in order to achieve syntactic consistency, which allows users to develop strategies in common terms that can be directly applied to each application. Without syntactic consistency, users may develop strategies using common concepts but they will usually have to use different commands for similar strategies and for different applications. Physical consistency is also important in simplifying the use of multiple applications. The sequence of interactions should be used to request the same functions in order to achieve physical consistency.

- *Direct manipulation*

The provision of dragging and picking operations for all graphical objects enables direct manipulation. Physical actions included as button presses or mouse actions whose results appear immediately on the screen are the techniques used in direct manipulation. The basic idea is to avoid syntactic forms of command languages, even the simpler ones, and to acquire only simple training. The results of the actions are obvious and are easily reserved thereby making the error messages unnecessary in many cases.

- *Rapid prototyping*

An application for the development of a user interface should pass through several iterations of prototyping, testing, evaluation and redesign before achieving the quality and usability requirement. A tool that assists the design and construction of user interfaces should therefore provide methods that allow changes to take place quickly by shortening the time required for each phase of the interaction.

Table 1. Approaches to interface design (Wallace and Anderson, 1993)

	Craft approach	Enhanced software engineering	Cognitive approach	Technologist approach
Philosophy	Craft-oriented Design through skill experience	Incorporate HCI into software engineering	Apply the psychological knowledge base to achieve optimal design	Quantify and automate the design process
Character	Monolithic evolutionary	Structured transformation	Structured transformation	Black box generation
Focus	Specification design	Specification	Specification Evaluation	Implementation
Role of practitioner	Craftsman/artist Multi-disciplinary collaboration	Traditional analyst broadening the scope of software engineering	Psychologist ergonomist human-factors specialists	Software tools developer
Tools	Brainstorming prototyping user evaluation	Software engineering methods CASE tools	Task analysis methods GOMS, CCT, KAT	UIMS mathematics formal grammars

- *Semantic feedback*

The feedback that depends on information given by the application is semantic feedback. In the applications using semantic feedback, the flow of dialogue is usually controlled by the user interface, limiting the way in which the application can intervene to provide feedback. The decisions concerning output and feedback cannot be directly affected since the application is only called in response to a user action.

Many of the aforementioned guidelines were developed when full screen, character-based systems were the norm (Kinshuk, 1996). Even though, such guidelines embody good design principles and are appropriate for graphical user interfaces they may not be suitable for representation of newer technology for example virtual reality environments.

Wallace and Anderson (1993) suggested four identifiable approaches to user interface design in a theory-based review of the user interface design practice. The first is the *craft approach*; where each design project is viewed as unique solutions evolve under the guidance of a skilled human factors expert to suit the circumstances. The objective of the design is to find the most appropriate features, based mainly on practical and economical considerations.

The second approach is to *enhance software engineering*; in this attempts are made to introduce human computer interaction (HCI) techniques into the range of traditional systems engineering which incorporate human factors such as user characteristics and task analysis for example the waterfall model. However, the main focus is on usability and the desire to serve the user effectively (Shneiderman, 1993).

The third approach that is the most theoretical approach to interface design is called *cognitive engineering*. According to Barnard (1991), this approach attempts to apply cognitive psychology to the problems facing designers to facilitate best design such as theories of information processing and problem solving to interface design. This approach is characterized by an attempt to measure the user's performance time and memory load for a given task, to identify prerequisite and acquired knowledge for a task, and to describe the user's mental models and mental processes for performing a task (Kinshuk, 1996).

Lastly, the *technologist approach*; this approach focuses on quantifying and automating the design process. The design process is based on user interface management systems and the idea that good interfaces can be extracted from the user (Wallace and Anderson, 1993). Each of these approaches is briefly summarized in Table 1.

The aim of each of the user interface design approaches should be to improve the quality of both the completed design and the design process according to Wallace and Anderson (1993).

EFFECTIVE USER INTERFACES - GRAPHICAL USER INTERFACE

In general graphical applications are becoming more popular nowadays in human interaction intensive areas. Marchisio *et al.* (1993) stated that graphical interfaces were known to motivate user creativity and increase user productivity. These interfaces are successful because of simple-to-use input devices, icons, windows, buttons and menus. Therefore it makes sense to move away from hard-to-remember, command-line-only interfaces to these, more attractive interfaces, which have dramatically widened the explicit communication channel between the user and computer (Kinshuk, 1996).

According to Kinshuk (1996), graphical user interfaces (GUI) provide natural and easy means for users to communicate with computers. However, Kinshuk stressed that the construction of GUI software requires complex programming that is far from being natural. Because of the 'semantic gap' between the textual application program and its graphical interface, the programmers themselves must theoretically maintain the correspondence between the textual programming and the graphical image of the resulting interface. For example, Miyashita *et al.* (1992) proposed a programming language environment by the *programming by visual example* (PBVE) scheme, which allows the GUI designers to 'program' visual interfaces for their applications by 'drawing' the example visualization of application data with a direct manipulation interface.

Various software tools have emerged in the last few years that assist in the development of graphical user interfaces. According to Kinshuk, (1996) most of these tools, however, concentrate on the implementation of GUIs rather than on their design. In addition, some of the tools, which assist in the GUI design process, do not encourage the practice of participatory design, which attempts to make interfaces more usable by involving users in the design process (Kinshuk, 1996). Kinshuk (1996) stated a few examples namely NeXT Interface Builder (Myres, 1992) and WindowsMAKER (BlueSky, 1991). Miller *et al.* (1992) developed TelePICTIVE, which is an experimental object-oriented software prototype designed to allow naive as well as expert users to work together in designing GUIs. TelePICTIRE is designed to support designers with diverse expertise in the collaborative design of GUIs.

SUGGESTIONS FOR BETTER USER INTERFACE DESIGN

Iannella (1992) suggested some guidelines for better user interface design, which are more related to the screen design rather than design methodology.

Screen layout:

- Functional areas (titles, help text, buttons) should be agreed on and consistently followed throughout the screen displays.
- Try not to include too much information on the screen, instead use lines and boxes and group similar information components and provide an aesthetically pleasing layout.
- Early prototyping with screen layout on paper and experimenting with different layout configurations will lead to a consistent and functional design.

- Keep the designs simple, clear, distinctive, with emphasis on critical information.
- Ensure that any textual information to be displayed on the screen is legible by using appropriate typestyles and typesets. For high-resolution graphical displays, Helvetica, Times, Garamond, and Courier are appropriate fonts.
- Never use more than three types of different fonts and point sizes on the same screen.
- A mixture of upper and lowercase characters is also more legible than all uppercase characters.
- Liberal use of white/blank space is an effective screen layout tool.

Windows:

- Provide a simple way for the user to navigate in a multi window system. For example provide a menu (or list) of currently open windows, which allows user to bring any of the windows to the forefront layer.

Menus:

- Menu item that represents some option (e.g. font size), then some feedback is required to indicate to the user which option is currently selected, for example by displaying a check mark (like ✓ or ◆) next to the items name in the menu list.
- Keyboard shortcuts (accelerators) should be provided on most (if not all) of the menu selections. This is useful for experienced users who prefer to use the keyboard instead of the mouse for menu selections. Again, display the keyboard accelerators sequence next to the item name in the menu list.
- If a menu option is unavailable, then provide feedback to the users as to its state by disabling the menu option (for example, graying out the words). Do not remove the option from the menu, as this will be inconsistent for the user.
- Menu options that performs some drastic change to the system, should ask the user for approval first, e.g. if the user selects the function 'Quit' and has not saved the document.
- Hierarchical menus (submenus) are an efficient way to offer options to each menu item. It is essential to provide a clear indication that there is a submenu available to each menu item.

List Selection:

- If the system expects an infinite number of options, then provide both options to the user. For example if the user is asked to select how many degrees a figure should be rotated (a popup menu of the numbers 1 – 360 could be a bit excessive) a text field coupled with a menu showing choices of angles could be provided.

Icons:

- Icons give visual clues to the user as to the function they perform, which should relate to a real world task. For example a pencil icon in a graphic program should allow the user to freehand draw on the application window.

Feedback:

- If some process takes longer than 2 seconds, provide feedback to the user including which task is currently being performed and some indication of the percentage complete. For example a message appearing that states "Copying file – 85% completed". In more graphical terms some type of moving bar could be included to show the percentage complete.

Help:

- Online help should be context sensitive. For example if a user is unfamiliar with certain terms say *momentum* used in the system, then display the help information with explanations of the terms in a glossary of commands table.

Errors:

- Error messages should be constructive and avoid negative phrasing and obscure codes and should offer the user some constructive help in suggesting a remedy for the situation.

Colour:

- Proper use of colour can be an effective tool to improve the usefulness of the user interface design. Related items on screen with similar colour backgrounds should be grouped together. In addition strong contrasting colours to focus attention on critical information and bright colours for danger signals and attention getters should be used.

Input devices:

- If the system is designed to use the mouse as an input device, say a multi function mouse, then it should follow some real world consistency. For example the left most button can be used for selection of shapes (draw) and the rightmost button for dragging.

Ambler (1998) proposed a list of user interface design tips and techniques. Some of the design tips and techniques were found to be suitable for the packages implemented in this study for example, making the user interface work consistently by placing buttons in consistent places on all the application windows, using the same text in labels and messages and consistent color scheme throughout. Diaz (1999) stated that it is important to maintain consistency when learning so that students' are not distracted, for example too many different fonts, sizes, and colors should be avoided. Additionally colors and backgrounds used should be consistent and not unnecessarily changed because these may draw away attention from the intended message (Diaz, 1999). A framework for developing multimedia systems by Carter (2002) suggests that when using multimedia for engineering education, each possible task has to be recognized such as users and chunks of content that might be involved.

Deubel (2003) provided a list of guidelines for designing interactive multimedia instructional design based on behaviorist and cognitive approaches, some key points suggested in her investigation were considered in improving the packages developed in this study. User-centered design (ISO 13497, 1999) expects that a system should meet the needs and characteristics of members of each unique group of users for example group of information providers (often the engineering educators) in addition to the more traditional groups of users who act as information consumers (who are generally the engineering students) (ISO 14915-1, 2002). ISO 14915-2, (2002) suggests that the design of a multimedia system involves identifying various navigation structures that allow the contents of different presentation segments to be used to meet the needs of different learning and exploration tasks. After identifying the presentation segments, they can be further designed by identifying suitable media objects to use to implement them (ISO 14915-3, 2002).

INTERFACE DESIGN FOR LEARNING ENVIRONMENTS

For a learning environment, whether it is conventional CAL or other forms of computer based learning aids, the interface is the way that the learner has access to the functionality of the package. According to Kinshuk (1996), the underlying functionality of the package (not the particular implementation, but the actual functional behavior) cannot be completely separated from the user interface. The author further added that functionality 'leaks through' into interaction. The development of the learner's model of the system and the domain being taught are shaped by what the learner-as-user can see, hear and do via the interface (Kinshuk, 1996). To exploit full functionality of the interface, interaction in a coaching package, Chen (1995) proposed a model of interface, where a portion of user interface penetrates into package components and be distributed into the entire coaching package. Apple Computer (Hefley, 1995) has explored the use of anthropomorphic (applying human qualities to inanimate objects) agents as a part of interface to lead the user through an application.

O'Malley (1990) argued that careful attention must be paid to three types of representations at the interface. The first is the representations of the semantics of the domain to be learned. The second is the representation of the system's functionality. This refers to the actual operations the system performs in representing the domain. It is treated as a separate type of representation, since it may not map completely onto the domain semantics. The third type of representation concerns what the user is expected to do by interacting with the system. This can be regarded as the 'ability' the system offers for interaction (Kinshuk, 1996).

O'Malley (1990) further noted that the interface is also the only way in which the coaching component has access to the activities of the learner. The implication of this is that the coaching or guidance system must have some representation of the learner and the machine interaction in order to model the development of the learner's understanding. This applies to as much to active diagnosis (assessment) as it does to passive diagnosis (modeling behavior). This issue of what needs to be represented even for a sufficiently accurate model (as opposed to an ideal model) is difficult since the learner interface machine is not bounded by the computer screen alone but includes the physical and social context in which the learner machine system is embedded (Kinshuk, 1996).

ROLE OF THE INTERFACE IN EDUCATIONAL ENVIRONMENTS

This section discusses the role of the interface in educational environments.

- *Interface as knowledge representation*

The design of the user interface to a piece of educational software is of vital importance for its educational effectiveness. This implies whether one is designing micro-worlds, simulations, intelligent tutoring systems or any other form of computer-based learning packages.

The two sources of knowledge representation in coaching systems were distinguished by O'Malley (1990) namely as an internal representation of the domain, and an external representation at the interface. The interface does more than map meaning onto a set of symbols whereas the external representation competes with the internal representation of the source of domain knowledge. The external representation can be the driving force of design of a coaching system and does not always play the supplementary role traditionally associated with user interfaces. For example, the design of *Guidon* (Clancey, 1983) took the traditional path from the construction of an internal model (domain representation), to augmenting the system with visual displays in Guidon-Watch (Richer and Clancey, 1985), whereas the design of Steamer (Hollan *et al.*, 1984) went from an external representation (using qualitative process theory), which was specifically designed to incorporate content of the visual display into the internal representation (Kinshuk, 1996).

The learning environments, according to O'Malley (1990), need to be designed 'outside-in' rather that 'inside-out' in order to ensure that the resulting design is usable to the extend that the interface 'disappears' for the users so that they concentrate on the domain itself. Another reason quoted by Kinshuk (1996) was to ensure that the representation by the coaching system of the learner-as-user is sufficiently accurate.

- *Making the Interface transparent*

If people are using computers to learn about something, their task should not be made more difficult by having to struggle with learning and using the computer system. Educational interfaces should be easily learnable and usable, but the argument is a little more subtle than this. The attention of the learner is drawn away from the domain being taught, if the learner has to concentrate on using the interface.

REPRESENTATION OF THE STUDENT IN THE SYSTEM

According to Kinshuk (1996) the system needs to model the processes whereby the students learn from interacting with the system so as to provide guidance in discovery learning environments. Even if the theories of learning can be developed that are detailed enough for student assessment modeling, they could not be applied to specific situations and adapted for particular learners, without being attached to domain semantics. Therefore, it is important to understand how learning takes place via interaction with the system. Kinshuk (1996) cited the following statements to be useful in representing the system to the user:

- *Representing domain semantics*

It is necessary to find out what student already knows in order to teach something new. One approach is to represent the domain to be taught in the form of simulation. An example is *Steamer* (Hollan *et al.*, 1984).

There are two main advantages of using metaphors in the interface firstly; the sensible use of metaphors allows users to draw analogies with their own experience, which enhances the learnability of the system because the concepts and techniques necessary for using the system are already familiar. Secondly, using a consistent and uniform metaphor at the interface enables users to make correct predictions about behavior of the system.

- *Representing system functionality*

The choice of appropriate metaphors to represent domain semantics (whether those domains are concrete or abstract) and the way in which the interface represents those metaphors are two separate issues. The constraints of software and hardware technology often require an extra translation step though ideally, there should be a direct mapping between the two. For example, file deletion concept can be represented by using the metaphor of throwing a document in the wastebasket, as in Windows 2000 interface.

Usually graphical or iconic means are used in the design of metaphors such as the 'desktop'. Metaphors can be conveyed by command-based systems that use textual interaction. Moreover, the visual modality is not the only source for presenting the metaphor. Auditory icons, such as the sound of glass shattering to convey deletion of a file, can also be powerful sources of metaphors.

Icons do not simply represent domain semantics; they also represent to a greater or lesser degree the functionality of the underlying system. For example an icon representing a folder in the Windows 2000 interface also suggests (via the user's understanding of the semantics being represented, which may be more or less successful) those items such as documents can be placed in or removed from the folder.

Successful icon representation depends on the types of representation used in mapping from the symbol to referent, such as whether it is an abstract or metaphorical relationship. Interpretation may also depend on mappings between icons since the icons that are used to make up the metaphors such as the desktop involve sets of symbols that are related in some way. The directness of the visual link between an icon and its referent, the extend to which the references can be differentiated within a set of icons, and the degree of capability between the relations of graphics elements and the common characteristics of the referents are the three interacting factors which contribute towards effective design of iconic interfaces as suggested by Rogers (1988).

REPRESENTATION OF THE USER IN THE SYSTEM

This section discusses the representation of the user in the system:

- *Defining the boundaries of the interface*

Kinshuk (1996) stated that in the HCI literature different people have viewed *user-computer interface* in different ways. For some, it is restricted to the visual display (i.e. what the user sees), or it includes the hardware (e.g. input devices), and for others it includes the social context in which the computer is being used.

The user computer interface is probably best considered as a distributed system. It is partly represented in the mind of the user i.e. during the course of the interaction that may or may not agree with what the designer had in mind. It is partly represented in the functionality of the software, partly in the documentation accompanying the system, and so on (Kinshuk, 1996).

- *Implications for assessment of student*

Use of uniform and consistent metaphor in designing interfaces can help in learning and using the system because they help provide constraints within which the user can predict the effects of the actions. However, the ability of a guidance system to collaborate with the user about the interaction presupposes that the system has some representation of the interface with which to reason and collaborate (Kinshuk, 1996).

In this respect, discovery-learning environments pose several problems for student assessment. Under the 'model-world' metaphor, although the learner's interaction with the system is constrained to prevent the errors, it is still flexible if compared to traditional CAL. Although simple and self-contained micro-worlds may be fairly easy to design, designing the functionality of a much more general and powerful environment is not an easy task. These issues are sharply focused when the aim is to provide adaptive learning environments with some sort of guidance built into the system even though they are not specific to the learning environments (Kinshuk, 1996).

The data-driven nature of interaction with graphical and direct manipulation interfaces implies that the system should be able to reason not only about its own state at any moment, but also about how that state appears to the user. In some sense, a representation of what's happening on the screen at any moment, as well as what's happening in terms of the domain being represented are the requirements of the system (Kinshuk, 1996).

STUDENTS PREFERENCES ON INTERFACES IN LEARNING ENVIRONMENTS

According to Kinshuk (1996) the most important consideration in interface design is how well it simplifies access to the program, how 'natural', or 'intuitive' it makes computer use. 'Ease-of-use' was not such a concern, when primarily engineers and technicians used computers. The user interface must easily allow novice, non-technical, users a hassle free interaction with the system if computers are to be used in higher learning institutions.

An evaluation study of Graphical User Interface vs. Command Line Interface on 102 undergraduate students enrolled in multiple sections of a Technical Writing course offered at the University of Southern Carolina, USA was conducted by Hazari and Reaves (1994). Students working on the command line interface used DOS based Microsoft Word 3.0 word processing software while students working on graphical user interface used Microsoft Word 4.0 on the Macintosh platform. These authors found that students preferred Graphical User Interface (GUI). Hazari and Reaves concluded that the GUI has greater potential for enhancing teaching and learning, not because of its short term effect on writing quality, but because it helps students use the technology easily and gives them more time to focus on learning. GUI users do not have to memorize complicated keystroke commands since the mouse input is used to automate routine tasks. The users do not have to go through a new learning curve while learning different application programs because strict software development standards force programmers to use a consistent design for dialogue boxes, buttons, tool palettes, and menu structures, (Kinshuk, 1996).

In summary this Chapter discussed the design of user interface, problems associated with traditional interface design methods, various design principles and recommendations. Interface requirements for learning environments were also discussed, and various features affecting the learning environment interfaces have been identified.

As the Windows environment was used to develop the TAPS packages, iconic metaphor used by Windows is considered better than command language based "conversational metaphor". Since the packages are to be used by students majoring in engineering courses, pictorial representation, text, animations, 2-D graphics and 3-D geometric models were used to represent information.

New technologies such as multimedia and desktop virtual reality (DVR) were the main technologies considered while developing the interfaces of the TAPS. However it was found that the immersive virtual reality technology is at its development phase and requires high specification hardware and software, which is not available in the student laboratories. Therefore, it was envisaged that while there is a possibility in future for inclusion of these technologies in the hardware components, in current situations, it is not feasible to use high-memory demanding real-time graphics, in introductory TAPS components.

However, while there exists a number of different approaches of user interface design, only a few of them focus primarily on the learning process and the user. The existing approaches could be practical but are not firmly rooted in the theory of learning, or too complicated to be useful for practitioners of interface design. Moreover, no approach has been found in the literatures that is specific to learning packages and the instructional strategies and methods that are relevant to this field. It is believed that by introducing appropriate methods and design principles, more researched guidelines and standards could help in the design of better user interfaces i.e. interfaces that are easier to learn and use.

REFERENCES

Aldrich, F., Rogers, Y., & Scalfe, M. (1998). Getting to grips with interactivity: Helping teachers assess the educational value of CD-ROMs. *British Journal of Educational Technology*, *29*(4), 321–332. doi:10.1111/1467-8535.00078

Ambler, S. W. (1998). *User Interface Design Tips, Techniques, and Principles*. Retrieved from http://www.ambysoft.com/essays/userInterfaceDesign.html#Principles

Barnard, P. (1991). Bridging between basic theories and the artifacts of human-computer interaction. In J. M. Carroll (Ed.), *Designing interaction: Psychology of the human-computer interface*, (pp. 103 –127).

Blue Sky Software Corporation. (1991). *WindowsMAKER*. Las Vegas, NV: Author.

Carter, J. (2002). A framework for the development of multimedia systems for use in engineering education. *Computers & Education Journal*, *39*, 111–128. doi:10.1016/S0360-1315(02)00029-5

Chen, J. W., & Chen, M. P. (1991). Toward the Design of an Intelligent Courseware Production System Using Software Engineering and Instructional Design Principles. *Journal of Educational Technology Systems*, *19*(1), 41–52.

Clancey, W. J. (1983). GUIDON. *Journal of Computer-Based Instruction*, *10*(1), 8–14.

Deborah, J. M. (1997). Managing the Design of the User Interface. In *Proceedings of CHI 97 Electronic Publications*. Retrieved May 25, 2003 from www.acm.org/sigs/sigchi/chi97/proceedings/tutorial/djm.html in 2002

Deubal, P. (2003). An investigation of behaviorist and cognitive approaches to instructional multimedia design. *Journal of Educational Multimedia and Hypermedia, 12*(1), 63–90.

Diaz, D. (1999). *The Pedagogies of Multimedia Presentation*. Cuesta Community College. San Luis Obispo, CA. Retrieved from http://www.ltseries.com/LTS/pdf_docs/mmpedago.pdf

Furnes, T. A., & Barfield, W. (Eds.). (1995). *Virtual environments and advanced interface design*: New York: Oxford University Press.

Hazari, S. I., & Reaves, R. R. (1994). Student preferences toward microcomputer user interfaces. *Computers and Education Journal, 22*(3), 225–229. doi:10.1016/0360-1315(94)90003-5

Hefley, W. (1995). Helping users help themselves. *IEEE Software*, 93–95. doi:10.1109/52.368272

Hollan, J. D., Hutchins, E. L., & Wetzman, L. (1984). STEAMER: An interactive inspectable simulation-based training system. *AL Magazine*, *5*(2), 15–27.

Hutchins, E. L. (1987). *Metaphors for interface design*. ICS report no. 8703, Institute of Cognitive Science, University of California, San Diego, California.

Iannella, R. (1992). Designing 'Safe' user interfaces. *Australian Computer-Journal, 24*(3), 92–97.

ISO 13497. (1999). *Human centered design process for interactive systems*. International Standard.

ISO 14915-1, (2002). *15-1. Software ergonomics for multimedia user interfaces – design principles and framework*. Final Draft International Standard.

ISO 14915-2, (2002). *15-2. Software ergonomics for multimedia user interfaces – control and navigation*. Final Draft International Standard.

ISO 14915-3, (2002). *15-3. Software ergonomics for multimedia user interfaces – media selection and combination*. Final Draft International Standard.

Kinshuk, R. (1996). *Computer Aided Learning for Entry accountancy students*. PhD. Thesis, De Montfort University, UK.

Landseadel, P. (1995). Methodologies in user interface design. *IEEE Aerospace and Electronic Systems Magazine, 10*(8), 15–20. doi:10.1109/62.406817

Marchionini, G. (1991). *Psychological dimensions of user-computer interfaces. Syracuse University*. (ERIC Document Reproduction Service No. ED 337 203).

Marchisio, L., Ronco, E., & Saacco, R. (1993). Modeling the user interface. *IEEE Communications Magazine, 31*(5), 68–74. doi:10.1109/35.212424

Miller, D. S. Smith, J. G. & Muller M. J. (1992). TelePICTIVE: Computer Supported Collaborative GUI design for designers with diverse expertise. *Proceedings of the ACM Symposium on User Interface Software and Technology*, Monterey, California, (pp. 151 –160).

Miyashita, K. Matsuoka, S. Takahashi, S. Yonezawa, A. & Kamada, T. (1992). Declarative programming of graphical interfaces by visual examples. In *Proceedings of the ACM Symposium on User Interface Software and Technology*, Monterey, CA, (pp. 107 – 116).

Myres, B. A. (1992). State of the Art in User Interface Software Tools. *Advances in Human-Computer Interaction, 4*, 1–31.

O'Malley, C. (1990). Interfaces issues for guided discovery learning environments. *Guided Discovery Tutoring* (pp. 5 – 18). London: Paul Chapman.

Pangalos, G. J. (1993). Designing the user interface. *Computers in Industry, 22*(2), 193–200. doi:10.1016/0166-3615(93)90066-A

Plass, J. L. (1998). Design and evaluation of the user interface of foreign language multimedia software: *A cognitive approach. Language Learning and Technology Journal, 2*(1), 40–53.

Plesis, J., Van Biljon, J. A., Tolmie, C. J., & Wollinger, T. (1995). A model for intelligent computer aided education systems. *Computers and Education Journal, 24*(2), 89–106. doi:10.1016/0360-1315(95)00001-3

Richer, M. H., & Clancey, W. J. (1985). GUIDON-WATCH: a graphical interface for viewing a knowledge-based system. *IEEE Computer Graphics and Applications, 5*(11), 51–64. doi:10.1109/MCG.1985.276331

Rogers, Y. (1988). *Pictorial representations of abstract concepts in relation to human computer interaction.* PhD. Thesis.

Shneiderman, B. (1993). *Designing the user interface: strategies for effective human-computer interaction* (2nd ed.). Reading, MA: Addison-Wesley.

Swaine, M. (1992). Multimedia and the art of user interface design. *Dr. Dobbs' . Journal, 17*(7), 113–117.

Wallace, M. D., & Anderson, T. J. (1993). Approaches to interface design. *Interacting with Computers, 5*(3), 259–278. doi:10.1016/0953-5438(93)90010-Q

Waterworth, J. A. (1992). *Multimedia interaction with computers: Human factors issues*. Chichester, UK: Ellis Horwood.

Chapter 4
Computer Aided Learning and Multimedia

DEFINITION OF CAL

In general, Computer Aided Learning (CAL) is the term given to software applications in which a computer is used to partially replace the function of a human instructor in the education or training of a learner/student. CAL is not limited to a particular field in education or constraint to a specific subject matter. The primary goal of CAL is to convey pre-defined theory/concepts to student so as to allow him/her to understand and apply gained knowledge at work place.

A CAL application offers the student a structured method of obtaining information as well as using the computer as navigational and information retrieval medium. CAL can therefore be thought of as a front end to a large information database. Early CAL packages offered the user information in the form of pages of text only. As technology evolved, applications started to present information using a wide range of media formats, including high-resolution graphics, narration, and even interactive video (Marshall, 1988). Hence the combination of one or more electronic media is subsequently known as multimedia.

Present CAL applications offer numerous advantages. Most importantly, CAL facilitates the implementation of effective training packages that can be made available to anyone who requires it without imposing any time constrain in learning. In addition, the CAL tutoring packages do not rely on the

DOI: 10.4018/978-1-60566-764-5.ch004

availability of skilled human instructor and is not influenced by the number of students requiring training (Dean and Whitlock, 1983). There are many benefits of employing CAL in the education sector as discussed in Section 4.

Although there are numerous benefits inherent in CAL, a major disadvantage with it is the way in which information is presented to the student. Conventional CAL packages present information at a pre-determined tutoring level and follow a set of structure. These packages do not take the student's basic knowledge or learning style into account and therefore lack the ability to adapt intelligently to the student's specific learning requirements (Vasandani *et al.*, 1989). The only form of student adaptation that is occasionally implemented is the pace at which the course material is presented (Sclechter, 1991).

Over the past few years, CAL packages have been designed to incorporate multimedia to allow learners to perform multi task simultaneously during a tutoring session. For example a learner can read text and be narrated by displaying a video clip to explain certain concepts of the subject matter. CAL in its simplest form does not cater for the individual student. Information is presented in a predetermined sequence, regardless of how knowledgeable the student is at the beginning of the learning activity, or how quickly or slowly the learner absorbs and understands the course material (Rickel, 1989). The incorporation of multimedia in CAL, on the other hand, provides the learner with the opportunity of exploring information in various media formats in addition to conventional text and graphics which focus on presenting information in a way that maximizes the student's learning process. In addition multimedia can be programmable i.e. gives the possibility of engaging the learner in activities, i.e. reacting, or responding, to selections made by learners (Cairncross, 2002).

IMPLEMENTATIONS OF INTERACTIVE MULTIMEDIA IN CAL

In general, the production of an effective interactive multimedia CAL package requires subject expertise, computing, authoring and modeling, and teaching skills. In the early days, developers had limited computing/programming skills and hence the resulting packages were often useful for teaching and testing factual knowledge but not good enough to promote learning and understanding. Attempts to overcome these shortcomings have been resource intensive and there is a need for a cost effective way of developing interactive multimedia CAL packages. Although there are a wide variety of interactive multimedia CAL packages in the market, there is considerable discussion and concern on the suitability of these packages in various education settings (Cairnscross, 2002).

Gardner (1990) argued that the most routine and simple tutorial (question and answer) or multiple choice style learning package can be effective in motivating students to study. The author inferred that it is going to be difficult to design a good piece of simulation or modeling package that challenges students to think and learn. Some researchers have developed learning packages and revealed its effectiveness; for example McAteer (1996) conducted a study on integrating two simulation packages on animal physiology in the University of Glasgow's Institute of Biomedical Sciences. The laboratory exercises aimed to complement the lectures by giving practical experience of scientific principles, to illustrate the techniques and procedures involved in practical aspects of physiology, to give hands on experience of investigative experimental work and to provide real data for handling, analysis and interpretation. From the observation of students engaged in their tasks across the laboratory room, McAteer noted that the simulation station was very much "one of the labs" rather than specifically "a computer assisted learning exercise". The author also realized that emphasis on learning and understanding the subject material was made very clear to most of the students when they were working through the simulation lab. The results showed that about 67% of the students favored the use of simulation package and that the students would use it for revision purposes.

In another study, Magin and Reizes (1998) tested students using different approaches to heat exchanger design and to examine the divergences between predicted and actual performance. The authors constructed a model-building package, which allows these issues to be investigated by simulation. They were concerned that their students would fail to gain a "real world perspective" when using the simulation package. Indeed, Magin and Reizes found that the students accepted the results blindly to the extent of accepting order of magnitude errors without questioning. This problem could have been avoided by linking the explorative powers of simulation with practical experiences on real situations.

Recent research includes *Intelligent 3-D Practice Environments* (Janet, 2002) which is a tutoring system that offers learning by doing and learning by discovery in a realistic practice situation, where the user will do problem solving on a set of sequential activities that are accomplished in a 3-D practice environment. The qualities of the *Intelligent 3-D Practice Environment* include the provision of a knowledge-based system that improves learning opportunities with dynamic advice and feedback. An expert system monitors user interactions to provide dynamic advice for the expected sequence of user activities and updates objects in sequence in a 3-D environment.

In engineering education, numerous studies have been carried out to implement CAL packages that incorporate multimedia and associated technologies. A list of such multimedia CAL packages reported in the literatures is presented in Table 1.

Although CAL has great potential in the education sector, it is not widely employed in classroom teaching. According to Hammond *et al.* (1992) the reasons for this include the poor quality of the available materials and the educational context in which the technology was used. The instructional goals embedded within a piece of learning package often failed to match the goals of the course it claimed to support. Yet, other reasons were concerned with the organizational and political issues, to certain extent financial constraints and whether by intent or default, all of which may give little support to those wishing to exploit innovative approaches (Kinshuk, 1996).

In spite of the aforementioned facts, interactive multimedia CAL has its own benefits and limitations over traditional teaching and learning (Gallagher and Letza, 1991, Cairncross, 2002).

BENEFITS OF INTERACTIVE MULTIMEDIA CAL

While much of the previous educational technology was designed to increase the power and effectiveness of lecture style presentations, more recent innovations have often been centered on the ability to promote self-directed and effective learning (Cairncross, 2002). Traditional techniques have gathered a number of students together in one place at one time, whereas interactive multimedia CAL approach to teaching has encouraged students to learn at their own pace, time and place. In general interactive multimedia CAL and its associated technologies offer various benefits to both the learner and the instructor as summarized below:

- Integration of computers in subject curriculum helps in preparing the students for their careers by familiarizing them with information technology and PCs (Shaoul, 1989).

Table 1. Multimedia CAL packages developed for engineering education

CAL Packages	*Educational Objectives*	*References*
Tutorial Shell (TS)	To help students to overcome their conceptual difficulties on understanding phase diagrams of engineering materials	Bailey *et al.* (1995)
GT-VIBS Multimedia Package	To teach engineering vibrations in a problem solving context through the use of simulation and visualization	Hmelo *et al.* (1995)
Computer Problem Class System (CPCS)	To improve student learning by providing tools for visualizing and solving problems in engineering dynamics subject	Scott (1996)
Computer Based Simulation (CBS)	To develop a substitute for laboratory experiment for the investigation of the operating characteristics of a centrifugal pump	Norrie (1996)
Multimedia Based Laboratory (MBL)	To prepare students for conducting their own experiments in the physical laboratory	Lynn *et al.* (1997)
Multimedia Tutorials Package (MTP)	To replace traditional lecture part of a course in engineering materials with studio exercises that could help students visualize 3-D evolving processes that cannot be presented effectively using static illustrations	McMahon (2000)
Interactive Educational Engineering Software (IEES)	To provide analytical, graphical and data management tools which allows the student to construct and control dynamic gas turbine simulations by manipulating graphical objects on the computer screen	John and Abdollah (1998)
Real Time Design (RTD) VRML Environment	To allow students to conduct design analysis of real world objects such as shafts, gears and pulleys in a 3-D environment	Karthik and Gramoll (1999)
Multimedia Based Environment (MBE)	To provide adequate MBE tutorials on engineering mechanics to motivate students in learning	Noronha *et al.* (2000)
Interactive Multimedia Intelligent Tutoring System (IMITS)	To permit individual tutoring of students, i.e. allow students to solve assignments in any manner the student chooses and is able to determine the student's ability to apply and understand basic concepts of engineering	Brian (2000)
Amoco Computer Simulation (ACS)	To innovate teaching of engineering design by developing problem solving skills	Mackenzie *et al.* (2001)
Computer Graphics Simulation (CGS)	To provide multimedia dynamic solid models for visualization	Wilson (2001)
Multimedia Mechanics of Materials Laboratory	To familiarize students with testing equipment, data acquisition, testing procedures and reporting results	Salvatore (2001)
Interactive Virtual Tutor (IVT)	To guide and provide helpful hints to students in an intelligent and interactive manner in solving problems in engineering mechanics statics	Gupta (2002)
Intelligent Practice Environment (IPE)	To allow the user to explore complex relationships through discovery and problem solving activities in a 3-D environment	Janet (2002)
Interactive Tutoring Components (ITC)	To help students understand the engineering mechanics dynamics subject with the development of multiple interactive tutoring components	Shang *et al.* (2005)
An Interactive Multimedia E-Learning System (IMELS)	To provide students with a comprehensive problem-based learning environment for the discipline of industrial engineering.	Lau *et al.* (2006)
Augmented Reality and Web 3D in Engineering Education	To allow students to interact with 3D web content using virtual and augmented reality. Learners of this interface can view multimedia content, to support a lecturers traditional delivery, either locally or over the Internet and in a table-top augmented reality environment.	Liarokapis *et al.* (2007)

- The use of multimedia and virtual reality technologies in CAL attracts learner regardless of age and gender, and captures the interest of unenthusiastic learners (Lambart, 1990).
- Learners are allowed to spend more time in areas with which they have difficulty. Students may concentrate on specific areas without holding up the rest of the class (Gallagher and Letza, 1991).
- CAL is flexible so the students can learn in their own free time without the need of constant human coaching. They can skip sections or topics of which they already have sufficient knowledge (Gallagher and Letza, 1991).
- CAL forces active participation of student. It works on one-to-one basis so that students must interact such as by answering questions before being allowed to proceed to the next stage. This is not the case in a classroom scenario (Gallagher and Letza, 1991).
- CAL ensures a more reliable course delivery and reduces the need to cope with different coaching styles and personality clashes between instructor and student (Gallagher and Letza, 1991).
- Because of the size of computer disks, more information can be stored virtually (soft copy) and the space required to store hard copies of text and books on shelves is not needed (Jensen and Sandlin, 1992).
- Small departments can benefit from CAL by reducing certain costs, e.g. simulations can replace expensive laboratory experiments and the need to have constant supervision of human experts (Gladwin *et al.,* 1992).
- CAL provides a suitable alternative for all teaching staff in dealing with the increasing number of students, and the higher pressure and workload imposed on the lecturers (Gladwin *et al.,* 1992; McDonough *et al.,* 1994; Darby, 1992).
- CAL allows the students to work at their own pace. Different students may have difficulty with different concepts and would require more time to understand. (McDonough *et al.* 1994).
- When CAL is used to replace traditional teaching, its cost is justified due to high student usage and reusability in various classes (McDonough *et al.,* 1994).
- The pedagogical strength of multimedia is that it uses the natural information processing abilities that humans already possess (Fred, 1995).
- In comparison to learning from a book or in a classroom setting, CAL can offer the student a more active role. On the emotional side, multimedia presentations can motivate and challenge students who like technological innovations and design (Guttormsen *et al.,* 2000).
- Computer simulations are relatively inexpensive compared with the cost of building and maintaining expensive experimental equipment (Mackenzie *et al.,* 2001).
- Interactive multimedia can be used to create an integrated learning environment, which combines explanations with illustrative examples. It can provide on-line assessment and feedback and opportunities to practice and experiment (Cairncross, 2002).
- 3-D CAL environments give the opportunity to imitate in virtual reality the experiments that are difficult to perform (Amon, 2002).
- Dynamic CAL multimedia models of a given problem often present challenges to students beyond what they have learnt in the traditional way (Liang, 2002).
- The availability of multimedia and virtual reality techniques with simulation initiates a new appearance for learning applications - real time presentations of 3-D data (Klett, 2002).
- Interactive multimedia CAL packages provide comprehensive coverage combining full-motion video, audio, 2-D and 3-D animated graphics, presentation slides, templates and text.

- CAL and multimedia simulations can provide a rich experience for the student. They can be a powerful resource for teaching i.e. providing access to environments, which may otherwise be dangerous or impractical due to size or time constraints; and facilitating visualization of dynamic or complex behavior (Thomas and Milligan, 2004).
- Multimedia technologies can provide a rich interactive learning environment that can be used in scheduled classes (Sieber *et al.*, 2004).
- With simulations and online materials, students are able to take a more active role in learning (Jesica and Tara, 2005).
- Animations and virtual simulations that require technological support provide the most recent and sophisticated implementations of multimedia learning (Antonietti and Giorgetti, 2006).
- Dynamic representations enable more efficient communication of complex concepts (Hennessy *et al.*, 2007).

LIMITATIONS OF INTERACTIVE MULTIMEDIA CAL

Despite the benefits mentioned earlier, several drawbacks of using interactive multimedia CAL can be summarized as follows:

- The effort and cost of developing an interactive multimedia CAL package may be about four times that required for instructor led courses (Wehr, 1988). The cost of developing CAL package not only involves salaries and training of personnel as well as purchase price of assisting software and hardware but may also include hidden costs such as student fees, distribution fees, annual maintenance fees and the commercial marketing of the packages (Collins, 1989).
- The computer may result in nothing more than a highly expensive electronic "page turner" if the students are faced with page after page of hard to digest text (Gallagher and Letza, 1991).
- The best CAL package cannot imitate the subtle of a well-trained, ambitious, conscientious and gifted instructor (Villiers *et al.*, 1992).
- The availability of hardware is restricted on the utility of CAL (McDonough *et al.*, 1994).
- Most of the CAL programs do not take into account of students' state of motivation and attention, which is not the case in human teaching situations (Dixon, 1998).
- People who have a verbal learning style may not be able to take full advantage of multimedia presentation. For many people, book has a value that cannot compete with computers. Books are generally easy to transport, can be personalized, offer comfortable reading when compared to a screen, and have texture (Guttormsen *et al.*, 2000).
- Self-contained or stand-alone (PC-based) multimedia CAL applications do not provide the opportunity for ongoing discussions (Cairncross, 2002).
- Multimedia can distract trainers and learners from objectives and content if it is just used solely for entertainment (Antonietti, 2006).
- Development for multimedia contents may require contracting out for specialized skills (Liarokapis *et al.*, 2007).

Although there are many authoring tools available nowadays to aid in the development of interactive multimedia CAL packages and some are easy to use without requiring good programming skills, these

Table 2. CAL systems (Guttormsen and Krueger, 2000)

Micro-world	**Micro-world uses the computer to create a problem-solving environment. They are closely related to artificial environments.**
Cognitive tools for learning	These tools are based on the constructivist principle that learners need to construct their own understanding of new concepts. Expert systems and authoring tools can be used this way.
Productivity tools	These tools include applications such as word-processors, spreadsheets, databases, graphics, desktop publishing and presentation packages. Whilst these tools are not specific to learning technology, if used within a pedagogical framework, they can support learning by enhancing the quality of the learning process and by improving student productivity.
Communications tools	These tools take several forms including electronic mail, electronic conferencing, video conferencing and World Wide Web. These tools allow learners to share ideas and information, to co-operate, to collaborate on joint work, submission of students' assignments and of tutors' comments on students' work.
Hypertext and Hypermedia	Hypertext organizes text as a network of nodes (pages, cards, and so on) connected by links (hyperlinks). The links enable unstructured navigation through the text. Hypermedia is a multimedia style of hypertext in which nodes may contain graphics, audio, video and other elements in addition to text.
Information	The computer presents the users with information on a specific topic.
Distance Learning	Students take part in a study program by accessing the teaching material via network technology.

packages require very high specifications in terms of hardware. Also, the level of flexibility achieved at the programmer level is not the same as compared with developing interactive multimedia CAL packages from scratch. Cairncross (2002) pointed out that if the value of multimedia is to be realized then guidance is required on how best to incorporate interactivity into the learning applications. It can be concluded from the arguments stated above (benefits and limitations section of interactive multimedia CAL packages) that interactive multimedia CAL packages seem to have high potential in the future but at present human tutors are indispensable.

CLASSIFICATIONS OF CAL IN THE CONTEXT OF LEARNING CONTENT

The developers of CAL apply different learning theories and techniques in CAL systems. In the literature (Moonen and Gastkemper, 1983; Alessi and Trollip, 1985), there are a number of descriptions of the form that CAL may take. A traditional approach is to divide CAL into three main streams; *drill and practice* (or exercise programs), *tutoring* systems (or instructional programs), and *simulation based* (or a real life or imaginary situation) (Kinshuk, 1996). Additionally Jong *et al.* (1992) distinguished three more areas; *problem solving* (specific learning goal), *testing* (examine the knowledge), and *databases/ information retrieval systems* (extracting information).

Stoner (1996), Guttormsen and Krueger (2000) classified other CAL systems as listed in Table 2.

Another classification of CAL systems comes from the level of control that can be exercised over the sequence taken by the learner through the program (Guttormsen and Krueger, 2000). At one extreme it is the program that fully decides the steps to be taken through the learning package. The other extreme would be that in, which the learner may choose (navigate to) any part of the section in the learning package at anytime.

COMPUTER BASED LEARNING

Computer-based learning (CBL) environments are examples of the use of computers to assist students explore and develop their own learning style (Kinshuk, 1996). CBL is different from any forms of CAL. Hsu *et al.* (1993) distinguished the uniqueness of CBL environments which allow students to work towards objectives that are not specifically set and paced by a computer program. Bentley (1991) stated that CBL environments enhance the true ability of computer and link it with a human approach, which focuses on the psychology of learning. Bentley explained that there are four basic features in which CBL environments outwit other forms of learning environments. These features can be summarized as below:

- The first is the speed in which the computer can respond to individual learner's need;
- The second is the way that the computer can offer, and respond to, a wide range of learner interaction;
- The third is the potential to represent information in a wide scope of formats from text to video;
- The fourth is the opportunity to provide unlimited choice of learning paths.

The abovementioned features of CBL environments can only be replicated by the most extensive development of the traditional approach – and even then the choice is limited to one or two carefully pre-defined paths (Kinshuk, 1996).

Guttormsen and Krueger (2000) recognized a new trend in learning systems called the hybrid system. In a hybrid system different learning systems are combined into a single system. The learning system then functions as the base program and uses information from web pages. Hypertext and Hypermedia can be integrated in a CBL system. A hypertext book can serve as an underlying information base. Simulations and other multimedia techniques can illustrate and enhance understanding of certain aspects of the material being presented.

Combining a simulation system and a tutoring system can alleviate the disadvantages of having each system operating by itself. Simulation based systems run the danger of giving the user too much freedom. To avoid the same a simulation alone can be backed up with tutoring information in some form, either by traditional class teaching or by integrating the simulation in a CBL system.

Tutoring systems and simulations support different knowledge structures. Both type of systems deliver the same knowledge, but in different ways. The differences are inherent in the design of the systems and the teaching philosophies behind them. Simulation based learning encourages natural knowledge representations from the perspective of experts, whereas tutoring systems encourages representations more suited to the novice. Thus, the combination of the two learning systems could deliver the benefits of both to the learner (Guttormsen and Krueger, 2000).

Bentley (1991) mentioned that to create truly effective computer-based learning environments the designers have to be learners themselves. In addition, the designers of computer-based effective learning environments need to have access to 'subject-matter experts', but if the designers themselves are ignorant of the subject area, it is very difficult for them to take the position of the learners.

When people know a subject well they tend to make broad assumptions about what others know, or need to know. They find it virtually impossible to take the point of view of a complete novice. They do, therefore, usually take the position of a person who is already advanced in learning the subject. This prevents them from developing software's of exploration and discovery (Bentley, 1991).

TAPS PACKAGES

While many software packages have been developed and used for the purpose of student learning in engineering, these packages do not provide the user adaptability in particular to students experiencing difficulties in studying Mechanical Engineering, i.e. students who normally need more time to understand a particular concept in engineering. As a result, the packages fail to provide adequate feedback, as they do not guide students to solve the engineering problem in a step-by-step approach (answers are given immediately without showing the solution after a question is posed). Additionally students of such packages do not know if they have applied the appropriate formulas to solve the problem (some may use wrong formulas or working approach) even though the answer given by them could be correct.

TAPS packages are developed to include multimedia features and simple intelligent functions such as alerting a student by displaying messages (hints) on screen if a wrong formula is applied or a wrong answer given in solving the selected engineering problem. However, if the user still cannot solve the problem, the student could approach the TAPS package by clicking on "solve" button to aid the student in solving the problem. The solution is given in a step-by-step manner showing how the answer is obtained. Further explanation about TAPS packages is given in Chapter 6 of this book. Additionally, desktop virtual reality features can also be incorporated to encourage students to interact and engage with the TAPS package (as discussed in section six and seven). These efforts have focused on conveying technical knowledge to the student solving the engineering problem in such a way so as to support the acquisition of theoretical knowledge.

THE NEED FOR DEVELOPING MULTIMEDIA COURSEWARE FOR ENGINEERING

Section 3 and 4 of this Chapter discussed the capabilities and limitations of multimedia as a multimedia presentation tool. Besides being a powerful tool for making presentations, multimedia and desktop virtual reality environment offer unique advantages in the field of engineering education. For instance, before conducting a complicated experiment, a learner can now rehearse on a virtual engineering problem, that is, by using a computer simulation. By wearing a special pair of glasses, which produce the impression of a three-dimensional object, the student is able to manipulate the objects parts as he/she would in a real situation. Engineers and technicians increasingly can enter into and interact with, an artificial, virtual reality environment that resembles the real world. In addition the computer can provide sensory impulses for the eyes and ears, creating a near perfect spatial object. As such, in the context of engineering users of multimedia courseware can interact, manipulate and visualize the objects better. Therefore, multimedia enables the courseware developer to provide a means for learners by which they could experience their subject in a vicarious manner.

Under conditions of chronic under-funding, multimedia can provide an enhanced or augmented (include both virtual reality and real world elements) learning experience at an economical cost. It is here that the usefulness of multimedia can be unleashed to provide long-term benefit to education. Furthermore, multimedia enables learning through exploration, discovery, and experience (Usha, 2003).

Multimedia can enhance the process of learning and can become more participatory, flexible in time and space, unaffected by distances and tailored to individual learning styles, increase collaboration between teachers and students.

The pedagogical potential of multimedia is that it utilizes the natural information processing abilities that humans already posses. The eyes and ears, in conjunction with the brain, form a formidable system for transforming meaningless sense data into information. The phrase "a picture is worth a thousand words" often understates the case especially with regard to animated images, as the eyes are highly adapted by evolution to detecting and interpreting animation. For example, a photograph of a crank shaft (part of an engine which translates reciprocating linear piston motion into rotation) apart from being aesthetically pleasing to engineering students, can contain a wealth of information relating to design, engine configuration, bearing, piston stroke and rotary engine parts. Therefore it is clear that multimedia can be used to discern significantly semantic features of the crank shaft not obvious in a static image or textbook.

For the student, one advantage of multimedia courseware over the text-based variety is that the application looks better. If the courseware includes only a few images at least it gives it relief from screens of text and stimulates the eye, even if the image has little pedagogical value. More often than not, the inclusion of non-textual media into courseware adds pedagogical value to the application. For example, a piece of courseware describing about curvilinear motion, would be valuable to the student if included images of particles (3D robotic arms, rods and collars), diagrams illustrating the motion of particles moving from point A to point B, narration explaining how the velocity and acceleration of the moving particles are computed. In this respect, using the text only, even in a creative way, has obvious limitations as compared to the use of both text and pictures.

In summary interactive multimedia CAL have great potential and have been widely used across a wide range of courses to promote learning. However, due to the different nature of each field of study, the degree to which computers can be used in teaching varies greatly. This Chapter discussed various aspects of implementation of computer aided learning in an academic engineering environment and reviewed the benefits and limitations of its implementation. Various representations of CAL according to the classifications of the learning content were also discussed with an overview of emerging technique of computer-based learning environments.

REFERENCES

Alessi, S. M., & Trollip, S. R. (1985). *Computer Based Instruction.* Englewood Cliffs, NJ: Prentice Hall.

Amon, T. (2002). Teaching biology in primary and secondary schools with the help of the dynamic HTML and web virtual reality (web3d) projects. *Interactive Educational Multimedia Journal, 4*, 89–98.

Antonietti, A., & Giorgetti, M. (2006). Teachers' beliefs about learning from multimedia. *Computers in Human Behavior Journal, 22*(1), 267–282. doi:10.1016/j.chb.2004.06.002

Bailey, J. D. Hall, J. L. Reed, P. A. S. & Colbourn, C. J. (1995). CBL in Engineering: Student's use of a Learning Resource on Phase Diagrams. *Computers Education Journal, 25*(1 – 2), 75 – 80.

Bently, M. (1991). *Intelligent Tutoring. Management Accounting*, (pp. 14 – 25).

Brain, P. (2000). The Learning of the Interactive Multimedia Intelligent Tutoring System (IMITS). In *Proceedings of the 30th ASEE/IEEE Frontiers in Education Conference*, (pp. 11 –16).

Cairncross, S. (2002). *Interactive Multimedia and Learning: Realizing the Benefits*. PhD Thesis, Napier University, Scotland.

Collins, A. M. (1989). Processes in acquiring knowledge. In *Schooling and the acquisition of knowledge* (pp. 339 – 363). Hillsdale, NJ: Lawrence Erlbaum Associates.

Darby, A. (1992). Computers in teaching and learning in UK higher education. *Computers and Education Journal, 19*, 1–5. doi:10.1016/0360-1315(92)90004-O

Dean, C., & Whitlock, Q. (1983). A Handbook of Computer Based Training (pp. 133 – 155). London: Kogan Page.

Dixon, M. (1998). Can you really learn from computers? *Training Officer*, (pp. 16 – 18).

Fred, R. (1995). *Understanding IT: Developing Multimedia Courseware. ITTI* – University of Hull.

Gallagher, I. D., & Letza, S. R. (1991). An Evaluation of computer aided accounting learning – a student's perspective. *Paper presented at the conference Reviewing Accounting Courseware*, Manchester University, pp. 78 – 95.

Gardner, J. (1990). Computers in higher education teaching and learning. In *Proceedings from the symposium held at the Queen's University of Belfast*, June 1990, (pp. 1.1 – 1.3).

Gladwin, R. P., Margerison, D., & Walker, S. M. (1992). Computer assisted learning in Chemistry. *Computers and Education Journal, 19*, 17–24. doi:10.1016/0360-1315(92)90006-Q

Gupta, A. (2002). Content Development for eLearning in Engineering Education. *Interactive Educational Multimedia Journal, 4*, 12–23.

Guttormsen, S., & Krueger, H. (2000). Using New Learning Technologies with Multimedia. *IEEE MultiMedia*, 40–51.

Hammond, N., Gardner, N., Health, S., Kibby, M., Mayes, T., & McAleese, R. (1992). Blocks to the effective use of information technology in higher education. *Computers and Education Journal, 18*(1-3), 155–162. doi:10.1016/0360-1315(92)90049-B

Hennessy, S., Wishart, J., Whitelock, D., Deaney, R., Brawn, R., & Velle, L. (2007). Pedagogical approaches for technology-integrated science teaching. *Computers and Education Journal, 48*(1), 137–152. doi:10.1016/j.compedu.2006.02.004

Hmelo, C. E., Lunken, E. Y., Gramoll, K., & Yusuf, I. (1995). Multimedia Courseware for Teaching Dynamic Concepts: Assessment of Student Learning. In *ASEE/IEEE Frontiers in Education '95*.

Hsu, J. J., Chapelle, C. A., & Thompson, A. D. (1993). Exploratory learning environments: What are they and do students explore? *Journal of Educational Computing Research, 9*(1), 1–15.

Janet, F. J. (2002). Sequence Learning with Intelligent 3-D Practice Environments. *MITRE Technical Papers*, Paper 2.

Jensen, R. E., & Sandlin, P. K. (1992). Why do it? Advantages of new waves of computer-aided teaching/instruction. *Journal of Accounting Education*, 39–60. doi:10.1016/0748-5751(92)90017-Y

Jessica, M., & Tara, M. (2005). Educational Applets for Active Learning in Properties of Electronic Materials. *IEEE Transactions on Education, 48*(1).

John, A. Reed, & Abdollah, A. (1998). Developing Interactive Educational Engineering Software for The World Wide Web With Java. *Computers Education Journal, 30*(3/4), 183–194.

Jong, T. D., Andel, J. V., Leiblum, M., & Mirande, M. (1992). Computer assisted learning in higher learning education in the Netherlands: A review of findings. *Computers and Education Journal, 19*(4), 381–386. doi:10.1016/0360-1315(92)90085-J

Karthik, R., & Gramoll, K. (1999). Design Education over the Internet using VRML. *In Proceedings of the Society Engineering Education,* (ASEE) 1999 Conf., Session 2625.

Kinshuk, R. (1996). *Computer Aided Learning for Entry accountancy students.* PhD Thesis, De Montfort University, UK.

Klett, K. (2002). Improving Engineering Student Learning in a Web-Based Learning Space Due to Virtual Reality Techniques and Advanced Interactivity. In *Proceedings of the 2002 eTEE Conference Switzerland,* (pp. 195 – 101).

Lambart, C. (1990). The electronic tutor. *Harvard Magazine, 93*(2), 42–51.

Lau, H. Y. K., Mak, K. L., & Helen, M. (2006). IMELS: An e-learning platform for industrial engineering. *Journal of Computer Applications in Engineering Education, 14*(1), 53–63. doi:10.1002/cae.20067

Liang, W. (2002). Using Multimedia in Teaching Dynamics. In *Proceedings of the 2002 Conference for Industry Education Collaboration, American Society for Engineering Education,* Session 4424.

Liarokapis, F., Mourkoussis, N., White, M., Darcy, J., Sifniotis, M., Petridis, P., et al. (2007). Augmented Reality and Web 3D to Support Engineering Education. *Proceedings of the Engineering and Technology Education Conference, Greece,* (pp.24 -28).

Lynn, E. K., Lenly, J., Weathers, W. M., Ronal, K., Kevin, A. & Mike, S. (1997). A Multimedia Based Laboratory Course for Environmental Engineering Fundamentals and Process Design. *ASEE/IEEE Frontiers in Education '97.*

Mackenzie, J. G., Earl, W. B., Allen, R. M., & Gilmour, I. A. (2001). Amoco Computer Simulation in Chemical Engineering Education. *Journal of Engineering Education., 90*(3), 331–343.

Magin, D., & Reizes, J. (1998). Computer Simulation of Laboratory Experiments: An Unrealized Potential. *Computers and Education Journal, 14*(3), 263–270. doi:10.1016/0360-1315(90)90009-V

Marshall, D. (1988). CAL / CBT – The Great debate. *Computer Based Training (CBT),* (pp. 218 – 234). Bromley, UK: Chartwell-Bratt.

McAteer, E., Neil, D., Barr, N., Brown, M., Drapper, S., & Henderson, F. (1996). Simulation Software in a Life Sciences Practical Laboratory. *Computers and Education Journal, 26*(1-3), 101–112. doi:10.1016/0360-1315(96)00011-5

McDonough, D., Strivens, J., & Rada, R. (1994). University courseware development: differences between computer-based teaching by users and non-users. *Computers & Education, 23*(3), 211–220.

McMahon, C. J. (2000). *Tutorials for Introduction to Materials Engineering* [CD-ROM]. Merrion Media.

Moonen, J., & Gastkemper, F. (1983). *Computer Assisted Instruction.* Utrecht, The Netherlands: Het Spectrum.

Noronha, M., Bittencourt, T. N., Proenca, S. P., & Guello, G. A. (2000). *Multimedia-based Environment in Structural Engineering Education.* SIAE-project Report, Brazil.

Norrie, S. E. (1996). Evaluation of Computer Based Laboratory Simulation. *Computers and Education Journal, 26*(1–3), 123–130. doi:10.1016/0360-1315(95)00057-7

Rickel, J. W. (1989). Intelligent Computer Aided Instruction: A survey organized around system components. *IEEE Transactions on Systems, Man, and Cybernetics, 19*(1), 40–57. doi:10.1109/21.24530

Salvatore, A. M. (2001). Experience with the Introduction of Multimedia into Mechanical Engineering Technology, Mechanics of Materials Laboratory. *Proceedings of the 2001 American Society for Engineering Education Society*, (pp. 1 – 10).

Sclechter, T. M. (1991). *Problems and Promises of Computer Based Training, promises, promises, promises: History and Promises of Computer Based Training* (pp. 1 – 17). New York: Ablex Publishing.

Scott, N. W. (1996). *A study of the introduction of educational technology into a course in engineering dynamics: classroom environment and learning outcomes.* PhD. Thesis.

Shang, H. H. (2005). Chuin [Interactive Tutoring Components for Engineering Dynamics. *Exploring Innovation in Education and Research.* iEER.]. *S. C. & Cho, C,* L.

Shaoul, J. E. (1989). Sharing Accounting Courseware – Some of the issues and only a few solutions. *Paper presented at the conference Sharing Accounting Courseware,* Manchester University, (pp. 18 – 27).

Sieber, V., Haynes, R., Dobson, C., Holley, D., & Andrew, D. (2004). Using Interactive Elements between Disciplines. *Journal of Interactive Media in Education, 16,* 1–10.

Stoner, G. (1996). Implementation Learning Technology – Learning Technology Dissemination Initiative Report, UK.

Thomas, R. C., & Milligan, C. D. (2004). Putting Teachers in the loop: Tools for Creating and Customizing Simulations. *Journal of Interactive Media in Education,* 15.

Usha, V. R. (2003). "Multimedia as an Educational Tool" Educational Multimedia Book (pp. 4 – 9).

Vasandani, V., Govindaraj, T., & Mitchell, C. M. (1989). *An intelligent tutor for diagnostic problem solving in complex dynamic systems*, (pp. 772 – 777).

Villiers, C. D., Pistorious, M. C., Slexander, P. M. & Plooy, N. F. (1992). Constraints on computer – assisted instruction in a distance education environment. *Computers and Control Engineering Journal,* 11 – 13.

Wehr, J. (1988). Instructor-led of computer based: Which will work best for you? [US]. *Training and Development Journal*, 18–21.

Wilson, L. (2001). Using Multimedia in Teaching Dynamics. In *Proceedings of the 2002 Conference for Industry and Education Collaboration*.

Chapter 5
Hardware and Software for Multimedia Development

INTRODUCTION

According to Peggy (2008),

It's relatively easy to transfer raw content from paper to technology delivery. But it's much more challenging to put material into a format that helps people learn.

The aforementioned statement is essential to ponder when it comes to multimedia hardware and software consideration. Computers are now making it possible to blend or integrate together multimedia elements such as audio, video, graphics, images and animation into a single learning package. However, blending these multimedia elements together to develop a learning package does not mean that student's proficiency in the subject matter could be enhanced. Furthermore it is not necessary to convert the entire textbook into a full working multimedia package for students to learn. Selected problems that are difficult to explain to the students from the textbook could be more appropriate and beneficial to the students if shown in the form of motion. On the other hand selecting appropriate multimedia elements and authoring tools could be difficult tasks for a new multimedia author. This is because proper development

DOI: 10.4018/978-1-60566-764-5.ch005

design process and careful selection of multimedia elements should be used appropriately to develop a high quality and cost effective package that could engage learners in their learning. Understanding the overall development process is an essential part for a multimedia learning package.

In general, five components are crucial for the development of a multimedia package, namely:

- Hardware (the multimedia computer)
- Authoring software (tools for developing multimedia learning packages)
- The multimedia author (the conceptual understanding and creative skills)
- The subject matter expert for example an engineer (if the domain is engineering)
- Students (the potential users of the multimedia package) as discussed in Chapter 2.

In addition to the above it is important to choose a suitable computing platform to run the final product. Although a number of platforms exist such as Silicon Graphics, Apple Macintosh, Sun Microsystems and Mainframes, this book confines only to Microsoft Windows platform. Windows-based system (or Windows operating systems) is selected as the choice for operating the TAPS (technology assisted problem solving) packages because it has a worldwide presence, availability and affordability. An operating system (OS) is the program that is responsible to manage all the other programs in a computer, once it is loaded into the computer's memory. The other programs are called the applications programs. The OS determines the distribution of time and order for multiple application programs running simultaneously. It also manages the sharing of memory among multiple applications. It communicates with the attached hardware devices about the condition of operations and errors that may have occurred (Usha, 2003).

The inventions in the field of hardware and software for multimedia are being upgraded and changing rapidly. The description in this Chapter is only informative (and does not necessarily claim to be exhaustive) for enabling the reader to familiarize with the concepts and capabilities of some of the software used in developing the TAPS packages which runs on Windows platform.

This Chapter attempts to describe the hardware and software which enable the developer to develop the end product called 'multimedia packages' or referred as TAPS packages for engineering in this book.

HARDWARE REQUIRED FOR MULTIMEDIA

Selecting multimedia hardware such as a computer often entails many conflicting issues and concerns. When considering a multimedia personal computer (PC) (a computer equipped with high main memory, CD-ROM, speakers microphone, PC camera, scanner and software tools for implementing multimedia applications) the main issue is usually related to budget. It is therefore recommended that a decent computer with adequate hardware is considered. Hardware interprets user commands into computer activity. When developing a multimedia application, a high speed computer and storage capacity is recommended. There are many more things that need to be considered such as which device makes a computer fast (processes information quicker), what is the device for storage, etc. The hardware devices thus can be divided into five categories namely system devices, memory and storage devices, input devices, output devices, and communication devices. This is further discussed in Table 1.

Table 1. Input and Output devices of a multimedia PC

Device	Descriptions
System devices	These are the devices that are the essential components for a computer. These include microprocessor, motherboard and memory. Microprocessor is basically the heart of the computer. A microprocessor is a computer processor on a small microchip. When the computer is turned on, it is the microprocessor, which performs some operations. The microprocessor gets the first instruction from the Basic Input/Output System (BIOS), which is a part of its memory. BIOS actually load the operating system into random access memory (RAM). A motherboard is a device in the computer that contains the computer's basic circuitry and other components. Motherboard contains computer components like microprocessor, memory, basic input/output system (BIOS), expansion slots and interconnecting circuitry. Additional components could be added to a motherboard through its expansion slot
Memory and storage devices	RAM (random access memory), also called primary memory, locates the operating system, application programs, and data in current use so that the computer's processor reaches them quickly. RAM is called "random access" because any storage location can be accessed randomly or directly. RAM is much faster than the hard disk; the floppy disk and the CD-ROM. RAM can be taken as short-term memory and the hard disk as the long-term memory of a computer. However, RAM might get slow when used to its limit. That is why, more memory is needed to work on multimedia applications. Today's PCs come with 128 or more megabytes of RAM. Users of graphic applications usually need 128 plus megabytes of memory. A hard disk stores and provides access to large amounts of data on an electro magnetically charged surface. Today's computers typically come with a hard disk that contains several billion bytes (gigabytes) of storage. The popular ones currently are 80 GB and above. Hard disk contains a part called disk cache which is responsible for improving the time it takes to read from or write to a hard disk. The disk cache holds data that has recently been read. The other type of hardware cache inside your computer is cache memory. Cache stores something temporarily e.g. Temporary Internet files are saved in Cache. A compact disk (CD) is a small medium that can store data pertaining to audio, video, text, and other information in digital form. Initially, CDs were read-only, but newer technology allows users to record as well. CD-ROM (Compact Disc, read-only memory) can store computer data in the form of text, graphics and sound. To record data into a CD, a CD recorder is needed. Normally this type of CD is either *CD-Recordable* (CD-R) or *CD-Rewritable* (CD-RW). For the latter the CD could be used as a floppy disk write, erase and again write data into the same disk. In the CD-R, once the data recording is completed, it becomes a CD-ROM and nothing can be deleted. Newer storage technology devices such as flash drives are gaining more popularity. These universal serial bus (USB) port connector device are more compact, faster, hold much more data, are more reliable for lack of moving parts, and have a more durable design. Currently the thumb drives comes in 8 GBs of storage size and above. Other option includes the external hard disk which comes in 40 GB and above.
Input devices	A keyboard is the primary text input device for a computer. It was very popular when DOS was the popular operating system. After the emergence of Windows, its role became limited to dealing with text and for some commands only. The keyboard contains certain standard function keys, such as the Esc key, Tab, Cursor movement keys and control keys. A mouse is also a primary input device but it is not suitable for dealing with text. A mouse is a small device that you move across a pad in order to point to a place on a display screen and thus execute a command by clicking it. The mouse is an integral part of any personal computer. A cable connects the mouse to the computer. Microphone is another input device that can interpret dictation and also enable us to input sound like the keyboard is used for text. A digital camera records and stores photographic images in digital form that can be fed to a computer as the impressions are recorded or stored in the camera for later loading into a computer. The digital cameras are available for still as well as motion pictures. A PC camera could also be used to record digital movies or photos directly to the computer.
Output devices	A printer is a device, which on receiving the signal from computer transfers the information to paper. Earlier the dot-matrix printer was a popular low-cost personal computer printer; now printers have taken its place. Dot-matrix printer strikes the paper a line at a time while inkjet sprays ink and laser printer uses a laser beam to attract ink (also called toner). A monitor is a device for display. It is just like a television set and is measured diagonally from two opposing corners of the picture tube. The standard monitor size is 14 inches. Very large monitors can measure 24 inches diagonal or greater. An amplifier is an electronic device that increases the power of a signal. Amplifiers are used in audio equipments. They are also called power amplifiers. Speakers with built-in amplifiers have become an integral part of the computers today and are important for any multimedia project.
Communication devices	A modem modulates digital signals going out from a computer or other digital device to analogue signals for a telephone line and demodulates the analog signal to convert it to a digital signal to be inputted in a computer. Most new personal computers come with 56 Kbps modems. Modems help the computer to connect to a network.

CONFIGURATION OF A MULTIMEDIA COMPUTER

A good multimedia system should have a Pentium 1.6 Ghz (or the one with similar capabilities) onwards processor, at least 4 GB of RAM, 160 GB onwards hard disk drive, 1.44 MB Floppy drive, 22 inch onwards SVGA monitor, 32MB AGP card, 52 X CD-ROM drive, a 32 bit sound card, high wattage sub-woofer speakers, 104 PS/2 keyboard, PS/2 mouse and 56K fax data voice modem. If you wish you can add a CD-recorder, scanner, printer, digital camcorder and a video-capture card. Remember, there is no set rule to define the exact hardware combination of a good multimedia computer. The combination is dependent on the nature and contents of the multimedia project you are dealing with. Fortunately, there exist hardware tools for performing almost any action; the need is to use only that hardware, which suits your purpose. For a ready reference, see Table 2.

SOFTWARE REQUIRED FOR MULTIMEDIA

According to Usha (2003), multimedia is making a difference by providing ways of delivering learning materials that are less expensive and more convenient. The key to any learning process is that it must be relevant and it must keep the learner engaged. Educational multimedia is no exception. This can be proved after seeing the growing use of graphics, illustrations, animations and sound in educational multimedia. It is therefore essential to choose that software which enables the multimedia author to execute the project with the minimum possible effort and maximum possible productivity. Multimedia software have unlimited features. The multimedia author can choose among several hundred colors, dozens of fonts, a wide variety of color-coordinated templates and many other incredible options.

However, today entire suites of integrated production tools are now available. The need is to use them judiciously to create good projects. Powerful features are continuously being added to the software that allows multimedia authors to work more smoothly and conveniently between applications. Emergence of these integration features has resulted in collaboration and unison of multiple tools. The integration has enabled us to use your graphics from a previous work and save time on rebuilding it (Usha, 2003). The options available in choosing multimedia software are enormous. All that the multimedia author has to do is to choose the right hardware and software to complete the multimedia learning packages. The next sub sections discuss software required for multimedia in brief.

Painting and Drawing Tools

In general graphics gives a great impact to multimedia presentation in influencing the students. It is the graphics that would create the first impression of the multimedia project. These tools are, therefore, very useful in giving the multimedia author the desired capability in terms of drawing and painting. Painting and drawing tools generally come with a graphical user interface with pull down menus for quick selection. It can be used to create almost all kinds of possible shapes and resize them. These tools have the capability to color with paint and clip arts. The multimedia author can use brushes of different sizes and shapes according to the need. In addition the multimedia author can use layers to give different treatment to each element. Most of these tools come with built-in plug-ins for performing different tasks. The completed drawing can be imported or exported in many image formats like .gif, .tif, .jpg, .bmp, etc. A good example of a drawing tool is Corel Draw®,

Table 2. Components of a multimedia PC

Component	Description	Standard
The Power Supply	It converts AC current into DC current as all computer components operate on DC current.	Any standard ATX Cabinet.
The System Board	All of the parts inside the computer are assembled on the system board.	Prefer to buy genuine board for the processor. Intel web-site has a motherboard selection feature.
Central Processing Unit (CPU)	The CPU is the brain of the computer. Pentium is a popular chip presently.	Pentium 4 processors, Intel, Celeron, AMD etc, with 1.5 GHz speed onwards.
RAM (The Main Memory)	Random Access Memory (RAM) is critical for multi-media. The more memory the better off the computer is.	Minimum 256 MB.
Pen / Flash Drives	Is a portable storage device for data files	Between 1 GB – 8GB (> 2 GB is recommended)
Hard Drives	Hard disk drive stores software and data. More storage is better for large projects.	80 GB onwards.
CD Drive (read/write)	CD drives can store what hundreds of floppy disks together can. Maximum capacity of CD is about 800MB presently.	ROM - 52 X RW 48X x 16X
Modem	Modem enables communications between your computer with other computers, the Internet and the World WideWeb.	56 kbps onwards.
Sound Card	Sound Cards allow conversion of digital sound to analog sound and vice-versa.	Sound Blaster e.g. Creative Live Value Card
Keyboard	The keyboard sends typed information to the system board.	Multimedia Key Board.
Monitor	Monitor is a display device. Choose how many colors they can display and about their resolution.	17"
Mouse	Used as a pointing device.	Scroll Mouse.
Printer	Inkjet printers have the ability to turn out good-looking output, including graphics at a lower cost than laser printers. Laser printers produce the best quality, but their cost is high.	Choose as per your requirements.
Scanner	Scanners are used to digitize photographs, artwork and documents.	Choose as per your requirements.
Digital Camera (Still and Movie)	To capture pictures and prepare movie.	Choose as per your requirements.
Video Capture Card	To capture analog video and convert into digital format.	Choose as per your requirements.
Graphics Card	To view graphics on the screen clearly.	32 MB Minimum

CorelDraw® and Illustrator® can be used to create illustrations from scratch. These tools have wide-ranging features to handle text and to create drawing with precision. It can also be used to improve clip art, pictures and photos. It is an ideal tool for any design project such as technical drawings, advertisements, logos, etc. It can be used in creating full-color illustrations for multifarious drawings and graphics for any designing project. It has lot of clip arts and high-quality drawings, which can be inserted into the under development multimedia project. CorelDraw® can also be used to generate drawing for an animation sequence. Other software tools such as MS Gif Animator, Animator-9, MotionBuilder, DAZ Studio and Bryce can also be used for the same. More painting and drawing tools are available at http://www.allgraphicdesign.com/graphicsblog/2008/05/16/online-drawing-tools-free-online-painting-sketching-tools/.

Image Editing Tools

While Painting and Drawing tools let the multimedia author create a drawing from scratch, Image editing tools are used to edit existing bitmap images and pictures. However, these tools are similar to painting and drawing tools as they can also create images from scratch. The image editing tools are capable of converting any image data type file format. Image editing tools are primarily used for reinventing and recreating the image, which make them an important tool for designing a multimedia project. Adobe Photoshop®, Illustrator® and Paint Shop Pro® are two good examples of image editing software.

Adobe Photoshop® is a cutting-edge image processing software package that enables the multimedia author to create and edit images on computers. Paint Shop pro® is also an exceptional drawing and painting utility that yields professional-quality effects. With both these tools the multimedia author can edit an image in almost any desirable way. One can add elements in layers; edit text and use effects filter to make the existing image look even inferior to the edited one. It can mix and manipulate colors at a click of a mouse button. The multimedia author can manipulate the images with special effects and techniques. Images can be imported and exported across programs in any format. These tools have been used to edit and create images for motion pictures, animations and for artwork. More image editing tools are available at http://www.creatingonline.com/image_editing/.

Sound Editing Tools

Sound editing tools allow the multimedia author to hear sound as well as visualize it. The multimedia author can also cut/copy and paste sound and edit it with great accuracy. In addition it can be used to integrate sound into the multimedia project very easily by using sound editing tools. One such software is CoolEdit®. CoolEdit® can be used to record music, voice, or any other audio. It can be used to edit, mix the sound with any other audio and add effects to it. CoolEdit® can record from a CD, keyboard, or any other sound played through your sound card. One good feature of this software is that it can read and write MP3, which is the hot sound format in the present times. Once the multimedia author is done with the sound file, it can help the multimedia author in converting the file to any desired format. In other words, there is a similarity in these editing tools - what Photoshop® can do to images; CoolEdit® can do for sound. Sound Forge® is another professional quality sound editing tool that is used in multimedia work. More sound editing tools are available at http://www.snapfiles.com/freeware/gmm/fwaudioedit.html.

Video Editing Tools

Animations are graphic scenes played back sequentially and rapidly. These tools adopt an object-oriented approach to animation. These tools enable the multimedia author to edit and assemble video clips captured from camera, animations and other sources. The completed clip with added transition and visual effects could be played back. Adobe Premiere® and Media Shop Pro® are two good examples of these tools. Adobe Premier® is a powerful tool for professional digital video editing. It is primarily used to produce broadcast quality movies. It has excellent editing tools that enable the multimedia author to work with complete flexibility. This software first digitizes the sound and video and then allows the multimedia author to edit them to preserve picture quality. It can edit video and multimedia movies in AVI (audio video) as well as MPEG (moving picture experts group) format. It can create titles and graphics and then add them to the multimedia project. It uses digital filtering for incorporating special effects. This software has applications in film editing and movie making. Another good example is Media Studio Pro®. For creating animations Macromedia Flash® is the industry standard tool. A file created in Flash is called a movie. A movie in Flash occupies very less file size, and hence is more popular for the Web. More video editing tools are available at http://www.snapfiles.com/freeware/gmm/fwvideoedit.html.

3D Modeling and Animation Tools

With the help of 3-D modeling and animation tools the objects that appear in perception in the project can look realistic (things can be depicted in the way they actually are). It has become conventional to use 3-D modeling tools in multimedia design. These tools offer features like multiple windows to view the design in each dimension. They have drag and drop menus from where the modeler (multimedia author) can drop shapes into the design and combine them to create complex designs. A good 3-D modeling tool is Alias Maya®. More 3D modeling and animation tools are available at http://www.dooyoo.co.uk/animation-3d-modeling-rendering/.

Alias Maya® is a tool for making 3D models and designs that can be converted into 3-dimensional animations. The modeler can virtually lead the imagination to go wild and visualize any object easily with the help of this tool. It has applications in creating web pages; designing advertisements; making cartoon films and in creating multimedia based training programmes. The modeler can also give special effects to the design especially in terms of sound, lighting and animation. Other good examples of 3-D modeling tools are 3-D Studio Max®, Cinema 4D®, and Lightwave®. However, the new user (modeler) of such tool may need sometime to get familiarized to design professional looking 3-D objects and effects. More 3D modeling and animation tools are available http://www.your3dsource.com/which-animation-software-to-use.html

Desktop Virtual Reality Tools

The use of visual technologies for teaching and learning in higher learning institutions has produced dramatic extensions of the once traditional lectures, demonstrations, and hands-on experiences. From the introduction of color photography to full-motion video to computer-generated presentations with graphics and animations, visual technologies have enhanced the preparation of workforce specialists and instructors by bringing into classrooms and laboratories a breadth and depth of realism that has enhanced comprehension, increased learning performance, and reduced teaching and training time. Occasionally, however, there arrives a training technology that causes a realization that "this changes everything." (Lynna and Floyd, 2004). Such a technology is known as virtual reality (VR). Lynna and Floyd further stated that the capabilities and possibilities for VR technology may open doors to new vistas in industrial and technical instruction and learning, and the research that supports them.

Table 3. Integrated design software

Software tools	Description
Page-based tools	These tools organize elements as pages of a book. These tools are used when the content of the package consists of elements that can be viewed individually. These tools organize them in a user-defined sequential form.
Icon based tools	These tools organize elements as objects. These tools display the flow diagrams of activities along with branching paths.
Time based tools	These tools organize the elements along a time-line. These tools play back the sequentially organized graphic frames at user-set speed and time.
Object Oriented tools	These tools organize the elements in a hierarchical order as related objects. These tools make these objects perform according to properties assigned to them.

In general VR can be divided into two categories namely desktop virtual reality (DVR) and immersive VR. Immersive VR employs more sophisticated hardware than DVR such as 3D goggles, head mounted helmet, wired data gloves and clothing or Cave Automatic Virtual Environments (CAVE) (projection of stereo images on the walls and floor of a room-size cube). At present, it is difficult to bring this technology into the classroom because it is expensive to be set up and requires large space with proper room temperature.

As VR has continued to develop, applications that are less than fully immersive have developed. These non-immersive or DVR applications are far less expensive and technically daunting than their immersive predecessors and are beginning to be employed by the educational sector in general. Desktop VR focuses on keyboard strokes and mouse clicks, joystick, or space/sensorball-controlled navigation through a 3D environment on a graphics monitor under computer control (Lynna and Floyd, 2004). One early application of DVR is the Virtual Reality Modeling Language (VRML). VRML applications can be distributed via Internet. Other tools that support DVR environments include Cortona™ and Cosmo™ viewer, Cult3D™, Pulse3D™, ViewPoint ™, Macromedia™ and Shockwave. The features of these DVR tools enable the user to move along any direction on the screen and have the object displayed continuously and updated instantaneously. Therefore, the user could gain a greater understanding of a given problem. More information about these tools is available at http://www.cult3d.com/;http://www.karmanaut.com/cosmo/player/;http://www.cortona 3d.com/cortona.

Integrated Design Software

Multimedia authoring tools or integrated design software are tools which organize and edit multimedia packages. These tools are required to design the user interface (as discussed in Chapter 3) for presenting the package to the learner. In other words, these tools are used to assemble various elements to make a single presentation. The multimedia author can compose comprehensive videos and animations with these tools. According to Usha (2003), there are four basic type of authoring tools viz. *Page based tools* (such as Tool book®, Visual Basic®), *Icon based authoring tools* (such as Authorware®), *Time based authoring tools* (such as Macromedia Director®) and *Object Oriented tools* (such as Media Forge®). More specifically, these tools can be summarized as following (Table 3).

A brief description of Macromedia Director® (time-based tool) is given next since most of the TAPS packages shown in this book have been authored using this tool (Director).

Macromedia Director is a multimedia authoring application capable of producing animations, presentations and movies. It provides a wide range of possibilities for integrating different multimedia elements. It supports inputs from programs such as Shockwave®, Photoshop® and Premiere®. It has applications in building professional multimedia presentations. The multimedia author can also integrate Real Audio and Real Video in Director projects. Compatibility of Director with other packages means that the multimedia author can use his/her favorite tools and software to create content for the package and then bring that content into Director for authoring and editing.

The newer version of Macromedia Director MX® improves productivity even more, as it can be used to construct highly interactive and quality materials that work perfectly with wide and narrow bandwidths, regardless of the resolution of the monitors. It also facilitates the use of high quality video files and support for 3-D geometric models designed using Alias Maya® (a 3-D modeling tool).

In summary, the development of effective multimedia packages that can be used by students in their learning requires the exploitation of a variety of software. These multimedia packages could be difficult to develop and could give rise to a number of design issues. In addition the use of authoring tools such as Director®, Flash® and Authorware®, requires the multimedia author to have a comprehensive knowledge and creativity on the capabilities of such tools under different design considerations. However, the incorporation of multimedia in engineering learning packages such as TAPS packages could provide added advantage to engineering students and instructors (Manjit and Ramesh, 2006). In addition multimedia technology has great potential to assist learning as well as to enhanced learner visualization and understanding of concepts in mechanical engineering.

Another essential component of interactive multimedia packages is user interface design. A good interface is clear, elegant, transparent and is easy to follow. However, the ability to interact with computers relies on buttons, links, instruction and dialogues that make up the interface. This was discussed in Chapter 3.

REFERENCES

Lynna, J. A., & Floyd, B. A. (2004).Desktop Virtual Reality: A Powerful New Technology for Teaching and Research in Industrial Teacher Education. *Journal of Industrial Teacher Education, 4*(41). Retrieved June 10, 2008 from http://scholar.lib.vt.edu/ejournals/JITE/v41n4/ausburn.html

Manjit, S. S., & Ramesh, S. (2006). Multimedia Learning Packages: Design Issues and Implementation Problems. [MOJIT]. *Malaysian On-line Journal of Instructional Technology*, *1*(3), 43–56.

Peggy, T. (2000). *The Role of Instructional Design in Multimedia Development.* Retrieved May 6, 2008 from http://www.learningcircuits.org/2000/feb2000/Troupin.htm

Usha, V. R. (2003). *Multimedia as an Educational Tool* (pp. 4 – 9). Educational Multimedia Book.

Chapter 6
Technology Assisted Problem Solving Packages:
A New Approach to Learning, Visualizing, and Problem Solving in Engineering

INTRODUCTION

Steif (2003) pointed out that,

The potential of the computer to offer new kinds of problem solving/learning experiences is only just being uncovered.

This is especially true in the domain of engineering where such technology needs much attention. Technology assisted problem solving (TAPS) packages are specialized computer programs developed to work as stand-alone (PC Based) or with Web servers that can supplement student learning; for revision, laboratory experiments, and self-study. In this book the term TAPS is used to represent interactive multimedia CAL in which the student is engaged with a computer tutor in the problem-solving task of the subject matter. TAPS packages offer similar pedagogic values as an experienced human tutor, with the added advantage of guiding students to solve engineering problems on a more flexible mode i.e. a student has the freedom of working on the problem at his/her own pace, repeat all or certain steps, spend more time at each or particular step until they are able to understand, and solve the problem. The objec-

DOI: 10.4018/978-1-60566-764-5.ch006

tive of these TAPS packages is to improve student's understanding of the selected engineering problems by guiding and presenting the problem solving steps accordingly. The ultimate goal is to instill a sense of independent learning, encouraging critical thinking, and to promote deep learning. When tutoring a student on solving an engineering problem, a human tutor is expected to gauge the student's background knowledge, deliver relevant course material at the correct level of detail, and clarify student's misunderstandings.

TAPS packages include the use of the computer to provide most aspects of instruction, which a classroom instructor could provide such as tutorials, questioning, feedback, contingent on answers, analysis and testing. The TAPS packages developed for this project has been customized to anticipate student needs, and have various interactive features built in to allow delivery control, navigation, and feedback. More specifically, the packages are designed to assist the student in learning, visualizing, and problem solving in a step-by-step approach.

The TAPS packages also employ a variety of multimedia elements such as text, 2-D animated and still graphics and 3-D animated and still geometric models, audio, video and animations, stereoscopic images, and simple artificial intelligence techniques to develop individualized computer based learning environments in which the student and computer tutor can have a flexibility that closely resembles to what actually occurs when a student and a human tutor communicate with each other. Such suppleness is important because without it, the package cannot be fully adaptive to the individual student's on-going learning and problem solving needs during instruction.

There are numerous difficulties with the implementation of realistic TAPS packages. The major problem with TAPS package development is that most of the features that are commonly found in non-computer-based tutoring packages are difficult to implement on the computer. In addition, many aspects of the tutoring process are taken for granted by the students. These include direct verbal feedback, visual and audio interaction, and an extensive knowledge base. When a student does not understand a concept, the norm is to ask a human tutor to provide a simpler explanation or to apply the concept to an everyday situation. This feature is difficult to implement in any computer based-tutoring package, because the computer does not have sufficient intelligence to understand and interpret the course material.

Based on these arguments, it is envisaged that an ideal TAPS package would be difficult to develop and implement. It is therefore necessary to identify key concepts that constitute a TAPS package and decide the best way of implementing similar forms of each of these concepts in a way that makes tutoring and problem solving environment as realistic and pedagogically effective as possible.

KEY CONCEPTS IN TAPS PACKAGE

There are a number of key concepts that can be applied in the development of a TAPS package. Some of these are similar to intelligent tutoring systems (ITS) whereby a computer tutoring system incorporates aspects of intelligence, in particular an assessment model (used to monitor the performance of the student), and domain knowledge representation. In TAPS package, these concepts can be divided into three main categories, namely learning scenarios, knowledge representation, and assessment modeling.

Learning Scenarios

A learning scenario is a situation in which the student's learning takes place. When implementing a TAPS package, the criterion for determining the most appropriate learning scenario is based on the interaction required between the student and the computer. The learning scenario selected will therefore be dependent on the type of information to be delivered to the student during the tutoring session, the amount of knowledge the student is expected to gain from completing the problem solving tutorials, and to a certain extent, on the knowledge base of the TAPS package.

In general, most computer based tutoring packages are implemented using one of the three learning scenario categories. The most common learning scenario category to be implemented is the explanation of theoretical concepts to the student. In this scenario, the TAPS package must convey pre-defined knowledge to a student in ways that maximizes his/her understanding of concepts being taught. This is the simplest learning scenario to implement, as the main challenge of developing the TAPS package is ensuring that more precise information is presented at the correct level of detail for students to comprehend and learn.

The second learning scenario that is commonly employed in computer based tutoring packages is the simulation of real-world tasks on a computer. These tasks include the detail operation of a specific component or the simulation of the process that the student is expected to perform in the future. In any event, the learning scenario must deal with simulating the appropriate real-world properties as accurately as possible on computer. This is a difficult requirement to implement successfully, as the TAPS package must both simulate the process as realistically as possible, as well as have the pedagogic ability to explain the process to the student in the best possible way.

In general, the most difficult learning scenario category to be implemented in a computer based tutoring package is the discovery of knowledge through investigation and exploration. In this third learning scenario, the student is required to actively participate in the learning experience, by manipulating the package and observing the direct response to the student's actions. An example of this scenario is a package to teach students Engineering Dynamics. For example, in the case of a perfectly elastic impact (Hibbeler, 2001), the total energy of two particles, as well as their momentum, is conserved. If mass A has velocities of V_{A_1} and V_{A_2} before and after a perfectly elastic impact, respectively, and mass B has velocities V_{B_1} and V_{B_2} before and after the impact, respectively then, we have:

Conservation of Kinetic Energy

$$\sum(\text{Kinetic Energy})_{\text{before impact}} = \sum(\text{Kinetic Energy})_{\text{after impact}}$$

$$\frac{1}{2}m_A\left(V_{A_1}\right)^2 + \frac{1}{2}m_B\left(V_{B_1}\right)^2 = \frac{1}{2}m_A\left(V_{A_2}\right)^2 + \frac{1}{2}m_B\left(V_{B_2}\right)^2 \tag{5.1}$$

Conservation of Momentum

$$\sum(\text{momentum})_{\text{before impact}} = \sum(\text{momentum})_{\text{after impact}}$$

$$m_A V_{A_1} + m_B V_{B_1} = m_A V_{A_2} + m_B V_{B_2} \tag{5.2}$$

While equations (5.1) and (5.2) can be completely understood from a mechanics point of view, it is particularly confusing to weak engineering students. For instance, in reality, after the ball is dropped to the ground, it will not rebound to the same height even without air resistance acting on it. This means that neither the total energy nor the total momentum is conserved during the impact. Does this mean that some non-elastic deformation or plastic yielding has occurred to the ball or the ground? Because of the uncertainty of the student's actions, this type of learning scenario is very difficult to implement successfully. The TAPS package must make allowance for any response made by the student and must be able to act accordingly; providing him/her with sufficient explanation for all decisions or conclusions made leading from the problem statement through a series of steps and solution.

Irrespective of the category of learning scenario implemented, the relevant course material will be memorized more effectively if the student is an active participant in the learning process. The tutoring environment may be further enhanced by involving most of the student's senses during the tutoring session. This allows the student to combine the knowledge acquired in the course with actual experience and application.

Knowledge Representation

The knowledge representation component of a computer based tutoring package can be divided into two categories, namely (a) domain knowledge and (b) pedagogical knowledge (Burns, 1991). Domain knowledge involves issues in the representation of knowledge and refers to the facts, figures, and interrelationships between the various objects in the domain. Pedagogical knowledge is the sequence of instructions that a computer tutor uses to carry out various tasks in operating a system. Pedagogical knowledge therefore involves the finding of techniques to solve particular problems. The knowledge of computer based tutoring packages contains definite information content and structure, as well as procedures for accessing and utilizing the information (Chu *et al.*, 1989).

Domain Knowledge

One of the major limitations with conventional computer based tutoring packages is that they have poor structure of knowledge of their domain in the database. The tutoring session typically consists of the presentation of information, problems with which to test the student's knowledge, answers to these problems, and at best, pre-specified branches based on the student's results obtained in the test.

Rickel (1989), noted numerous disadvantages with these computer based tutoring (CBT) packages as summarized below:

- The CBT package is unable to adapt to the requirements of the student
- There are no facilities with which to assess the student's true misunderstandings
- The pre-specified branches prevent the CBT package from handling unanticipated answers
- Pre-specified answers leave the CBT package with no criteria for judging student responses other than correct or incorrect

If human tutors are expected to possess a great deal of domain competence, this should be the ultimate challenge for TAPS packages.

The domain knowledge of a TAPS package should incorporate the necessary information so as to correct the above limitations. This implies that the TAPS package should contain an extensive knowledge database and have the ability to filter out the course material that is not directly relevant to the student. In addition, the TAPS package should have the ability to interact with the student in the same manner as a human tutor. Interaction is perhaps the most difficult part of TAPS package design. A good TAPS package should be able to answer course-related questions asked by the student, as well as present summaries and overviews whenever these are required. Furthermore, the TAPS package should know when and how to present the student with information and should be able to determine immediately whether the student has understood this information or not. A good TAPS package should constantly monitor the student and have the ability to automatically offer explanations to match the student's current level of understanding.

The domain knowledge component of a good TAPS package requires a great deal of intelligence and effort to implement successfully. In general, it is for this reason that to date, no computer based tutoring package has been commercially developed and fully accepted by learning institutions. Even if such packages exist, these packages may be used only for a short period of time. Therefore, it will probably take many years of research in the field of Artificial Intelligence before an intelligent computer based tutoring package could be successfully developed. Currently, when developing a TAPS package, it is necessary to compromise on various aspects of the domain knowledge, to ensure that a simplified advanced computer based tutoring package, can be physically realized.

Pedagogical Knowledge

Pedagogical knowledge is an essential component of a TAPS package. Although the domain knowledge is responsible for filtering useful information from the vast knowledge base, the pedagogical knowledge is responsible for relating this information to the student. The pedagogical knowledge decides how to interact with the student, when to interrupt the student, and how to address the student while he/she is using the tutoring package.

Although the pedagogical knowledge is burdened with many responsibilities, the most important of these is determining a strategy to deal with student errors. Since students are seldom consistent, a computer tutor cannot simply provide correct answers to a student's mistake. When a student makes a mistake, the TAPS package must select between ignoring the error, pointing it out, correcting it, or somehow guiding the student towards recognizing the error and correcting it without the explicit help of the computer tutor.

There are numerous trade-offs in correcting a student explicitly, trying to entrap the student into discovering the error without the help of the computer tutor, or simply allowing the student to view the consequences of any mistakes made.

THE ASSESSMENT MODEL

Introduction

The assessment model is a dynamic model of the student's knowledge and capabilities, maintained and constantly updated by the computer-based package. Its purpose is to evaluate and account for the student's actions and responses. Human tutors do an excellent work on moderating student's answers in the context of their assumed level of understanding and past learning behavior, thus effectively adapting their instruction to the student's competence and abilities. Although adaptation to the student is almost second nature for human tutors, it is an extremely difficult characteristic to implement in TAPS package.

The function of the assessment model is to provide the student with feedback by comparing the student's actions to those prescribed by the TAPS package. This feedback is used to inform the student which actions are correct and which are incorrect. In this way, the student receives tuition while interacting with the TAPS package.

Conventional (classroom) assessment could occur through a variety of methods, for example quizzes, exams, oral test, or homework. However, the most common technique used for assessing the student in computer-based tutoring package is the assessment of the number of correct and incorrect answers upon completion of a course topic. Adaptation to the student level of understanding is usually limited to the presentation of a pre-specified course material, based on the student's response to the questions of the test. Most available conventional computer tutoring packages do not have the ability to keep track of the student's insufficient knowledge, except at a very basic level. For current tutoring packages, the assessment model will have to be greatly simplified so that it may be practically realized.

Requirements for Assessment Modeling

The student assessment model represents an overview of the student's capability level. There are a number of fundamental rules that should be adhered to when developing an assessment model in a TAPS package. These can be summarized as follows:

- The model must be able to represent knowledge, concepts and skills.
- The model must include the knowledge that the student has acquired, and that which the student has been exposed to and shown some understanding.
- The model must be able to represent the student's misconceptions.
- The model must be able to include a history of the student's problem-solving performance.

Student Diagnosis

Student analysis plays an important role in assessing and correcting the student's misunderstandings. It is important that whenever a student makes a mistake, the computer tutor points out the error, offers an explanation and guides the student effectively in solving the problem. If the computer tutor only tells the student that he/she is incorrect, it has not performed its teaching task, but instead shifted the problem back to the student.

The aforementioned types of misconceptions in a student's knowledge that may be analyzed are dependent on the knowledge represented in the tutoring packages. Because each tutoring package has a limited domain and pedagogical knowledge component, an extensive student analysis is a difficult facility to incorporate into a TAPS package. There are occasions where the student has the correct answer but expresses it in a different way from that recognized by the tutoring package. Similarly, a student may appear to be missing certain skills when instead, he/she is employing a totally different strategy that is not programmed in the assessment model.

Misconceptions in the Students' Understanding

The assessment model describes what a student should know and do in a particular situation. When the student's actions do not match those suggested by the assessment model, the reasons can be attributed to a number of causes, for example the lack of knowledge stored in the assessment model, the use of inappropriate knowledge to tutor the student, student's inability to apply this knowledge to the present scenario, or an incorrectly defined model (Sleeman, 1985). Evaluating a student's misconceptions involves determining the probable cause for the student's incorrect behavior or action.

After evaluating a student's misconceptions, a computer based tutoring package will provide instructions for correcting these misconceptions and improve the student's problem-solving skills. The selection of the appropriate instructions is guided by the domain knowledge tutoring strategies (Sleeman, 1985).

Tutoring Strategies

Tutoring strategies are means of the computer tutor to impart the knowledge to the student. There are two tutoring strategies that affect the assessment model directly. These are (a) the adaptation of the tutoring package to the student and (b) the limitation of the number of interrupts allowed to the student.

Adaptation to the Student

One of the most important aspects of assessment modeling is the ability of the computer to adapt to the requirements of the student. The following strategies offer useful guidelines to the successful adaptation in response to the behavior of the students (Galdes and Smith, 1990).

- It is useful to look at the cause of a definite error in addition to the type of error made. Before correcting the student, it is important to assess whether the error was one of a careless nature, or whether it resulted from flaw in the student's understanding.
- When a student appears to be struggling with the concept or problem and has not specifically asked for help, it is advisable to allow the student extra time before interrupting. The student's attitude should determine some of the parameters in the tutoring process.
- In specific cases, where the student's error is fundamental, it is useful to teach procedures for error detection instead of just correcting the error.

Limit the Number of Interrupts to the Student

An important tutoring strategy is to ensure that the student is not offered excessive assistance on a topic, unless it is required. The TAPS package should function as a passive observer, offering guidance only where necessary/requested. This method is essential to ensure that the student discovers knowledge through problem solving and experimentation. The following strategies could be adopted (Galdes and Smith, 1990).

- The computer tutor should apply a 'pause' strategy whenever possible for correcting errors that arise when a student is performing a task.
- It is not recommended to interrupt the student for every action that cannot be fully explained, especially if the student is likely to return to the correct solution path at a later stage.
- If the student appears generally confused and is likely to request assistance in the near future, it is advisable to wait for the student to initiate the dialogue, as the student's initiative may give insight into the specific problem.

CONTRIBUTING TECHNOLOGIES

There are various technologies that have, and will, contribute significantly in the development of TAPS packages presently, and in the future. It is assumed that with the current level of progress in the field of computer hardware and software, most of these contributing technologies could dramatically change the environment of TAPS packages. Although computer hardware is a contributing technology in itself, the technologies that are of greatest interest are those of data storage, multimedia, Virtual Reality, Artificial Intelligence, and user-interface design. In the following Section, each of these is briefly described.

Multimedia Attributes

Cairncross and Mannion (1999) described three main attributes of multimedia applications, namely multiple media, interactivity, and delivery control. These attributes can be further shown with its sub-functions or properties as depicted in Figure 1.

According to this model (Figure 1), a given piece of information could be delivered using one or more media element. For example, an image can be used to illustrate a text-based description. The information originally presented on screen can be supplemented by the use of audio, video, and pop-up boxes. Audio is useful as text can be minimized on the screen. Thus, multimedia has the ability to support multiple representations of the same piece of information in a variety of formats.

Figure 1. Key attributes model of multimedia

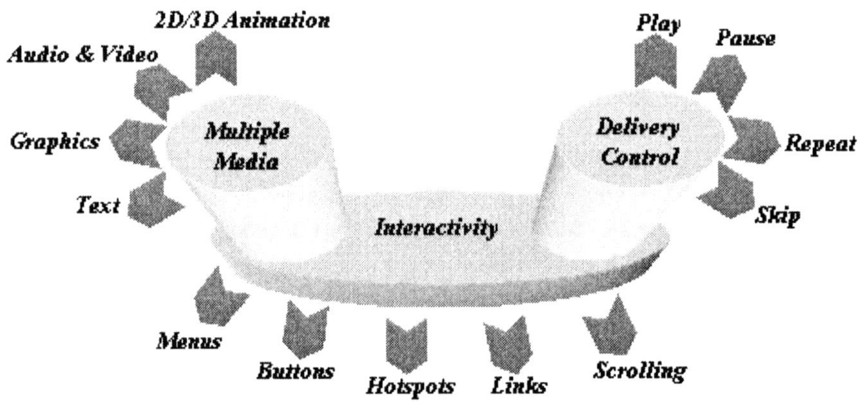

In general, multimedia applications demand some kind of delivery control. The non-linearity offered by many interactive CAL multimedia learning applications provides a learner/user greater navigational freedom. Users may go to any section in a multimedia tutorial and in any order (Cairncross and Mannion, 1999). Dynamic media such as video and audio can be controlled i.e. pausing, playing and repeating clips.

Another attribute is interactivity. Interactivity in multimedia assisted learning applications should go further than simply allowing a user to choose his/her path and, pointing and clicking at various menus and buttons. Most multimedia applications provide some interactivity in which it responds to user instructions. What makes the difference even in a simple educational software is whether the software allows the user to work at his/her own pace, in the order desired, repeating sequences at user's will, manipulate virtual objects on screens, and simulation of experiments or industrial processes (Cairncross and Mannion, 1999).

Data Storage

Data storage devices such as CD-ROM and hard disk are providing educators with better interactive opportunities. The capacity to store vast amounts of text and graphics, along with multiple audio and video tracks, for a reasonable price, makes these storage media ideal for TAPS packages implementation. Latest emerging storage devices such as handy drives and portable hard disks with more than one gigabytes of storage space can also be used to store multimedia data and is more convenient to be transported by students. This enables a wealth of multimedia data to be incorporated into transportable PC-based TAPS packages.

Virtual Reality Attributes

Virtual Reality (VR) is a remarkable technology that allows three-dimensional artificial worlds to be created on computer. What makes this technology unique is that it is possible to move about and interact within these artificial worlds in a way that allows all navigational and manipulative movements made by the user to be emulated in this computer-generated environment (Pimentel and Teixeira, 1993). This is accomplished using immersive VR input devices, such as the head mounted display (HMD) and data glove. Using this equipment, the user believes that he/she is actually immersed in this artificial world. A typical VR system, in general, consists of one or more input devices, several forms of output devices, and a computer to manage all the data.

The goal of VR is essentially to create a new and flexible form of communication between computers and humans. This requires a step-by-step analysis of why humans do what they do so well, and why computers do what they do so inflexibly, and hence finding methods to make the computer simulate what humans do (Lavaroff, 1992). On the other hand, there are also desktop virtual reality (DVR) applications that do not require expensive hardware equipments to be used. DVR requires a PC or laptop, some specialized hardware such as 3-D graphics card, 3-D sound card, a 6-D tracker, a joystick, and software that displays and permits navigation in virtual environments such as Cortona™ and Cosmo™ viewer, and Macromedia™ Shockwave. The delivery control features of DVR as shown in Figure 2 are an extension of the key attributes model of multimedia shown in Figure 1. These new features enable the user to move along any direction on the screen and have the object displayed continuously and updated instantaneously. Therefore, the user could gain a greater understanding of a given problem. In terms of cost, DVR is cheaper than immersive VR systems.

Virtual Reality would be one of the key technologies to influence TAPS packages in the near future. With constant developments and improvements in the field of computer hardware and software, it will eventually be possible to create immersive VR TAPS packages at an affordable price. Such packages could help students learn and understand better i.e. if an environment responds realistically to various inputs given by the student, the student can observe exactly how the system functions in reality.

Artificial Intelligence

The goal of Artificial Intelligence (AI) is to build computerized systems that can make intellectual decisions, comparable to those made by humans (Merrill *et al.*, 1988). One discipline of AI that is particularly relevant to TAPS packages is expert systems. An expert system is a computer-based system comparable of solving complex problems at the competency of a human expert. To do this, the computer usually obtains information from the user about a specific problem, by asking the user a series of questions. The expert system then assesses the user's response to help solve the specific problem (Crews, 1992). Expert systems are especially useful in the field of CAL, particularly in assessing the student, because they can be programmed to evaluate the response of the student and gauge his/her knowledge and understanding of the subject matter. Expert systems can also be used in TAPS packages to construct a dynamic model of the assessment that may be adjusted throughout the tutoring or problem solving session. More information on expert systems is given by Shute and Regian (1993) & Kinshuk (1996).

Figure 2. Extended key attributes model of multimedia and DVR

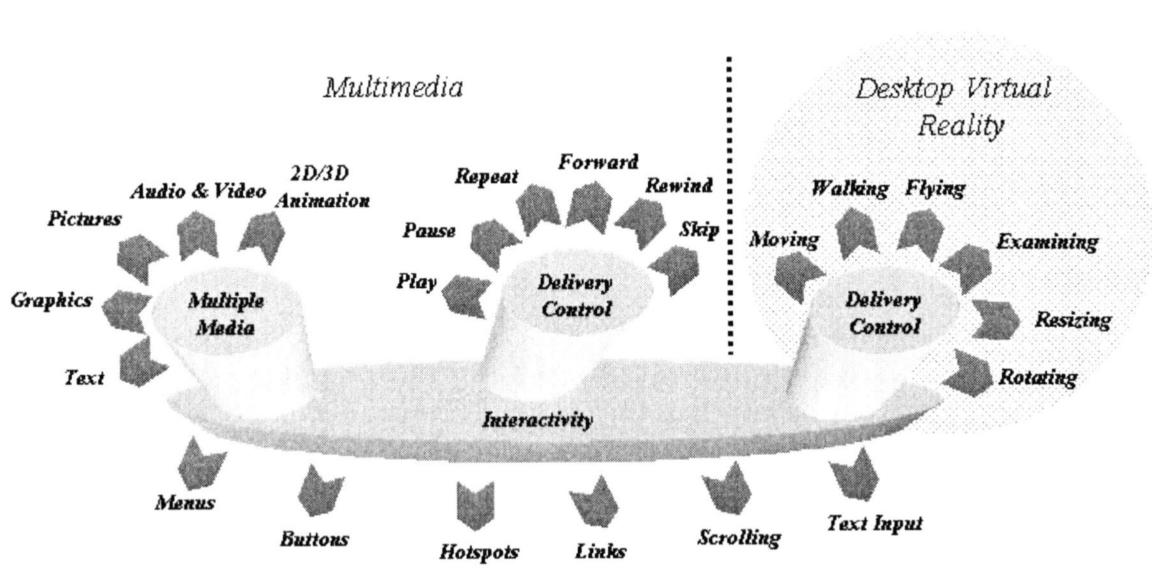

USER INTERFACE DESIGN

Although most components of a TAPS package can be considered as part of the user interface, it is useful to distinguish those aspects that are explicitly concerned with interaction between the computer tutor and the student. This includes the actual presentation of the information, as well as the acceptance of the user input. Since interaction with the student is primarily through the high level / natural language (e.g. English), the package must have abilities to understand the student's response and generate text of its own. There is a strong argument for user interfaces that include only text with this graphical environment (Swaine, 1992).

User interface design (discussed in Chapter 3) should be robust and at the same time allow flexibility to the student. The output should appeal to as many of the student's senses as possible. Future interfaces would likely be more intuitive and more realistic mainly due to emerging technologies, especially in the field of Virtual Reality.

DEVELOPMENT ASPECTS OF TAPS PACKAGES

In the late 1960's and early 1970's 'generative systems' or so-called 'adaptive systems' were employed in which meaningful problems could be generated and solved by using the computer. The intention was to eliminate the need for having pre-stored teaching material, problems, solutions, and associated diagnostics, but instead to generate them. This drastically reduced the memory usage and systems could generate and provide as many problems as the student needed to some desired level of difficulty. Uhr (1969) implemented a series of systems that generated problems in arithmetic that were 'custom made' to a student's performance. Suppes (1967) and Woods and Hartley (1971) produced systems with similar abilities. These programs were restricted to drill-and-practice type exercises in the domain as well as structure. They did not possess any real knowledge of the domain and they could not answer questions. The gap between the student's cognitive processes and the internal workings of the programs was far apart and only parametric summaries of behavior were used to guide problem generation, rather than an explicit representation of the student's knowledge (Sokolnicki, 1991). Yazdani (1986) stated that:

none of these computer-based tutoring systems has human-like knowledge of the domain it is teaching, nor can it answer the serious questions from the students as to "why" and "how" the task is performed.

Hawkes *et al.* (1986) found that computer-based tutoring (CBT) systems had many disadvantages such as:

- CBT systems attempted to produce total or almost complete courses rather than concentrating on building systems for more limited topics.
- CBT systems had no 'knowledge' or 'understanding' of the subject it tutored or of the students themselves.
- CBT systems had severe natural language barrier that restricted user interaction with them.
- CBT systems tended to be static rather than dynamic.

Figure 3. Integrated model of multimedia effects on learning (Hede and Hede, 2002)

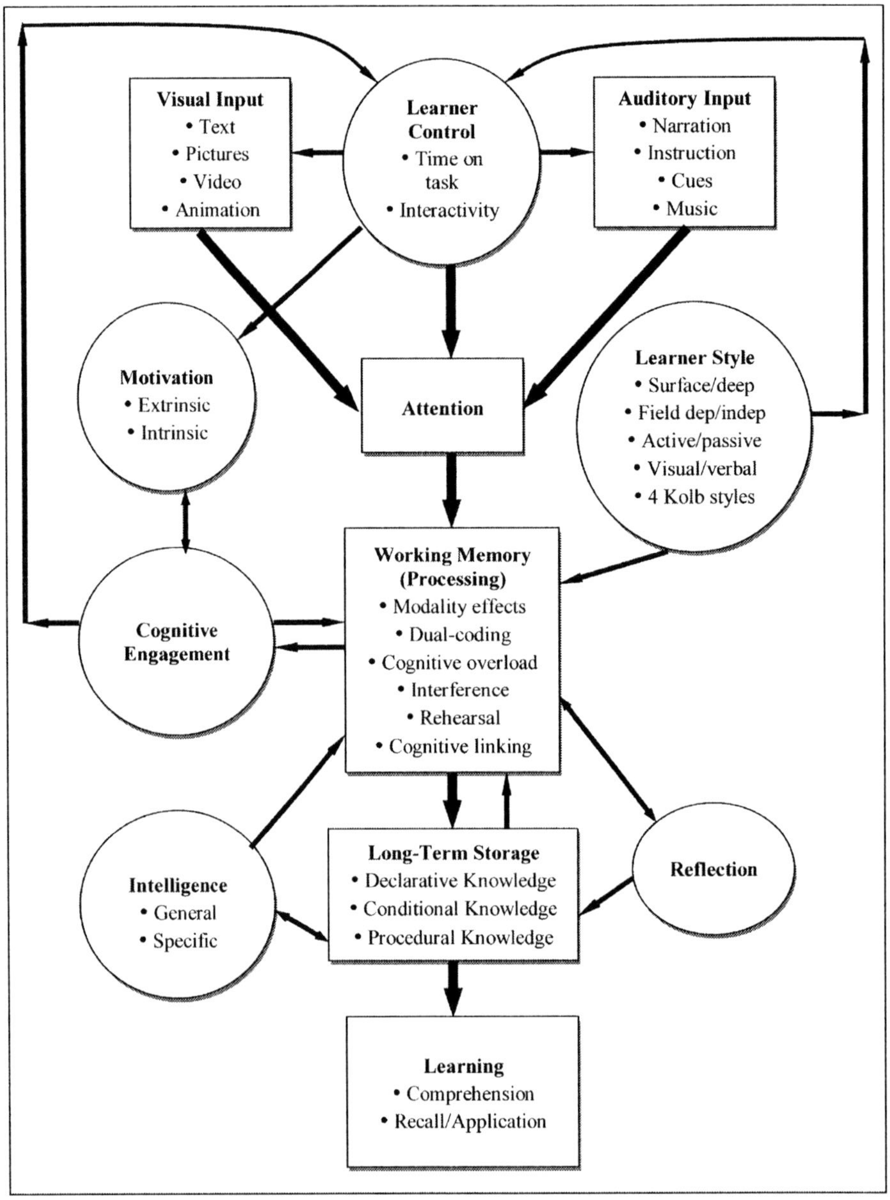

Self (1988) argued that a computer-based tutorial system should have a representation of what is being taught, who is being taught, and how to teach the student. As such, TAPS packages should dynamically analyze the solution and use principles to decide what to do next, rather than simply providing solutions. Providing a truly 'interactive' system was recognized to be a non-trivial task that needed experts from several disciplines. The major features of a TAPS package, therefore, should incorporate the following aspects:

- TAPS packages are suitable learning aids where the knowledge domain of a good human tutor can be maintained.

- TAPS packages provide more detailed analytical errors rather than simple drill and practice.

MULTIMEDIA EFFECTS ON LEARNING

In general, although various multimedia models that describe effects on learning have been proposed in the literature (Garcia *et al.*, 2001), there are currently no specific guidelines available regarding the general configuration of the models. Some of the sound pedagogical and psychological principles that predicate an effective multimedia design model to inform computer tutoring package developers on how to maximize the learning of intended users should be investigated further. According to Garcia *et al.* (2001), these models can help the developer to:

- Decide what multimedia elements are suitable to integrate in the tutoring package.
- Include the appropriate functions and services that should be provided by these elements.
- Decide the kind of communication and material that must flow between each one of these elements.

However, the development construction for delivering effective multimedia instructional materials rooted in engaging multimedia effects is guided by the integrated model of multimedia effects on learning proposed by Hede and Hede (2002) as illustrated in Figure 3.

The various conceptual elements designated by the boxes and ellipses in the model (Figure. 3) represent constructs at the theoretical level as well as the variables at the operational level. Most of the conceptual elements are, in fact, multi-dimensional as indicated by the dot-points, which also designate further constructs/variables. The arrows in the model indicate either a casual or an associative relationship between conceptual elements. The complete model comprises 12 inter-related conceptual elements and their complex interactions that multimedia and learning designers need to be aware of, to account for multimedia effects on learning. For ease of explanation, Hede and Hede (2002) has grouped the elements in the model as follows:

a) *Multimedia input* (three elements: visual input, auditory input and learner control);
b) *Cognitive processing* (two elements: attention and working memory);
c) *Learner dynamics* (three elements: motivation, cognition engagement, and learner style);
d) *Knowledge and learning* (four elements: intelligence, reflection, long-term storage, and learning).

Multimedia Input

The group of elements relating to input making the learners access to the content of the instructional material needs to be addressed in an integrated model (Figure 3). Vision and learning form the two primary input modalities where visual input takes the form of text, pictures, diagram, video and animation while auditory input takes the form of narration or commentary, instructions and music. The effective combinations of these inputs are used in developing multimedia.

According to Farrell and Moore (2000) and Tripp (2001) design features aid the learner in navigating through the various sources of information provided in the multimedia environment. Content-rich databases for more detail information can be made accessible to the learner using links and hyperlinks. Multimedia provides learners with varying levels of interactivity, which has been conceptualized in many different ways in the literatures (Sims 2000; Kettanurrak *et al.,* 2001; Cairnscross, 2002). McNeil and Nelson (1991) pointed out that there was evidence of learner control being less efficient than program control though it was assumed to be a positive feature of multimedia. This was further studied by Stemler (1997) who concluded that learner control in multimedia needs to be tailored to the learner's capabilities.

Cognitive Processing

The two elements for an integrated model i.e. attention and working memory for the next group of factors, which are involved in processing the information accessed through the input process. In his works conducted in 1976 and 1980 Hede came to the conclusion that attention serves to focus the learner's concentration on one input at a time though there is evidence that several inputs can be monitored simultaneously at a perceptual level.

The main processing takes place in the working memory and it is here that the real complexities on multimedia come into play. In 1992 Baddeley introduced the construct of working memory, which has widely been accepted by researchers like Niaz and Logie, (1993), Mayer *et al.,* (1996) and Mousavi *et al.,* (1995). Working memory comprises an executive processor plus two short-term stores, i.e. a "phonological loop" and a "visuo-spatial sketchpad." According to Baddeley in 1992 the verbal material (covering both text and narration) is retained beyond a few seconds by "sub-vocal rehearsal" in the phonological loop.

Factors affecting the way the working memory processes multimedia information are as follows:

1. Modality effects, which according to Penney, (1989); Mousavi *et al.,* (1995); and Tindall-Ford *et al.*, (1997) are a result of dual coding that enables both auditory and visual inputs to be processed simultaneously.
2. Cognitive overload occurs when input exceeds the limited capacity of working memory, i.e. when identical information from one source disrupts semantic processing of information from another source.
3. Retention of information depends on whether it is subjected to rehearsal.
4. Cognitive linking establishes referential connections between verbal and visual representations (Mayer and Anderson, 1991; Mayer and Sims, 1994; Mayer *et al.,* 1996).

Learner Dynamics

There are three conceptual elements relating to learner dynamics. According to Taylor *et al.,* (1997), the key variable in learning is the first motivation. In 1998, Najjar studied that extrinsic motivational factors such as the design features of a multimedia package are difficult to provide initial incentive for learners to access the material but sustained effort occurs only when they encounter intrinsic motivational factors provided by interesting and challenging content which according to Stoney and Oliver (1999) leads to cognitive engagement (a process where by learners become motivated to take full control of their learning). The integrated model sees the various motivational factors impacting on learner control, where, the time and effort learners devote in engaging with multimedia are considered specifically.

There are a number of ways of classifying learner style, which influence the way they access multimedia. Dillon and Gabbard (1998) reviewed three approaches to learner style. The first distinguishes between field dependence and field independence that determines the extent to which a learner relies on the context in which information is presented. The second approach classifies learners according to whether they are surface processors or deep processors of information, the former relying on memorization and rehearsal whereas the latter using content structuring techniques which seemed to be more effective in multimedia environment. The third approach is based on the activity versus passivity of learners – different feature of multimedia presentation will be more appropriate for active and passive learners. Smith and Woody (2000) adopted another learning approach that distinguishes between visual and verbal processors. They found that multimedia is best suited for learners with highly visual style. Finally, a number of recent studies have examined multimedia in terms of Kolb learning style inventory based on (a) diverges, (b) assimilators, (c) converges and (d) accommodators (Karakaya *et al.*, 2001; Kettanurak *et al.*, 2001, Kraus *et al.*, 2001). These different approaches to learning style need to be accommodated by an integrated model as discussed in Chapter 2.

Knowledge and Learning

Four elements that form the final group of factors are as follows: (a) intelligence, (b) reflection, (c) long-term storage and (d) learning. In 1998, Fetherson advocated the view that intelligence is multi-faceted involving seven different intelligences. According to him the more of these that are stimulated by a multimedia package, the more effective it will be. In 1997, Taylor *et al.*, stated that the process of reflection relates to self-directed learning and entails learners thinking critically about their current knowledge and their learning strategies.

The long-term storage is where one's knowledge is stored. Long-term storage receives processed information from working memory but also supplies working memory with the basis for cognitive linking whereby connections are established between new content and what is already known. In 2001, Yildirim *et al.*, felt the need to distinguish between declarative, conditional and procedural knowledge all of which are involved in the learning process. Research done by Kalyuga *et al.*, (1998); (2000), (2001a); Kalyuga *et al.*, (2001b) has shown that the relative effectiveness of different multimedia strategies varied with the level of learner knowledge and experience. The final element in any model of multimedia effects is, of course, learning which comprises of the immediate level of comprehension of material accessed through multimedia plus the ability to recall and apply one's acquired knowledge.

Cairncross (2002) deduced seven functions of how multimedia can be used to support key aspect of learning and teaching. A summary of the functions is shown in Table 1.

EFFORTS AND APPROACHES OF CAL PACKAGES DEVELOPMENT IN ENGINEERING

Since the introduction of CAL in 1960's and the advancement in newer instructional delivery systems, a number of efforts have been made in the development of CAL packages for teaching and learning, with particular emphasis in engineering. Some of these efforts are relatively comprehensive, aspiring to provide almost complete teaching systems for students to adapt in their learning. Other efforts are more focused for example, within Mechanics of Materials subject, such as to construct shear force and

Table 1. Interactive multimedia support for teaching and learning (Cairncross, 2002)

Learning and Supporting Activities	Associated Teaching Function	Interactive Multimedia Support	Illustrative Examples
Setting the Scene and Maintaining Interest	Orienting	Multiple media User control over delivery	• Text explanations with audio guidance and links to help.
	Motivating		• Audio overview with text bullet points and graphics. • Personalisation through getting user to log-on and remembering details and using name.
Conceptualisation	Presenting	Multiple media User control Over delivery	• Use of most appropriate medium for information type e.g. animation to illustrate dynamic processes. • Use of redundancy (where appropriate) to support individual preferences e.g. text description with graphical representation • Content structuring to support different levels of interest. • Provision of different navigation routes through applications.
	Clarifying		• Hyperlinks to provide pop-up glossary or context sensitive help • Hyperlinks to concrete examples • Hyperlinks to supporting material when questions answered wrongly • Interactive dialogues.
	Elaborating		• Thematic links to associated material.
Construction	Consolidation	User activity	• Interactive simulations or experiments.
	Confirmation		• On-line assessment
Internal Reflection	Consolidation	Multiple media User control over delivery User activity	• Decision making i.e. where to go next. • Providing alternative viewpoints
	Confirming		• Self-assessment questions.

bending moment diagrams, within Mechanics Statics subject such as analysis of structures, and within Mechanics Dynamics subject such as the study of projectile, curvilinear, and rectilinear motions. Some early developments of CAL packages in engineering include:

- The work by Grammoll and Abbanat (1996) contains about forty-three creative real-world problems that cover a typical dynamics course. The problems were illustrated using text, graphics, animations and audio. Each problem contained a simulation that models it and gives the learner the power to dynamically manipulate physical systems and explore engineering concepts.
- The work by Vaughan (1998) in the introduction of Fluids Mechanics course contains ten modules covering topics ranging from fluid statics to boundary layers. In addition to those modules, a laboratory simulation was also developed. This simulation consisted of six sections with experiments covering measurements of basic fluid properties, pressure and velocity measurements, applications of Bernoulli equation, applications of momentum equation, and pipe friction. By using an active learning approach, instructional technology can benefit students of varying backgrounds and skill levels. The students are able to view the same information from several perspectives, strengthening connections and transferability.

- The work by McMahon (2000) on Introduction to Materials Engineering CD-ROM multimedia-based course serves as both a comprehensive one-semester material education course for non-materials science and engineering majors. The course employs computer-presented tutorials that can be utilized in several ways; as a replacement for classroom lectures in courses where the class meetings are devoted to studio-type active learning, as a supplement to classroom lectures that are illustrated by the animations provided, as a basic course in institutions where faculty from other branches of science or technology serve as coaches for specialized subjects such as materials science and engineering, or as a self-study course for those who must pursue the subject on their own.
- The work by Callister (2003) on an Introduction to Materials Science and Engineering CD-ROM contains eight dynamic learning modules where students can view and manipulate 3-D projections and activate animations that bring these concepts to life. In addition, students have the opportunity to improve their problem-solving skills and at the same time evaluate their progress.

In general, most CAL packages have the capability to control the presentation of multimedia types of information including text, sound, 2-D and 3-D graphics and geometric models, interactive video, and even animation of virtual reality, for instruction. This technological advancement provides a better and great potential for CAL packages to be more effective than human tutors since human tutors do not have the ability to generate or control the presentation of multimedia information when using conventional teaching aids such as the overhead projector or the white board. Technology such as virtual reality has also influenced CAL packages and is being researched for its effectiveness in education (Nicole and Tracy, 2003).

A more recent research by The Intelligent Systems Application Center (ISAC) (2004) includes the development of applications that are used to enhance educational delivery. A principle activity of ISAC is developing educational software that is interactive, intelligent, and can be used for training in specific professional areas that provide expert consultations including engineering problems. One such application is the interactive multimedia intelligent tutoring system (IMITS). A unique quality of this tutoring system is that it is concerned with delivering basic principles, reasoning skills, as well as encouraging creativity. The result is a set of tutoring systems that have the capability of dynamically organizing and modifying a single lesson plan and a series of lesson plans based on the ability of the students.

Cairncross (2002) noted that multimedia brings with it many benefits over other methods of presenting information. As such multimedia is making significant contributions in variety of domains, including job training, scientific and medical field, help desks and other areas. Powerful, sophisticated systems can now be created on less expensive microcomputers. However, in order to be successful, designers or developers of computer learning packages need to work closely with human tutors in order to avoid repeating problems associated with conventional CAL when attempting to integrate interactive multimedia based learning into the classroom. Additionally, Cairncross and Mannion (2001) stressed that it is necessary to re-examine the key features of multimedia and their use to enhance learning.

The bottom line of high costs of development, delivery, and maintenance schedules are considered to be major set back of CAL packages. The coach/learner interaction that tries to anticipate everything ahead of time is going to be costly, if not a hopeless goal (Kinshuk, 1996). CAL packages methods are beginning to be explored to reduce the high costs of development, delivery, and maintenance. One way to reduce cost is to develop modular components that can be reused (Janet, 2002). For example, the system interface component and the rule base for processing procedural steps and goals could be reusable components in a learning package.

A great deal of attention has also been paid to the interface design of TAPS packages due to the fact that the interface provides the user with all the functionality of the software.

In summary, the design of a TAPS package requires various key components to be successfully integrated. These include the appropriate choice of learning scenario, a comprehensive domain and pedagogical knowledge, and a dynamic student assessment model. With these components firmly in place, the TAPS packages have the ability to present relevant course materials at a level of detail ideally suited to the individual style of learning. In addition, the TAPS packages should be able to constantly assess the capabilities of the student and provide adequate feed back throughout the problem solving process. A good TAPS package demands a great deal of computer intelligence to be incorporated into the problem-solving package. Presently, it is not possible to represent all the characteristics of a human tutor in TAPS packages. Consequently, it is permissible to compromise on the knowledge representation of a human tutor to a certain extent to produce a tutoring and problem-solving package that can teach students in a more effective manner than other already existing CAL packages.

REFERENCES

Baddeley, A. D. (1992). Is working memory working? *Quarterly Journal of Experimental Psychology*, *44A*(1), 1–31.

Burns, H. (1991). Intelligent Tutoring Systems. Evolutions in Design. *The Evolution of Intelligent Tutoring Systems: Dimensions of Design* (pp. 1 – 11). Mahwah, NJ: Lawrence Erlbaum Associates.

Cairncross, S. (2002). *Interactive Multimedia and Learning: Realizing the Benefits*. PhD Thesis, Napier University, Scotland.

Cairncross, S., & Mannion, M. (1999). How multimedia functions in engineering education. *Engineering Science and Education Journal*, *8*(3), 100–107. doi:10.1049/esej:19990302

Cairncross, S., & Mannion, M. (2001). Interactive Multimedia and Learning: Realizing the Benefits. *Innovations in Education and Teaching International*, 156–164.

Callister, W. D. (2003). *Student Learning Resources. Materials Science and Engineering an Introduction* [CD-ROM]. Chichester, UK: John Wiley & Sons Inc.

Chu, R. W. Mitchell, C. M. & Govindaraj, T. (1989). *Characteristics of an ITS that evolves from the tutor Operator's Assistant*, (pp. 778 – 783).

Crews, P. (1992). Instruction: An unfulfilled promise of Expert System Technology. *Engineering Systems Design and Analysis*, *4*, 257–263.

Dillon, A., & Gabbard, R. (1998). Hypermedia as an educational technology: A review of the quantitative research literature on learner comprehension, control and style. *Review of Educational Research*, *68*(3), 322–349.

Farrell, I., & Moore, D. (2000). The effect of navigation tools on learners' achievement and attitude in a hypermedia environment. *Journal of Educational Technology Systems*, *29*(2), 169–181. doi:10.2190/BJW2-XE9H-HMPM-VWRN

Fetherson, T. (1998). A socio-cognitive framework for researching learning with IMM. *Australian Journal of Educational Technology*, *14*(2), 98–106.

Galdes, D. K., & Smith, P. J. (1990). Building an Intelligent Tutoring System: Some guidelines from a study of human tutors. *Proceedings of the Human Factors Society 34th annual meeting*, (pp. 1407 – 1411).

Garcia-Cabrera, L. Rodriguez-Fortiz, M. Jose, & Parets-Llorca, J. (2001). A Cognitive Model for Adpative Hypermedia Systems. In *The 1st International Conference on Web Information Systems Engineering, Workshop on WWW Semantics* (Vol. 2, pp. 29-33). Washington, DC: IEEE Press.

Gramoll, K., & Abbanat, R. (1996). *Multimedia Engineering Dynamics*. Reading MA: Addison-Wesley.

Hawkes, W. L., Sharon, J. D., & Kandel, A. (1986). *Fuzzy expert systems for an intelligent computer based tutor.* Technical Report No. 86-5. Learning Systems Institute, Centre of Educational Technology, Florida State University.

Hede, A. J. (1976). *Selective attention as a perceptual process in vision and hearing*. Unpublished Doctoral Dissertation, University of Sydney Australia.

Hede, A. J. (1980). The pre-post cueing technique in the study of selective attention. *Acta Psychologica*, *46*, 115–128. doi:10.1016/0001-6918(80)90004-9

Hede, T., & Hede, A. (2002). Multimedia effects on learning: Design implications of an integrated model. In S. McNamara and E. Stacey (Eds), *Untangling the Web: Establishing Learning Links Proceedings ASET Conference 2002*, Melbourne, (pp. 7-10). Retrieved January 23rd, 2003 from http://www.aset.org.au/confs/2002/hede-t.html

Hibbeler, R. C. (2001). *Engineering Mechanics: Dynamic* (9th Edition). Upper Saddle River, NJ: Prentice-Hall.

The Intelligent Systems Application Center (ISAC) (2004). Retrieved May 20, 2005 from http://www.temple.edu/imits/isac.htm

Janet, F. J. (2002). *Sequence Learning with Intelligent 3-D Practice Environments*. MITRE Technical Papers, Paper 2.

Kalyuga, S., Chandler, P., & Sweller, J. (1998). Levels of expertise and instructional design. *Human Factors*, *40*(1), 1–17. doi:10.1518/001872098779480587

Kalyuga, S., Chandler, P., & Sweller, J. (2000). Incorporating learner experience into the design of multimedia instruction. *Journal of Experimental Psychology*, *92*(1), 126–136.

Kalyuga, S., Chandler, P., & Sweller, J. (2001a). Learner experience and efficiency of instructional guidance. *Educational Psychology*, *21*(1), 5–23. doi:10.1080/01443410124681

Kalyuga, S., Chandler, P., Tuovinen, J., & Sweller, J. (2001b). When problem solving is superior to studying worked examples. *Journal of Experimental Psychology, 93*(3), 1–10.

Karakaya, F., Aincough, T. L., & Chopoorian, J. (2001). The effects of class size and learning style on student performance in a multimedia-based marketing course. *Journal of Marketing Education, 23*(2), 84–90. doi:10.1177/0273475301232002

Kettanurak, V., Ramamurthy, K., & Haesman, W. D. (2001). User attitude as a mediator of learning performance improvement in an interactive multimedia environment: An empirical investigation of the degree of interactivity and learning styles. *International Journal of Human-Computer Studies, 54*, 541–583. doi:10.1006/ijhc.2001.0457

Kinshuk, R. (1996). *Computer Aided Learning for Entry accountancy students.* PhD Thesis, De Montfort University, UK.

Kraus, L. A., Reed, W. M., & Fitzgerald, G. E. (2001). The effects of learning style and hypermedia prior experience on behavioral disorders knowledge and time on task: A case-based hypermedia environment. *Computers in Human Behavior, 17*, 125–140. doi:10.1016/S0747-5632(00)00030-3

Lavaroff, N. (1992). *Virtual Reality Playhouse* (pp. 15 – 27). Corte Madera, CA: The Waite Group.

Mayer, R. E., & Anderson, R. B. (1991). Animations need narrations: An experiment test of a dual-coding test of a hypothesis. *Journal of Educational Psychology, 83*(4), 484–490. doi:10.1037/0022-0663.83.4.484

Mayer, R. E., Bove, W., Bryman, A., Mars, R., & Tapangco, L. (1996). When less is more: Meaningful learning from visual and verbal summaries of science textbook lessons . *Journal of Educational Psychology, 86*(3), 389–401. doi:10.1037/0022-0663.86.3.389

Mayer, R. E., & Sims, V. K. (1994). For whom is a picture worth a thousand words? Extensions of a dual-coding theory of multimedia learning. *Journal of Educational Psychology, 88*(1), 64–73. doi:10.1037/0022-0663.88.1.64

McMahon, C. J. (2000). *Tutorials for Introduction to Materials Engineering* [CD-ROM]. Merrion Media.

McNeil, B. J., & Nelson, K. R. (1991). Meta-analysis of interactive video instruction: A 10-year review of achievement effects. *Journal of Computer-Based Instruction, 18*(1), 1–6.

Merill, P. F. Hammons, K. & Tolman, M. N. (1988). *Computers in Educations,* (2nd Ed.), (pp. 11 – 19). Boston: Allyn and Bacon.

Mousavi, S. Y., Low, R., & Sweller, J. (1995). Reducing cognitive load by mixing auditory and visual presentation modes. *Journal of Educational Psychology, 87*(2), 319–334. doi:10.1037/0022-0663.87.2.319

Najjar, L. J. (1998). Principles of educational multimedia interface design. *Human Factors, 40*(2), 311–323. doi:10.1518/001872098779480505

Niaz, M., & Logie, R. H. (1993). Working memory, mental capacity and science education: Towards an understanding of the "working memory overload hypothesis. *Oxford Review of Education, 19*(4), 511–525. doi:10.1080/0305498930190407

Nicole, S., & Tracey, H. (2003). National Center on Accessing the General Curriculum (NCAC) Report. Retrieved June 2003 from www.cast.org/system/galleries/download/ncac/

Penney, C. G. (1989). Modality effects and the structure of short-term memory. *Memory & Cognition, 17*(4), 398–422.

Pimentel, K., & Teixeira, K. (1993). V*irtual Reality – through the new looking glass* (1st Ed.), (pp. 19-252). New York: McGraw-Hill.

Rickel, J. W. (1989). Intelligent Computer Aided Instruction: A survey organized around system components. *IEEE Transactions on Systems, Man, and Cybernetics, 19*(1), 40–57. doi:10.1109/21.24530

Self, J. A. (1988). Student models: what use are they? *Artificial Intelligence Tools in Education,* (pp. 73 – 86).

Shute, V. J., & Regian, J. W. (1993). Principles for evaluating intelligent tutoring systems. *Journal of Artificial Intelligence in Education, 4*(2/3), 245–271.

Sims, R. (2000). An interactive conundrum: Constructs of interactivity and learning theory. *Australian Journal of Educational Technology, 16*(1), 45–57.

Sleeman, D. H. (1985). UMFE: A user modeling front-end subsystem. *International Man-Manchine Studies,* (23), 71 – 88.

Smith, S. M., & Woody, P. C. (2000). Interactive effect of multimedia instruction and learning styles. *Teaching of Psychology, 27*(3), 220–224. doi:10.1207/S15328023TOP2703_10

Sokolnicki, T. (1991). Towards knowledge-based tutors: a survey and appraisal of Intelligent Tutoring Systems. *The Knowledge Engineering Review, 6*, 54–95. doi:10.1017/S0269888900005622

Steif, P. S. (2003). Design and evaluation of problem solving courseware modules for mechanics of materials. *Journal of Engineering Education*, 1 – 30.

Stemler, L. K. (1997). Educational characteristics of multimedia: A literature review. *Journal of Educational Multimedia and Hypermedia, 6*(3/4), 339–359.

Stoney, S. & Oliver, R. (1999). Can higher order thinking and cognitive engagement be enhanced with multimedia? *Interactive Multimedia Electronic Journal of Computer-Enhanced Learning, 1*(2).

Suppes, P. (1967). Some theoretical models for mathematics learning. *Journal of Research and Development in Education, 1*, 5–22.

Swaine, M. (1992). Multimedia and the art of user interface design. *Dr. Dobbs' . Journal, 17*(7), 113–117.

Taylor, J., Summer, T., & Law, A. (1997). Talking about multimedia: A layered design framework. *Journal of Educational Media, 23*(2/3), 215–241.

Tindall-Ford, S., Chandler, P., & Sweller, J. (1997). When two sensory modes are better than one. *Journal of Experimental Psychology. Applied*, *3*(4), 257–287. doi:10.1037/1076-898X.3.4.257

Tripp, S. (2001). Cognitive navigation: Toward a biological basis for instructional design. *Educational Technology and Society*, *4*(1), 41–49.

Uhr, L. (1969). *Teaching machine programs that generate problems as a function of interaction with students*. Paper presented at the 24th National Conference of Education, Virginia.

Vaughan, R. (1998). The Use of Multimedia in Developing Undergraduate Engineering Courses. *Journal of Materials, 5*(5). Retrieved February 23, 2001 from www.tms.org/pubs/journals/JOM/9805/Voller/Voller-9805.html

Woods, P., & Hartley, J. R. (1971). Some learning models for arithmetic tasks and their use in computer based learning. *The British Journal of Educational Psychology*, *41*, 35–48.

Yazdani, M. (1986). Intelligent tutoring systems survey. *Artificial Intelligence Review*, *1*, 43–52. doi:10.1007/BF01988527

Yildirim, Z., Ozden, M. Y., & Aksu, M. (2001). Comparison of hypermedia learning and traditional instruction on knowledge acquisition and retention. *The Journal of Educational Research*, *94*(4), 207–214.

Chapter 7
Development and Usage of TAPS Packages in the Mechanical Engineering Course

INTRODUCTION

The Mechanical Engineering course is largely based on practical skills and requires the acquisition of basic skills and domain knowledge before applying them on real problems. In order to design and develop a technology assisted problem solving (TAPS) package particularly to guide students in learning and solving engineering problems, it is necessary to be acquainted with its development and its process of realization in practical terms in computer software. User interface design has been applied in learning environments as discussed in previous Chapter 3. Therefore it is informative to discover the extent to which they help engineering students in their learning and thereby be incorporated in TAPS packages. This examination includes an overview of good practice in the positioning and operation of navigational features, visual screen presentation, the nature of presentation, help and feedback and views on the role of the learner in using the TAPS packages. This Chapter discusses the need to learn practical Mechanical Engineering skills and reviews the tutorial and situational learning approaches. Additionally the Chapter provides an overview of TAPS packages and the approach adopted for problem solving and student learning.

DOI: 10.4018/978-1-60566-764-5.ch007

THE MECHANICAL ENGINEERING SUBJECTS

In general Mechanical Engineering subjects confound students with a combination of physical laws and engineering examples that must be interpreted together. The laws already very clear to the teaching staff do not make sense to the students until they are applied to realistic problems. These problems are not clear until segregated by appropriate use of the physical laws. The student needs experience in a range of related cognitive and social areas in order to succeed. However, not all students have the underlying psychological and cultural background to make use of the learning resources that are provided (Scott, 1996). Therefore it is necessary to evaluate if additional tutoring packages such as TAPS packages could help them learn and solve engineering problems better.

STUDENTS' CONCEPTIONS AND PROBLEM SOLVING IN ENGINEERING MECHANICS

Research has shown that in general students studying physics and engineering subjects encounter many difficulties in understanding the concepts of Engineering Mechanics. For example in kinematics topic, in a study of student understanding of two-dimensional motion, diagrams of trajectories of moving objects were shown to five students in an introductory university course and to five physics faculty (Reif and Sue, 1992). The participants were told whether the objects were speeding up, slowing down or moving with constant speed and were asked to draw the acceleration vectors at specified points. The novices did very poorly at these tasks; even the experts had some difficulties. A detailed analysis of how the two groups approached these tasks enabled the investigators to identify the underlying knowledge and skills required for successful performance.

Some investigations have focused on student understanding of the graphical representations of motion. A descriptive study that extended over several years and involved several hundred-university students helped identify a number of common difficulties encountered by students in making connections between the kinematical concepts, their graphical representations, and the motions of real objects (McDermott *et al.*, 1987). Another study identified that students have difficulties with the graphical representation of a negative velocity (Goldberg and Anderson, 1989).

On the other hand, the topic of dynamics and misconceptions about the relationship between force and motion has been extensively studied. Less well documented are difficulties students have in interpreting the relationships between force and more complex concepts, such as work, energy, and momentum. Some samples of investigations reported in the literature on student understanding of mechanics course can be summarized as follows:

- Prior to instruction, more than 100 students in an introductory university Mechanics course were given a short-answer test on concepts of force and motion (Champagne *et al.*, 1980). The test used a technique abbreviated as D.O.E. (demonstration, observation, explanation). The results revealed that the students, who had previously studied physics, had mixed ideas such as the following: a force will produce motion; a constant force produces constant velocity and the magnitude of the velocity is proportional to the magnitude of the force; acceleration is due to an increasing force; and in the absence of forces, objects are either at rest or slowing down. The results of another study also indicated that both before and after an introductory course in Mechanics, many students seemed to believe that motion implies a force (Clement, 1982).

- In a study involving curvilinear motion and trajectories of moving objects, about fifty undergraduates were asked to trace the path that a pendulum bob would follow if the string were cut at each of four different positions along its path (Caramazza *et al.*, 1981). Only one-fourth of the students gave a correct response.
- Other studies have examined student difficulties with situations involving gravity. A study of several hundred first-year university students in Australia involved the use of simple lecture demonstrations related to gravity (Gunstone and White, 1981). For example, students were asked to compare the time it would take for an equal-sized steel and plastic balls to fall from the same height. On this task, 75% of the students gave different answers.

Since the Engineering Mechanics problems mentioned above are common among undergraduates of many higher learning institutions and have been experienced by University Tenaga Nasional (UNITEN) instructors and students, it is viable to design and use TAPS packages to help engineering students to understand Engineering Mechanics concepts and to apply these concepts in solving problem. The primary focus of TAPS packages is directed to students experiencing difficulties with Mechanical Engineering.

TAPS PACKAGES IN MECHANICAL ENGINEERING SUBJECTS

Since engineering subject involves a simultaneous mix of mathematics and physics in a challenging way to students Vallim (2006), some instructors, in general, have made additional efforts to develop and use multimedia computer aided learning packages as described in Chapter 4 for teaching. The following are some difficulties experienced by the instructors in using conventional teaching methods in engineering, compelling them to turn to the multimedia packages as an additional learning aid:

- In the area of engineering, the traditional communication model follows a one-dimensional, linear path that focus on the instructor/lecturer as the most important element of a communication transaction. This model does not take into account the level of the learners. In addition, traditional learning methods could not engage the learners in visualization tasks and work on virtual experiments (Kahn, 1992; Janson, 1992).
- Engineering Mechanics Dynamics, like many other engineering subjects, is fundamentally about problems solving through the application of scientific principles. The engineering problems are often complex and relationships among the variables of an experiment can be difficult to visualize (Scott, 1996).
- Traditionally, problems in Engineering Dynamics are presented to the student as a combination of schematic diagrams and text descriptions. The shapes and lines that make up the schematic diagram have very specific engineering meanings, and the words accompanying the diagram also give rise to student error because critical information about the solution of the problem is often concealed in the text in unexpected ways (Scott, 1996).
- Theory oriented approach results in some disparity between text coverage and student comprehension (Ratan and Mitty, 1997).

- One of the difficult issues to deal with engineering within the curriculum at the introductory level is the process of abstraction of real and practical situations into mathematical models (Gramoll, 2001).
- Although many forms of learning aids have been used by educationalist to support them in their teaching (Fogler *et al.*, 1992; Squires *et al.*, 1992), there is a need to provide better-enhanced learning aids. For example multiple tools such as calculator, glossaries of words, and electronic notepad can be integrated in a single learning package that can perform multiple tasks simultaneously, is user-friendly, and caters learner's requirements and could guide the learner when reaching an impasse (wrong answers).
- In general, the feedback that students receive on their homework is relatively ineffective. Feedback usually comes too late; solutions are often made available to students after the week's homework is complete. By then, however, students are often focused on another course, on the next homework, or the posted solutions (Steif, 2003).
- Engineering Dynamics subject is difficult to understand from the textbook alone because there are many cognitive steps that lead from a problem through a series of steps to solution. Subsequently, this scenario creates additional educational difficulties, such as some learners lack the ability to translate mathematical word problems into the form necessary for effective computation and poor visualization of the problem that ultimately leads to lack of interest in the subject matter.

Although there are many conventional computer aided learning (CAL) packages available in the field of mechanical engineering, much of the efforts in the engineering CAL packages have attempted to replace the lecture and not focus on problem solving skills. Multimedia based technologies have the potential of providing a mean for dealing with the aforementioned issue in a dynamic (animated), provocative, and cost-effective manner that not only will increase the effectiveness of the educational program but will also increase the quality of the resulting students.

As the demand for more economical and effective learning packages increases, this study aimed to design and evaluate-teaching problem solving skills in engineering subjects with nontraditional approaches in the facilitation of student learning. These approaches include the use of computers with contributing technologies such as multimedia, simulation, desktop virtual reality, and visualization.

The next Section describes the approach adopted in developing the TAPS packages (selected engineering problems) and discusses how this approach helps in building up the essential problem solving skills while improving the theoretical understanding of learners.

DEVELOPMENT OF TAPS PACKAGES

This Section describes the approach taken to integrate computer-based technologies in problem solving learning environment, subsequently termed as technology assisted problem-solving (TAPS) packages. To assist students who need additional support in applying principles presented in lectures to problems in the subject matter, four design approaches were used in developing the TAPS packages namely 2-D graphics and animation (design approach 1), coach-based environment (design approach 2), 3-D virtual environment (design approach 3), and desktop virtual reality (design approach 4). These TAPS packages can be classified as cognitive tools for learning, problem solving, testing, and simulation as described in Chapter 2. These approaches are further explained through Sections 7.6 – 7.9. The reasons for employing TAPS packages can be summarized as follow:

- To use and store the knowledge of experienced instructors (human) and make the same easily accessible to the students;
- To develop a suitable user interface for simplifying the difficult engineering concepts;
- To help students who need additional support in applying principles presented in lectures to problems to acquire problem solving skills;
- To provide encouragement to students in independent learning by incorporating simple intelligence (expert system like rules) in the TAPS packages;
- As an attempt to improvise the limitations of the already existing computer based learning packages thereby making them more acceptable as effective learning aids in UNITEN.

The TAPS packages developed for this study used selected engineering problems that are difficult to understand by first year engineering undergraduates taking the Engineering Mechanics and Dynamics subjects at UNITEN. Since the information, diagrams and sketches are presented in a static way in engineering textbooks, multimedia and desktop virtual reality technologies were found to be a suitable alternative in delivering technical information to students in the subject matter. For example, each problem-solving step in the TAPS package can be narrated and shown in an animated form to help students understand the problem being presented.

The TAPS packages for this study were design using the ADDIE model (as detailed in Table 2, Chapter 2) and developed using commercial application systems such as Microsoft Visual Basic® and Macromedia Director®, as its main interface environments. The employment of these technologies is intended to enhance students understanding and visualization of mechanics problem that otherwise would be difficult to understand via the traditional method of learning from a textbook. These authoring environments provide tools to develop user interactions, enter text, perform animation in two and three dimensions, evaluate user input, and integrate multimedia attributes such as audio, video, animations, and graphics.

The Macromedia Director MX improves productivity even more, as it can be used to construct highly interactive and quality materials that work perfectly with wide and narrow band-widths, regardless of the resolution of the monitors. It also facilitates the use of high quality video files and support for 3-D geometric models designed using Alias Maya® (a 3-D modeling tool).

A complete list of authoring systems employed in the development of TAPS packages and other editing tools such as sound, video, and 3-D modeling is summarized in Table 1.

Director and Visual Basic have generally been the authoring environments of choice for this study because of the ease in which the TAPS packages can be constructed. Additionally, Macromedia Direc-

Table 1. Authoring Systems employed in developing the TAPS packages.

TAPS Package	Authoring Systems, 3-D Modelling, Sound, and Image-Editing Tools
1. 2-D Graphics and Animation	Macromedia Director®, Microsoft Visual Basic® and, Adobe Photoshop®
2. Coach Based Environment	Microsoft Visual Basic®, Adobe Photoshop® and, Cakewalk Sonar®
3. 3-D Virtual Environment	Macromedia Director®, Maya®, Adobe Photoshop® and, Cakewalk Sonar®
4. Desktop Virtual Reality Environment	Macromedia Director®, Maya®, Adobe Photoshop® Cakewalk Sonar® and, 3-D Producer 1.1®

tor authoring tool allows more control over the program in the package under development. Selected sample source codes are provided in Appendix B.

2-D GRAPHICS AND ANIMATION TAPS PACKAGE (DESIGN APPROACH 1)

This TAPS package examined various levels of interaction for a tutorial and problem-solving topic on Structural Analysis in the Engineering Mechanics Statics subject as shown in Appendix C. The aim of this package is to enhance learning and understanding of Engineering Mechanics concept based on the equations of equilibrium which analyze engineering structures composed of pin-connected members.

Apparently the method of teaching as shown in Appendix C can be tedious, difficult, time consuming and requires the instructor to repeat the entire exercise several times until the student understands. Thus, in order to overcome this situation, tweening technique was employed where animations were used to illustrate motions such as movement of the support and rotation of structure. Tweening is a technique that allows "in-between" images to be created between supplied key frames using linear interpolation. In tweening, key frames are provided and "in-between" frames are calculated to allow smooth dynamic movement of member. The equation for tweening in a linear interpolation (***P***) is given by equation (7.1) (Hill, 2001):

$\boldsymbol{P} = \boldsymbol{A}\,(\mathbf{1} - \boldsymbol{t}) + \boldsymbol{B}\boldsymbol{t}$ (7.1)where ***A*** is the initial location of the point, ***B*** is the final position of the point, and ***t*** is the time measured from 0 to 1 seconds.

In general, a number of techniques can be used to generate a 2-D animation such as cell-based, stop motion, rotoscoping, and path-based animation. Animation is done differently in 2-D and 3-D-based animation. However, they have some common properties such as the key-frame, which holds all the information about the state of the animation at that point in time. Key-frames are renditions of two or more points, usually the beginning and ending frames of a specific animation.

Since Mechanics Statics concepts are built in a linear fashion, the TAPS package is structured to present information sequentially. The tutorial of this package contains several sections that are made up of any number of pages/procedures. Each page builds at a time, so that a particular concept is illustrated/explained as the user clicks the "Continue" button. While it is intended that the learner will proceed through the content in a linear fashion, the capability to move back and fourth throughout the tutorial is also provided in the TAPS package making it more student oriented as sometimes they need to review the previous image to get a better understanding of the present stage of solving a problem. As each page builds, several elements such as text, equations, images, graphics, and animations are displayed and manipulated as typically shown in Figures 1 and 2.

At the click of the "Continue" button, the analysis to compute the unknown reactions are carried out in a step-by-step approach as described in the earlier section using the equations of equilibrium. As the text and figures are written and displayed on the screen, the TAPS package will prompt the students at various checkpoints to see if student understands the step that has been executed. At this point, if for some reason the student is unclear, the student can move back to the previous step. Similar approach is adopted to determine the force induced in each member of the truss by considering the respective joints and applying the equations of equilibrium. Finally, the solution to the problem in question is illustrated in an animated form as shown in Figure 3.

During the developmental stages of the TAPS package, it is vital to build a screen sequentially as this allows students to see the steps involved to obtain the solution. For example, as the user clicks the

Figure 1. Free body diagram of the truss showing forces as represented by the colored (red) arrows.

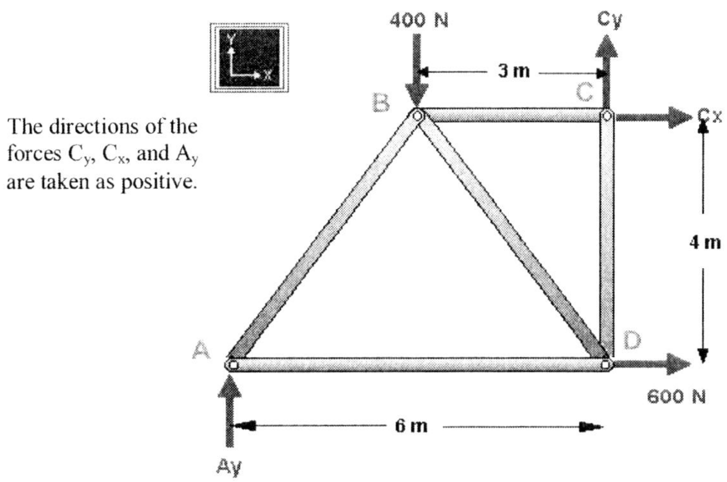

"Continue" button, relevant forces causing movements about *C* is written on the screen. During the event, the whole sequence is carried out dynamically between each of the screens.

The interface (shown in Appendix C – Figure 1) referred to as the TAPS package for (design approach 1), provides an interactive environment in which the students solve the Mechanics Statics problem. The menu of the TAPS package allows the user to repeat a procedure/step that is not clear.

To illustrate the use of tweening technique used in the TAPS package, a swinging rocker is shown in Figure 4, in-betweens are created from points ***A*** to ***B***. The resulting 2-D animation creates a dynamic rocker that swings from points ***A*** to ***B***. The effect of tweening beyond time, $t = 0.1$s, results in what is called interpolation. When $t > 0.1$s, the image results in the tweened points moving in the direction of

Figure 2. Illustrate animation steps involve in the computation of forces, Cx.

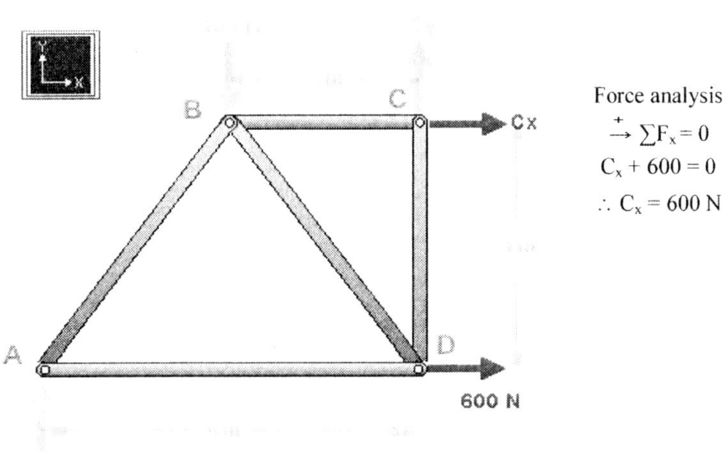

Figure 3. The correct free-body diagram showing all forces acting in each member is drawn on the screen.

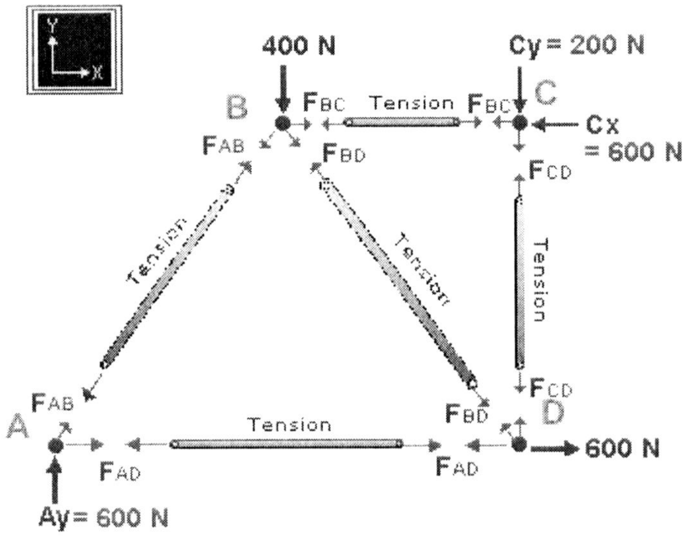

A to ***B***. Similarly the same technique is used to analyze the joints at ***A***, ***C*** and ***D***. A typical animation sequence for analyzing forces acting at one joint of the truss is shown in Appendix D.

In addition to tweening the image's path, the authoring tool used to develop the TAPS package was also used to tween the size, rotation, tilt, merge, and change the color of an image to semi-transparent. One advantage of the authoring tool is that it can tween all of these properties simultaneously. To make an image fade, the image merge settings can be made to tween and to make the image spin or tilt, the rotation settings can be made to tween. Similarly to create gradual shifts in color, the color settings could be made to tween.

Figure 4. Timeline animation of tweening an image

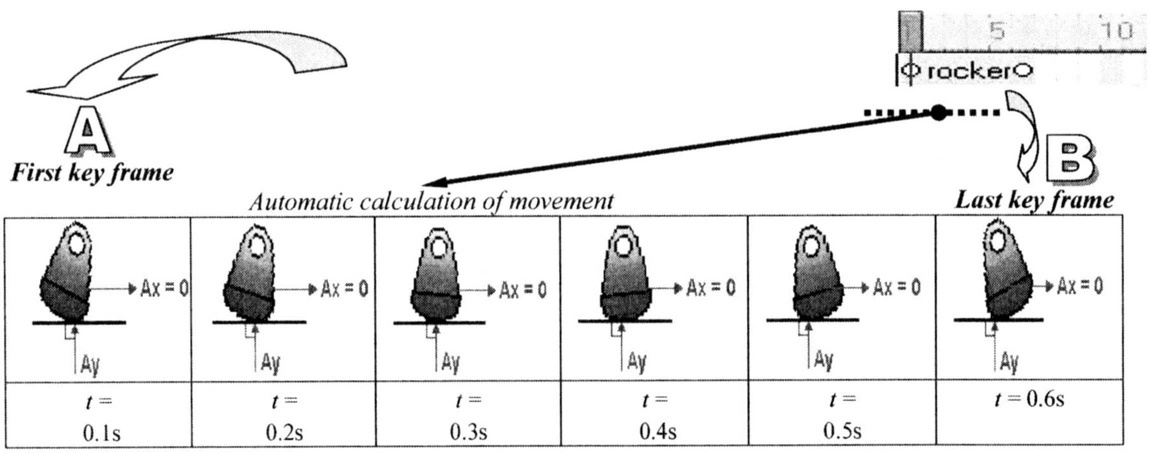

Figure 5. Video explaining engineering concepts

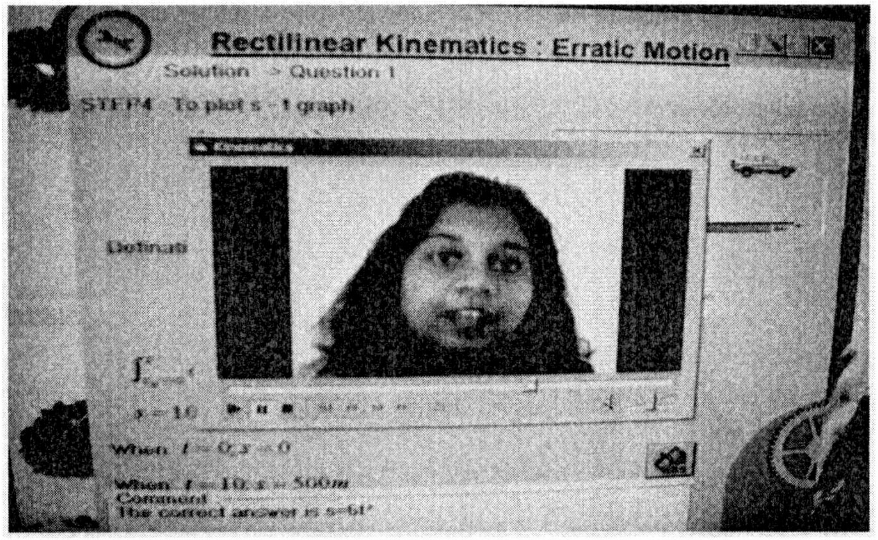

COACH BASED ENVIRONMENT TAPS PACKAGE (DESIGN APPROACH 2)

In this package, attempts were made to provide an environment that could help the student solve Engineering Mechanics problems without the help of a human instructor. The aim of this package is to investigate whether such TAPS package could enhance student's understanding and be effective as an unsupervised learning aid. Coach environment is defined as a TAPS package that could guide a user by providing a step-by-step approach to solve a task. This TAPS package has been developed to solve rectilinear kinematics, erratic motion, and structure problems. The Coach based TAPS package uses similar multiple media features of the TAPS package described in Section 7.6. In addition, it provides the user with more dynamic and interactive support than what has been provided in existing computer aided learning packages such as motion, and feedback response in the event when student makes mistake.

In coach environment, the user can approach the "Hint" and "Solve" buttons in the event if the user reaches an impasse (has no idea how to proceed and solve the problem or gives a wrong answer). Typical example in the coach environment TAPS package includes basic information, context-sensitive hints or tips, or procedural steps required to solve the problem. Since not all students are capable of solving engineering problems by attending a single tutorial, the TAPS package is designed to help and show the user how to solve a problem leading from the question and a series of steps to solution. Every task and step is shown in an animated process on the screen with audio or video narration more about the problem as shown in Figure 5.

The TAPS package also provides reasoning support (simple expert system rules are used) and explanations of complicated concepts while the user is trying to solve the problem presented in the package. Thus this approach can help students make decisions and complete tasks better and also provide explanation for reasoning, enabling continual performance and improvement. In addition, the TAPS package could further provide more classroom time for demonstrations and coverage of theoretical issues.

The coach environment TAPS package was designed in such a way that it would not deviate from the method of teaching adopted by the lecturer teaching the Mechanics Dynamics course. The package

Figure 6. A selection list of formulas and equations

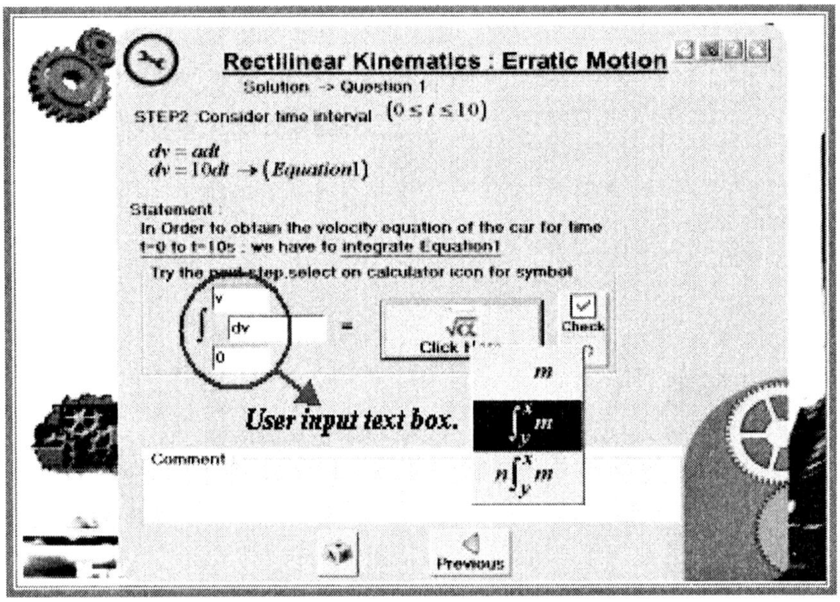

has striven to imbibe the same pedagogical philosophy that is used in the classroom. For example, in solving Mechanics problem, students are normally encouraged to draw free body diagrams and kinetic diagrams where appropriate, label all known forces, moments and other parameters, select appropriate datum or reference point, and use appropriate formulas and equations to solve the problem.

Past research by Gupta (2002) indicated that immediate feedback has a positive impact on learning, and thus the package was designed to provide immediate feedback, for example by displaying blue colored messages in the comment box on the computer display to indicate correct student actions, red to indicate incorrect ones, and black to indicate no value entered in the text input box. In addition, if a student inputs the wrong answer, the student may immediately be narrated and questioned prior to giving a hint by clicking the *hint button*. Students can activate the calculator (just like they would use scientific calculators during classroom test, quizzes, and examination) to perform simple calculations and the results/answers could be copied and pasted in the text input box.

In this TAPS package (Design approach 2), users are encouraged to select and define the correct formula from a drop list and then use them to enter symbolic equations, and finally to solve for the numerical answer as shown in Figure 6.

In order to eliminate poor algebraic skills from undermining the theoretical issues the student is trying to learn, the package further enables the student to select an equation, and click on a *solve button* which will then perform the algebraic/arithmetic manipulations necessary to simplify the equation. This feature is considered to be important as it can help learners understand the use of correct equations to solve a particular problem.

When solving engineering problem, students need to use and input special characters that are unavailable on the standard keyboard such as the "power of two". Thus a character map table is included in the TAPS package. Users can click the character map button as shown in Figure 7 that will display all the characters available for editing. Once the required character is selected, the user can copy and paste into the text input box, where appropriate.

Figure 7. The character map window displaying all the characters available for user to copy, edit and paste into text input boxes where appropriate

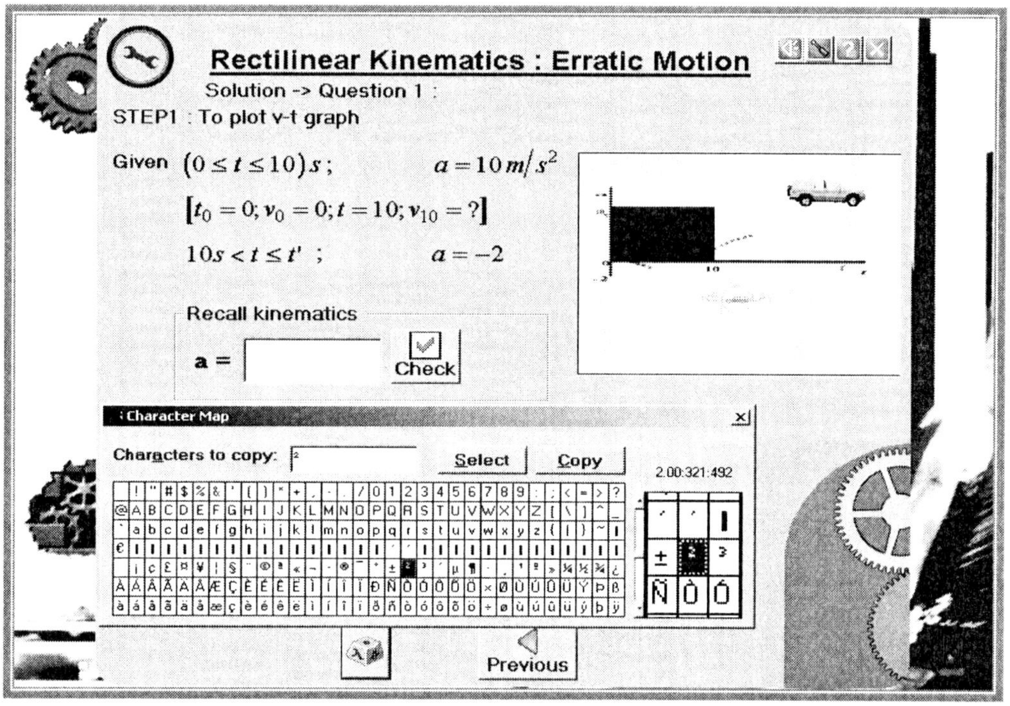

In addition a notepad (for students to make notes and print) and glossaries of commands (for students to understand the meaning of a certain engineering term) are also included in the TAPS package. These features are considered as essential components of multimedia learning packages (Cairncross and Mannion, 1999). Cairncross and Mannion (1999) believed that having control over the delivery of information could help promote a sense of ownership over the material, which in turns leads to engagement and active learning.

The interface of the TAPS package provides an interactive environment in which the user works the Mechanics Dynamics problem as shown in Figure 8.

The coach-based TAPS package has a user-friendly environment that is built on six major modules, namely the action interpreter, the assessor, the interface, the help, graphs to show (velocity, time and speed) *v-t* and *s-t,* and a database to store the student's progress score.

The system environment consists of the given conditions of a problem and a Dynamics problem-solving engine. The problem-solving engine contains approximately fifteen conditions comprised of decision-making rules. An example of each type of the rules is shown in Table 2.

The given conditions of a problem are used as input to the problem-solving engine. Outputs from the problem-solving engine consist of all the equations necessary to solve the problem. These equations are then used by the action interpreter and assessor to provide appropriate hints. If the student correctly solves a step, the screen will display the next subsequent step. The steps are iterated until the final step is completed.

The action interpreter module interprets the student's problem solving action in the context of the current problem and determines the type of feedback to provide. For example, if the student enters an

Figure 8. The interactive interface of the coach based TAPS package

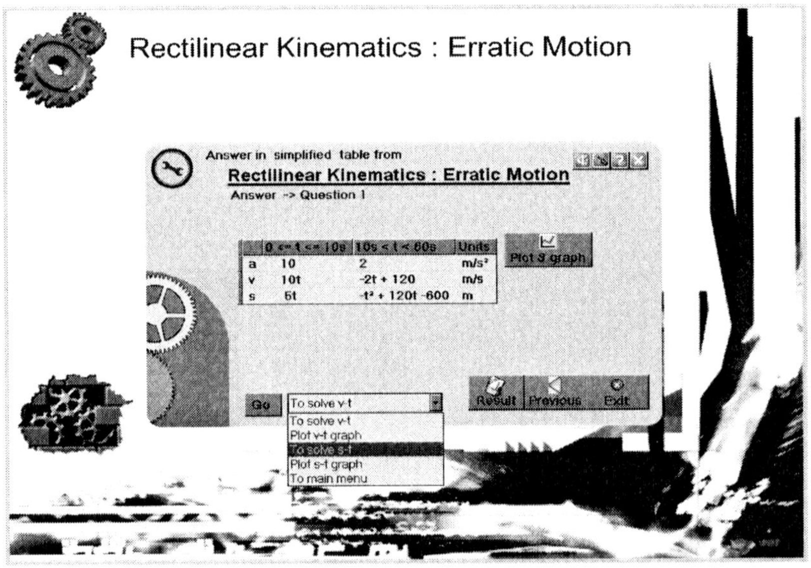

equation, it is compared to the set of equations produced by the problem-solving engine and if a match occurs, the message colored in blue indicating that the equation is right is displayed as shown in Figure 9. If there is no match then the message colored in red indicating incorrect equation is displayed. When the student has reached an impasse and has no idea how to proceed, the student can click the *hint button*; which may aid the student in solving the problem.

On the other hand, if the student has input the wrong formula, the student will be prompted if a hint is needed. If the student attempted to answer without approaching the hint button and still gives the wrong answer, a *solve button* will be visible as shown in Figure 10. The student can then click on the *solve button* to allow the TAPS package to guide the student in solving the problem. If the answer given by student is correct, the student may then proceed to the next step. If a complete solution has been ac-

Table 2. Typical TAPS package problem-solving decision making rules

STEP 1	*If*	
		the sub goal is to determine the time needed to stop the car correct kinematics formula applied
	then	
		create a sub goal to find out how far has the car traveled
STEP 2	***If***	
		correct symbolic equation selected correct values input in text boxes *and* correct value of velocity applied
	then	
		create a sub goal to determine the next subsequent step
STEP n		

Figure 9. Screen caption of the engineering problem being solved by user

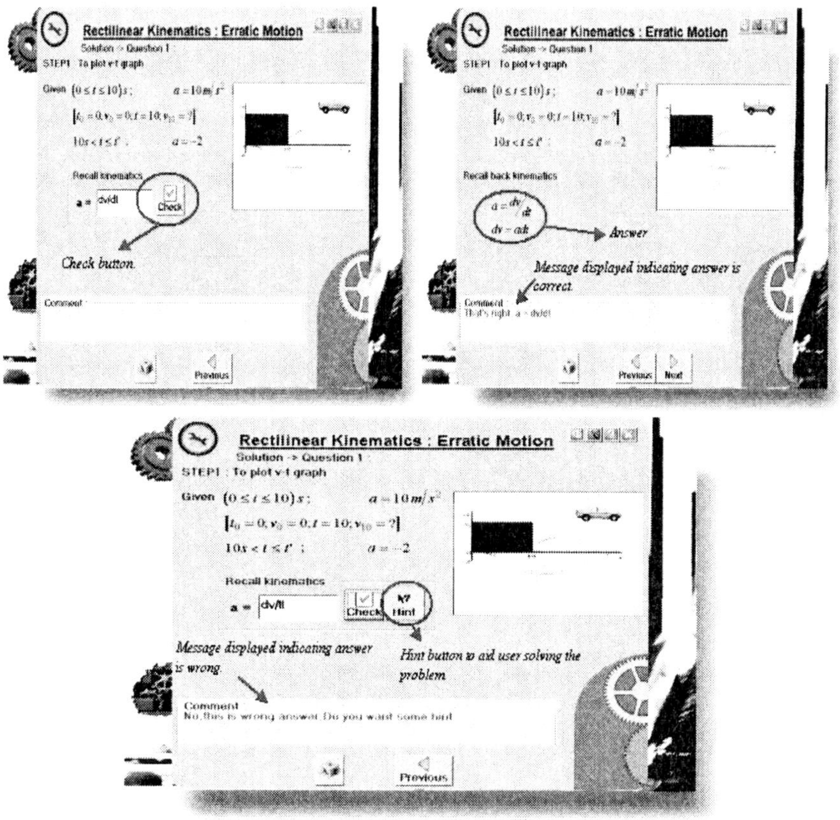

complished, except for numerical substitution, the student could choose the *solve button* for the TAPS package to do the appropriate substitution.

The assessment model of the TAPS package is developed to actively monitor the performance of the student. The assessment model is a simplified rule-based system that constantly tests and grades the student. The score of the student is stored in a database and updated accordingly.

In an ideal intelligent tutoring system (ITS), the knowledge representation component provides the student with expertise in tutoring, including the ability to answer specific problem-related questions asked by the student. The system must assume the role of an expert human tutor, possessing the requisite knowledge on the subject matter, and being able to deliver the knowledge in a way that maximizes the student's learning motivation and learning process. To do this the system must have extremely broad knowledge base, encompassing all the information to be delivered to the student, as well as detailed explanations, summaries and overviews of the more difficult concepts, examples to enforce these explanations and summaries to put this information into context.

Due to limitations of the availability of the current level of technology and resources at UNITEN, it is not economically viable to build a TAPS package that possesses all the features of an ITS. However, it is possible to extract key features from an ideal ITS and use these features to implement a good TAPS package. This is the process through which the TAPS package evolved.

Figure 10. Screen caption of the engineering problem being solved by the TAPS package

The knowledge representation component of the TAPS package is not influenced by any factor. The domain knowledge presents the information sequentially, thereby presenting the student with relevant information only.

3-D VIRTUAL ENVIRONMENT TAPS PACKAGE (DESIGN APPROACH 3)

In this package the incorporation of multimedia interactivity to study curvilinear motion of cylindrical components in a 3-D virtual environment was developed. The instructor teaching the Mechanics Dynamics subject at UNITEN found that the students have difficulty in understanding and solving the problem of curvilinear motion as shown in Appendix E. This problem was taken from Hibbler (2001) textbook. In the normal classroom lectures, the student's perception of the collar sliding outward along the rod is that the collar has a linear motion. As such, the rod and collar shown in Appendix E was modeled using a 3-D modeling software tool and animated to show the rotational movement of the rod and the motion of the collar sliding outward on the moving rod. This animation is to show the students that the collar experiences a curvilinear motion. As to generate and show the curvilinear path taken by the collar, a simple algorithm was constructed to plot the path in the 3-D virtual environment. The algorithm is used to show the path taken by the 3-D object (for example a robotic arm) from the starting point labeled ***A*** to the ending point labeled ***B***. The algorithm is further explained in Section 7.9.

In the example, in Figures 11, 12, and 13, the movement of the rod as well as the motion of a collar on the rod can be studied simultaneously with the incorporation of multimedia technology. The benefit is two fold; the student can firstly visualize the relative motion of the collar with respect to the moving rod and secondly, the path of the collar during a given time period e.g. $t = 0$ to $t = 5$s can be viewed in real-time and analyzed. This is illustrated in Figure 11 where initially when time $t = 0$ the collar is originally located on the rod at ***p*** and after a time interval of say 5 seconds, the collar has moved from its original position to a new position ***q*** as shown in Figure 13. Close observation indicates that the collar has taken a curve path and thus has a curvilinear motion.

DESKTOP VIRTUAL REALITY ENVIRONMENT TAPS PACKAGE (DESIGN APPROACH 4)

In this TAPS package, another selected engineering problem based on curvilinear motion was developed. Similar features of the problem presented in package 3 (Design approach 3) were used in this package. In addition progress was made to implement a 3-D problem-solving model that was tested in a desktop virtual reality (DVR) environment for greater interaction and visualization. Every effort was made to give clear explanations on linear and curvilinear motion in this package. In the brief tutorial of this TAPS package, 2-D animated examples illustrating the motions are displayed and narrated to the students as shown in Appendix F (Figures F-1 and F-2). Additionally to make the tutorial more

Figure 11. Linear motion of the rod and collar rotating about y-axis (t = 0s)

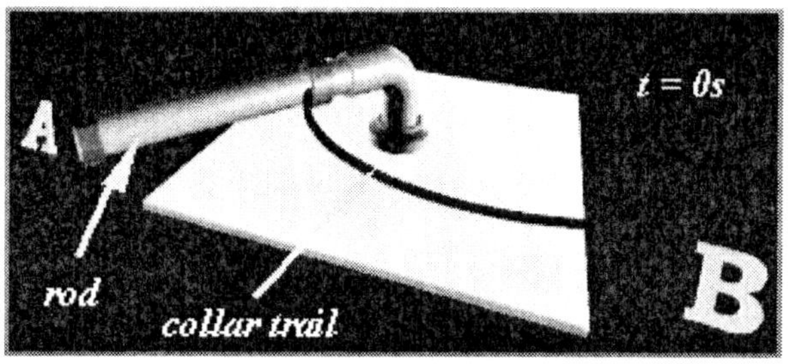

Figure 12. Linear motion of the rod and collar rotating about y-axis (t = 3s)

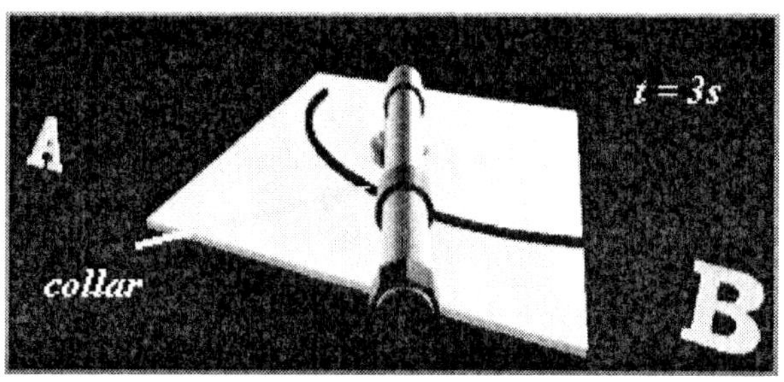

Figure 13. Linear motion of the rod and collar rotating about y-axis (t = 5s)

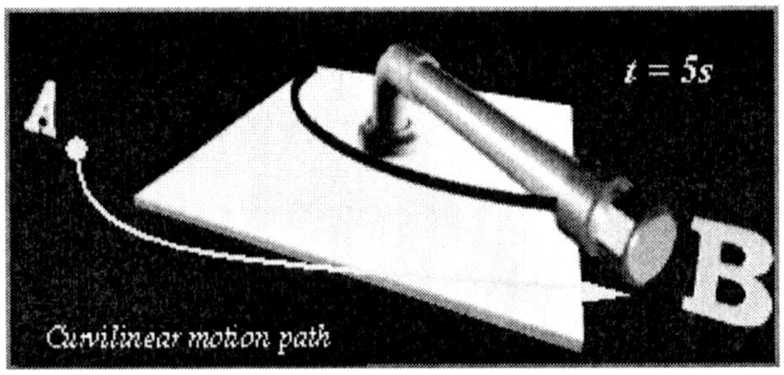

interesting, 3-D animated example models are used to explain the concepts of the motions as shown in Appendix G (Figures G-1 - G-3).

The student is then explained about the components of a particle that experiences curvilinear motion as shown in Appendix H (Figures H-1 - H-5). In this example when the vehicle (van) takes a corner, the magnitude of velocity is tangent to the motion of the van thus the direction of the animated arrow is used to depict the correct path of the velocity. If the velocity of the particle is known at any two instant points, the acceleration of the curvilinear motion could be shown as well.

In specialized systems, many different coordinates can be used such as cartesian and rectangular. However, to express a curvilinear motion, polar coordinates are used. Polar coordinates are useful in situations where information is most conveniently expressed in terms of distance from the origin. In the example shown in Appendix H (Figure H-4), the origin area of the particle is made to dim and shown by depicting a red colored animated arrow pointing at the origin of the particle. Thus the use of multimedia enhances the student's understanding whereby the student can clearly see and understand the origin point of the particle. From the origin point, the location and distance of the particle at any given point can be shown by animating and extending the red colored arrow. During this instance, additional text and formulas are displayed and narrated to the student explaining important concepts.

Once the tutorial explaining the concepts involved in curvilinear motion are clearly understood by the student, the student could proceed to solve a problem-solving question in the TAPS package as shown in Appendix I (Figure 1). Here the student could use the knowledge transferred from the tutorial to solve a real life problem in a virtual environment. The problem shown in Appendix J (Figures J-1 – J-4) is an example of an industrial robotic arm. The movements that could be carried out by the robotic arm such as arm extension, vertical and horizontal rotation are animated and shown to the student as depicted in Appendix J (Figure J-2). Although the robotic arm movements could be easily shown and explained by the human instructor, the instructor would need to demonstrate each movement at a time and this can be time consuming. On the other hand the TAPS package could be used to perform similar task simultaneously, repeatedly and narrated at the same time hence reducing the workload of the instructor.

After the student has understood the movements of the robotic arm, a short demonstration (see Appendix J - Figure J-3) of the robotic arm gripping an object (e.g. ball) while the arm is extending outwards at a constant rate is shown by the TAPS package. Here as shown in Appendix J (Figure J-4) the student is prompted with a question where the student has to determine the magnitudes of the velocity and acceleration of the ball. This is where the interaction between the student and TAPS package takes place and the knowledge of the student is tested. The student could approach the notepad tool provided in the TAPS package to type and show the calculation steps and use the calculator to obtain the final result. As for the special mathematical symbols, the student needs to approach the character map table, select the required symbol, then copy and paste it into the notepad. Although in normal circumstances, the student is not expected to use the calculator tool, but this tool is found to be suitable for learners who need extra assistance. Once the answers (value) have been entered in the text boxes, the student can click on the examine answers button to see if the answers are right. In the event if students get the wrong answer, the student has an option to repeat until the student gets the correct answer or simply by clicking on the solve button to see the solution steps.

Most conventional CAL packages provide only the answers to a solution but not the steps in solving the problem. In the TAPS package, the complete solution is provided and narrated leading from a series of steps to the answer as typically shown in Appendix K (Figures K-1 – K-4) and the animated motion of the robotic arm based on the answers is shown in Appendix L. As such the student is provided with the complete solution instantly instead of waiting for the answer paper to be marked by the instructor and returned and discussed a week later. Additionally the curvilinear motion of the robotic arm could be observed clearly and visualized by the student.

To further help students to visualize the curvilinear motion of the robotic arm based on different time say between 1 – 10 seconds, the student may approach the virtual lab option as shown in Appendix M (Figures M-1 and M-2). In user interaction with virtual worlds, consistent realistic behavior of objects is very important (Bowman & Hodges, 1997).

As such, it is desired that the objects can respond in a natural and predictable way to the actions.

The main interface (virtual environment) as illustrated in Figure 14 is a 3-D model of a robotic arm that can be viewed and interacted with in a 3-D environment. The 3-D interface provides an interactive environment in which the students visualize the Mechanics Dynamics problem. It allows the student to move, resize, rotate and interact with the robotic arm on the display screen. In addition the user can adjust viewpoints i.e. solid or wire-frame mode (without texture) of the 3-D robotic arm and change the display options such as changing the background colors from a color palette list.

The DVR TAPS package has a user-friendly environment that is built on five major modules, namely the action interpreter, the assessor, the interface, motion path generator, and a randomized multiple-choice questions quiz. The action interpreter and the assessor functions are similar to the ones in the 3-D coach based TAPS package described in Section 8. The problem-solving engine contains the object motion trail (path) generator algorithm to show curvilinear motion path/trail taken by the robotic arm, say for example from the start to the end point of a path. An example of the algorithm is shown in Table 3.

The path generated to depict the curvilinear motion exhibited by the virtual robotic arm is shown in Figure 15. The motion path algorithm was designed to construct a short sequence of intermediate motions to transform and rotate the robotic arm from point ***s*** to ***t*** as shown on Figure 15. These motions can serve to fill in the intermediate scenes between ***s*** and ***t***, thus such scene can greatly reduce the amount of work the instructor has to do in the traditional way.

The motion path algorithm designed for the 3-D robotic arm is a mechanism that is used to show the path taken by the robotic arm. For instance the initial 3-D robotic arm is given a starting position ***s*** in a virtual environment as shown in Figure 15a, with a desired ending position ***t*** as shown in Figure 15b. The movement path of the robotic arm is based on the time input by the student, in this example say the time input is 3 seconds. Therefore the robotic arm should rotate in a curvilinear motion path from point ***s*** to ***t*** in the given time interval by plotting the motion path taken by the robotic arm. In this scenario the generated motion path on the screen can enforce visualization in the sense that it can clearly show the

Figure 14. The main interface (virtual environment) of the mechanics dynamics problem of the DVR TAPS package

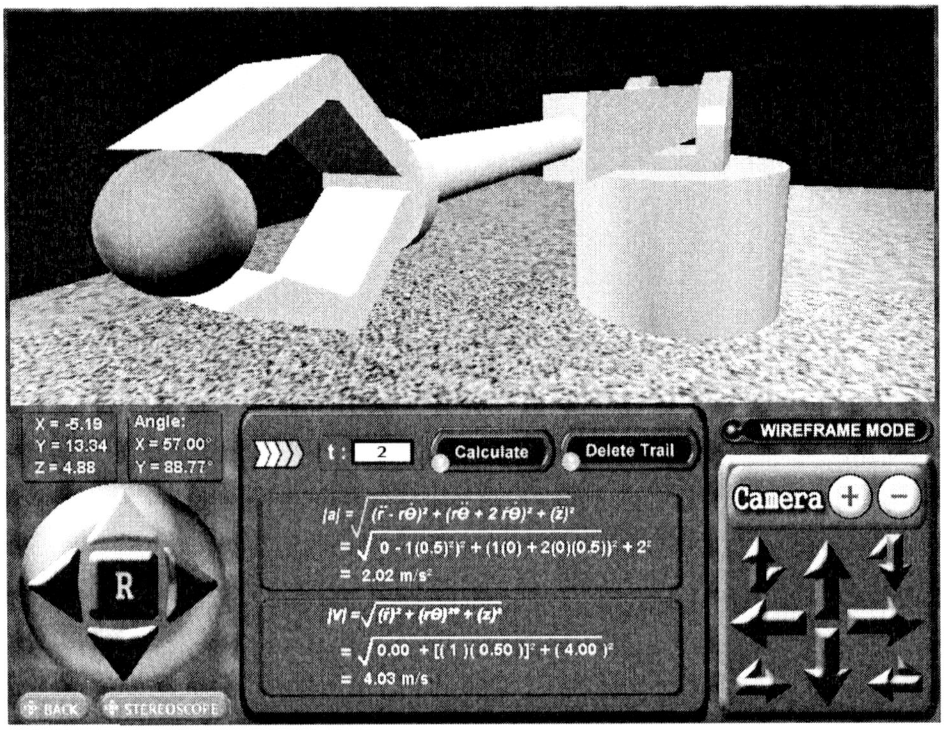

curve path taken by the robotic arm while moving from point ***s*** to point ***t***. This sort of motion could be difficult to explain in the traditional approach, for example from 2-D drawings or static images.

Once the student gets a clear understanding of the overall process in solving the problem, the student could try out the multiple-choice questions (MCQ) quiz that is provided to gauge their level of understanding about the topic learnt. Upon completion of the MCQ quiz, the TAPS package provides the student with feedback regarding the quiz results as typically shown in Figure 16. The TAPS package grades the student and informs the student of the number of correct answers, and the overall percentage achieved. A suitable comment is also presented to the student, recommending specific action, depending on the score obtained. For example, if a student obtains $\leq 75\%$ for a quiz, the TAPS package will recommend the student to revise the problem solving steps tutorial completely. It is then possible for the student to redo the quiz and improve the score. The MCQ quiz is design in such a way that the students can interact with the package and keep on trying until they obtain the correct answer to all the wrong ones only. However, if a student attempts a particular quiz question more than once, only the first attempt of each correct answer is taken into account and recorded by the DVR TAPS package.

The DVR part of the TAPS package is the stereoscopic image gallery as shown in Figure 17. A normal and complete stereoscopic animated sequence view of the robotic arm is shown in Appendix N (Figures N-1 and N-2). In addition, typical rotational sequences of the robotic arm and output showing the path plotted by the DVR TAPS package on time input by user are also shown in Appendix N (Figure N-3).

The stereoscopic views are available as static images and 3-D animations. However, to view the ste-

Table 3. The robotic arm path algorithm

1	*Accept time (valid 1 ~ 10 seconds)*
2	*Temporary transform and assign "object" to temporary variable and initialize to world position*
3	*Temporary variable name = model*
4	*Assign properties i.e. height, width and length to model*
5	*Set model height to 0.1*
6	*Set model width to 0.3*
7	*Set model length to 0.3*
8	*Temporary variable = (model)*
9	*Assign (Temporary transform) to world position*
10	*Model = model + 1*

11	*IF (glob_lift_limit = 180) THEN*
	normal_mode = FALSE
	reverse_mode = TRUE
	larger_angle = -1
	ELSE

12	*IF (glob_lift_limit = 0) THEN*
	normal_mode = TRUE
	reverse_mode = FALSE
	larger_angle = -1
	ENDIF

reoscopic images the user needs to wear an inexpensive pair of simple stereo glasses. The paradigm of desktop virtual reality (stereoscopic views) provides many benefits. Firstly the student can gain a better view of the path and motion of the 3-D robotic arm in the 3-D virtual environment. Secondly it enhances the learning process whereby the student can view the 3-D robotic arm from different angles.

Additionally the 3-D visualization and stereoscopic presentation of the robotic arm as well as the function to change the time (in seconds) of the robotic arm accelerating from the start to the end point of a path can be calculated and shown by the TAPS package. For example, when the user inserts the time in seconds (in the text input box), the velocity of the robotic arm based on the time input by the user is

Figure 15. Original position and path generated to depict the curvilinear motion exhibited by the virtual robotic arm

Figure 16. Typical student progress screen in the DVR TAPS package

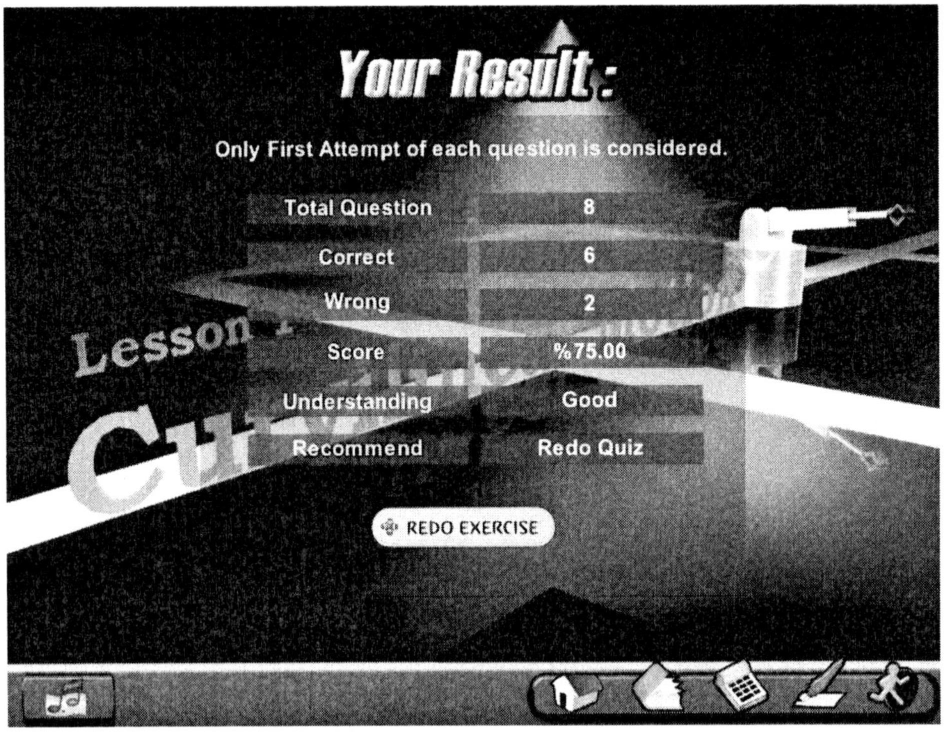

automatically calculated by a built in function and the solution is shown in steps on the display thus the problem solving process is enhanced by the TAPS package. As a comparison, a sequence of the animated robotic arm in a 3-D virtual environment and DVR animated mode is shown in Appendix N.

Fundamentally the TAPS package provides direct visualization of 3-D geometric model created using Alias Maya™ (a 3-D modeler). It allows the students to explore the model by using the built-in features in the package (i.e. moving, walking, flying, examining, resizing, rotating and changing viewpoints). With these features, the user can move along any direction on the screen and have the displayed 3-D robotic arm continuously and instantaneously updated. The interface of this TAPS package has been implemented using Macromedia Director™ (an authoring tool), it is a stand-alone (PC based) application, which is available on the CD-ROM and currently runs on Windows platform.

To further help the students to visualize the motion of robotic arm; extended polygonal views of the 3-D robotic arm are available with their correct positions and orientations in a wire-frame (the model without textures) representing the physical extends of the robotic arm. Figure 18 shows the image of a robotic arm along in a wire-frame mode and the motion path. The path is produced in a 3-D virtual environment to show users how the robotic arm has moved from one point to another based on the time input by the user. As such it can be clearly seen by students that the robotic arm has rotated in a curve thus has curvilinear motion.

Figure 17. The stereoscopic images of the robotic arm

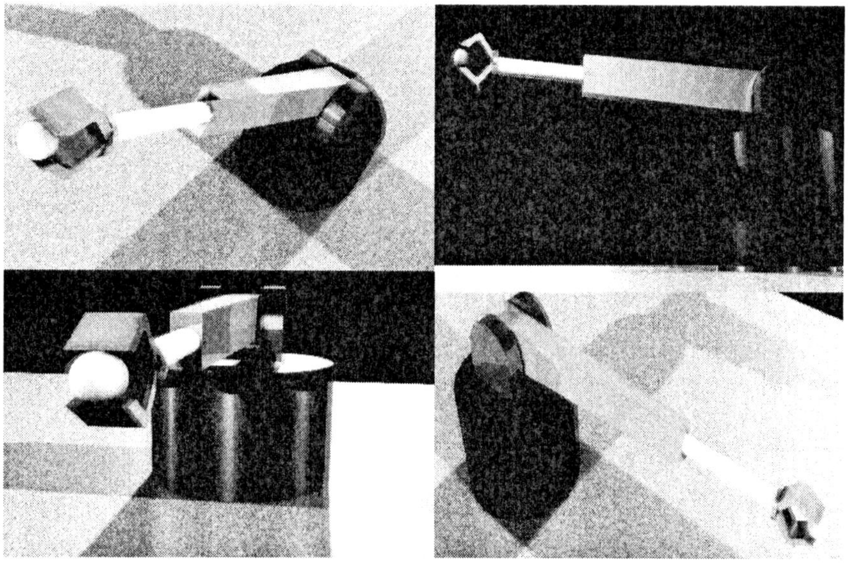

Figure 18. Wireframe mode exhibiting the motion path of the robot arm in DVR TAPS package

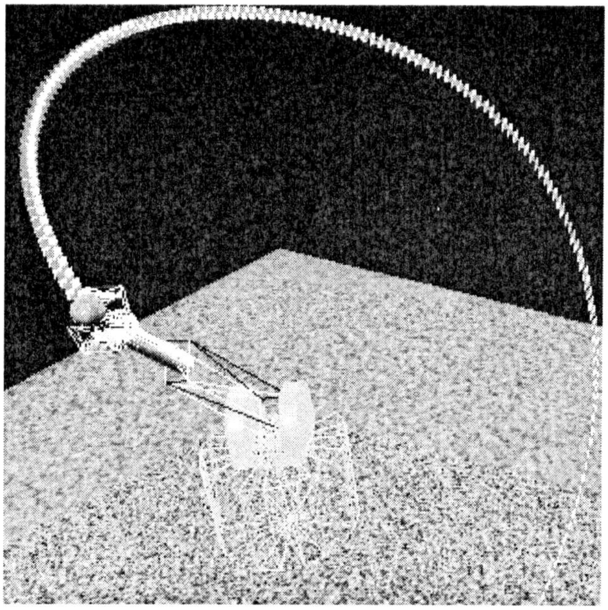

CONFIGURATION AND INTERFACE OF THE TAPS PACKAGES

The approach used for developing the interfaces of the TAPS packages described in Sections 6 – 9 could be classified as 'Craft Approach' (Wallace & Anderson, 1993) as described in Chapter 3. The TAPS packages were designed and evaluated by 60 students taking the Engineering Mechanics Dynamics subject at UNITEN. Many improvement suggestions were made which helped in improvising the interface in its acceptable form.

Some principles for good user interface design from the literature reviewed in Chapter 3 were applied while designing the interface. Efforts were made to provide a consistent interface of each of the TAPS package, such as semantics, syntax and others. Navigation was also made as simple and intuitive as possible.

Since a problem given in the subject of Engineering Mechanics Statics and Dynamics has to be solved in a series of steps/procedures leading from problem statement to the solution, a similar problem solving model and user tools were adopted for the TAPS packages described in Sections 6 – 9. However the solution of each selected engineering problem depends on the number of steps involved in solving the problem. An example of the problem-solving model and user tools designed for one of the TAPS package is shown in Figure 19. The same model is adopted by all the other TAPS packages stated in this book (only the steps in solving the problem differs which is based on the problem statement).

The organizational design flow of the problem-solving model shown in Figure 19 can be represented as a list of steps/procedures that must be executed in a serial order which lead from a problem statement to the solution. In the TAPS packages, the student will be given a brief objective (either in the form of text or narration) of the problem followed by an example. The student then follows the steps shown by the TAPS package by clicking the next button until the solution is given. At the end of the tutorial, the student could solve another problem presented in the TAPS package. Additionally students may use the user tools shown in Figure 19 to help them in solving the problem. The other TAPS packages were also structured to solve the engineering problems in a similar fashion.

ACQUISITION OF THE ENGINEERING CONCEPTS

Students gain knowledge in two phases. Firstly, the student grasps the understanding of engineering concepts with the help of precise textual explanations in Basic Theory (e.g. what is a truss, rectilinear kinematics, curvilinear motion) part of the package, and then obtains technical knowledge and problem-solving skills on the subject by solving the problems presented in the TAPS packages.

ARTICULATION OF THE CONCEPTS

After acquiring the concepts and representations of the technical knowledge Basic Theory part of the package, students could work on a problem presented in the package that is structured progressively into its theoretical components to introduce simpler concepts first. The problem presented in the TAPS package, attempts to guide the student to solve the problem in a single model. While the TAPS package is displaying the problem solving steps, some of the theoretical components are assigned values (numeric/formulas) and animated graphics, which are sufficient to address the whole problem. In the process, the TAPS package displays the steps in solving the problem, the TAPS package could prompt the student to answer or apply a formula, where the student has to derive the values associated with the remaining components applying the theoretical and technical knowledge gained from the Basic Theory part. A student who does not understand a certain part of the problem solving process may go directly

Figure 19. Problem-solving model and user tools adopted to develop one of the TAPS package

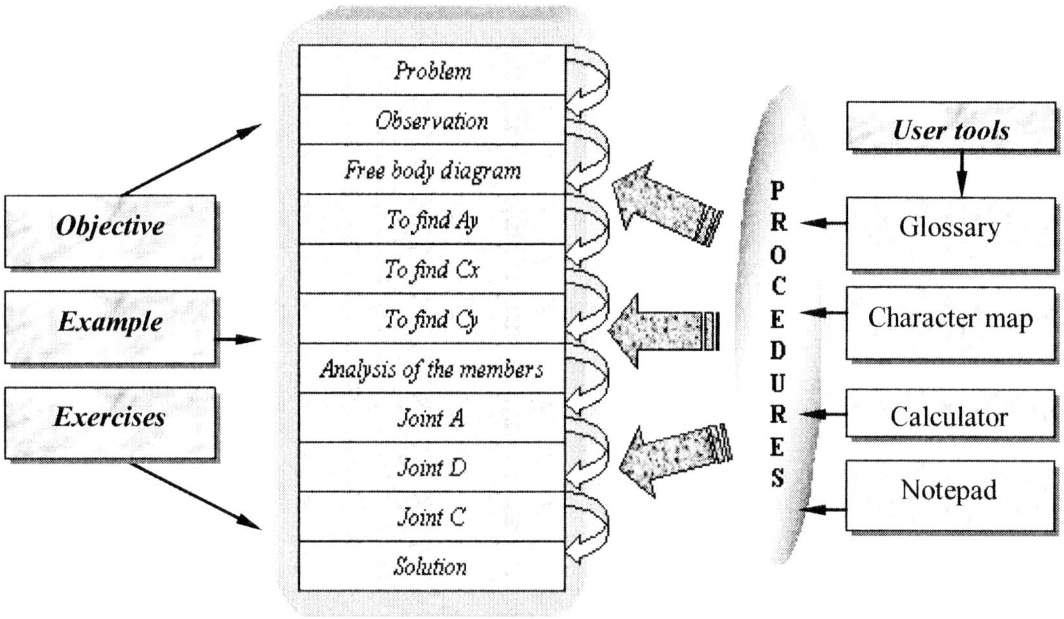

to the particular problem-solving step in the TAPS package to repeat or start the tutorial again from the beginning.

DISCUSSION ON KEY DESIGN FEATURES OF TAPS PACKAGES

In general, all four TAPS packages developed in this study were designed based on some of the design guidelines as described in Chapter 3. For example, the TAPS packages were developed by adopting graphical user interface (GUI) since it is generally accepted that GUI provide natural and easy means for users to communicate with computers (Kinshuk, 1996; Kinshuk and Patel, 2003). GUI was found to be suitable for engineering students to interact with the TAPS packages, since these students lack experience in using command driven packages where the user has to type and press the "*Enter key*" on the keyboard to process the task.

Direct manipulation design was used extensively throughout the design of TAPS packages because it is natural and fast. For example icons were adopted to represent the object of the task whereby the user uses a pointing device such as mouse or graphics pen to make selections and the results appears immediately on the screen. In addition, it is generally accepted that icons could provide a more logical way of interacting with the TAPS packages since most students are familiar with Windows platforms. Similar view was also stated by Pangalos (1993).

Design guidelines suggested by Iannella (1992) were also used to design the TAPS packages. For example, in designing the screen layout of the TAPS packages, suitable consistent titles, help, text and buttons were used. Screen designs were kept simple, clear, distinctive, and emphasized on critical information. For example, in designing menus, if the menu option is unavailable, the option is made to dim it by disabling the words or images. In designing the icons, suitable icons were adopted as it could give visual clues to the student as to the function they perform. For example, a pen icon in the TAPS package allows the user to invoke the windows notepad so that the student could type important notes. Design guidelines suggested by Diaz (1999) were also used. For example, the fonts, colors and sizes were maintained consistently throughout the design of the TAPS packages as to avoid students from being distracted. A more detailed summary of key features and differences of each TAPS package are given in Table 4.

In summary, the TAPS packages described in this Chapter were developed to see if the packages could help slow learners understand how to solve selected engineering problems. These problems were previously found to be difficult for slow learners to understand during normal classroom lectures. Therefore the TAPS packages were developed to familiarize slow learners with the concepts of solving Engineering Mechanics Dynamics problems in a step-by-step fashion. It guides the students to solve a problem presented in the TAPS package. Some good design principles for graphical user interfaces for CAL suggested by various researchers (Ianella, 1992; Ambler, 1998; Diaz, 1999; Carter, 2002) have been applied while developing the user interfaces of TAPS packages. Similar layout and icons are used throughout TAPS packages so that students could have an easy access to the information. Marchisio *et al.*, (1993) suggested that icons are suitable means of interaction to convey information in a package, which can be supplemented, with textual information to enhance understanding. In addition icons help students familiarize with the package consuming minimum duration of time and avoiding distraction.

The TAPS packages utilize similar Microsoft Windows-like facilities such as pop-up windows for displaying error messages and hints, symbolic icons and various visually interactive dialogue boxes throughout the problem solving process. Iannella (1992) pointed that a good interface provides feedback in a helpful manner so that the cognitive workload is minimum on the students. Since constant feedback is an important part of the TAPS packages, timely warning and guiding messages with appropriate colors helped students significantly in understanding and correcting their mistakes. Appropriate use of sound and video clips in the TAPS packages assisted students in understanding important concepts of engineering. Menu-based user interface was found by the students as a useful tool in solving the selected engineering problems in a step-by-step fashion. This new graphical user interface design approach provided students with an easier way of interacting with the TAPS packages as compared to the traditional approaches for developing computer based learning environments where alphabets and numeric were used as options to interact with the software. Usage of the TAPS packages for problem solving purposes in engineering substantially reduced the time taken by learners in their learning and enhanced their problem solving skills.

Kinshuk (1996) stressed that keeping information as little as possible on any part of a screen is a good design principle. Keeping this as a guideline, the TAPS packages only displays related information such as objectives of the problem, formulas to be applied, steps involved to solve the problem, and solution rather than providing lengthy theoretical explanatory text about the problem. The reason is to provide autonomy in the TAPS packages.

All features of the TAPS packages, from the interface, multimedia and DVR, have been deliberately designed to allow maximum control over their paths through the packages, while at the same time providing easy access of information on good problem solving skills, to allow learners to make best use of the control they have. Students must be able to concentrate on problem solving skill without losing time for learning to interact with the packages. As such the users are provided with a precise and easy to use and learn interface in each of the TAPS package. The content of engineering problems solving task are

Table 4. Summary of key features and differences of each TAPS package

Key features	2-D TAPS Package	Coach Based TAPS Package	3-D TAPS Package	3-D Desktop Virtual Reality TAPS package
Objective	• To examine various levels of interaction for a tutorial and problem-solving topic on Structural Analysis.	• To provide an environment that could help the student solve Engineering Mechanics problems without the help of a human instructor.	• To incorporate interactivity multimedia to study curvilinear motion of cylindrical components in a 3-D virtual environment.	• To study curvilinear motion of cylindrical components where the user could interact in a desktop virtual reality (DVR) environment for greater interaction and visualization.
Design approach	• Employed 2-D graphics/images, colors, and tweening technique to produce animations.	• Employed 2-D graphics/images, colors, sound, and tweening technique to produce animations. • Provide user with more dynamic and interactive support. • Provide user support tools such as calculator and notepad.	• Employed 2-D graphics/images, 3-D geometric model, colors, sound, and tweening technique to produce animations. • Provide user with more dynamic and interactive support in a 3-D virtual environment. • Provide user support tools such as calculator and notepad. • Allow user to plot curvilinear path of a moving object in a 3-D virtual environment.	• Employed 2-D graphics/images, 3-D geometric model, colors, sound, and tweening technique to produce animations. • Provide user with more dynamic and interactive support in a 3-D virtual environment. • Provide user support tools such as calculator and notepad. • Allow user to plot curvilinear path of a moving object in a 3-D virtual environment and real time interaction. • Provides animated and static stereoscopic images.
Learning/ Problem solving approach	• Provides learning objectives to user. • A sequence of steps and solution of the problem are presented to the student. The student moves forward to the next step or back to previous step or solution.	• Provides learning objectives to user. • A sequence of steps and solution of the problem are presented to the student. The student moves forward to the next step or back to previous step or solution. • Allows user to select a formula from a list provided and validates if the correct formula has been applied to solve a problem. • Allows user to make mistakes. • Employed simple expert system rules to coach user.	• Provides learning objectives to user. • A sequence of steps and solution of the problem are presented to the student. The student moves forward to the next step or back to previous step or solution. • Allows text, numeric and special characters as input data.	• Provides learning objectives to user. • A sequence of steps and solution of the problem are presented to the student. The student moves forward to the next step or back to previous step or solution. • Allows text, numeric and special characters as input data. • Employed simple expert system rules to coach user.
Association	• The user needs to be able to associate the virtual environment of 2-D elements and his/her interactions with related information such as images, textual resources, and the other data with the problem solving process performed by the user.	• The user needs to be able to associate the virtual environment of 2-D elements and his/her interactions with related information such as images, textual resources, and the other data with the problem solving process performed by the user.	• The user needs to be able to associate the virtual environment of 3-D geometric elements and his/her interactions with related information such as images, textual resources, and the other data with the problem solving process performed by the user.	• The user needs to be able to associate the virtual environment of 3-D geometric elements and his/her interactions with related information such as images, textual resources, and the other data with the problem solving process performed by the user.
Coaching	• Coaching is provided to enhance the user's problem solving experience while performing complex tasks in 2-D environment.	• Coaching is provided to enhance the user's problem solving experience while performing complex tasks in 2-D environment. • The built-in expert system rules managers the user's activities in the 2-D problem solving environment and provides dynamic coaching advice and feedback based on the user's activities in a 2-D coach based environment.	• Coaching is provided to enhance the user's problem solving experience while performing complex tasks in 3-D environment.	• Coaching is provided to enhance the user's problem solving experience while performing complex tasks in 3-D environment. • The built-in expert system rules managers the user's activities in the 3-D problem solving environment and provides dynamic coaching advice and feedback based on the user's activities in a 3-D coach based environment.

continued on following page

Table 4. continued

Feedback and Assessment	• For a positive experience, the user is given feedback and assessment to understand his/her progress.	• For a positive experience, the user is given feedback and assessment to understand his/her progress. • Allows text, numeric and special characters as input data. • User' progress, i.e. scores is stored in a database. • The problem solving environment provides timely feedback and assessment directly related to the user's interactions. The feedback helps increase the user's ability to reason and analyse the problem solving environment.	• For a positive experience, the user is given feedback and assessment to understand his/her progress.	• For a positive experience, the user is given feedback and assessment to understand his/her progress. • Allows text, numeric and special characters as input data. • Multiple choice questions and answers are randomized. • The problem solving environment provides timely feedback and assessment directly related to the user's interactions. The feedback helps increase the user's ability to reason and analyse the problem solving environment.
Text / Contents	• Minimum text of theory is used in order to make use of multimedia elements.	• Minimum text of theory is used in order to make use of multimedia elements.	• Minimum text of theory is used in order to make use of multimedia elements.	• Minimum text of theory is used in order to make use of multimedia elements.
Interactivity	• The student interacts and observes meaningful tasks i.e. such as movement of the support and rotation of structure.	• The student interacts and observes meaningful tasks i.e. such as rectilinear kinematics, erratic motion, and structure problems. • Provides the user with more dynamic and interactive support than what has been provided in existing computer aided learning packages such as motion, and feedback response in the event when student makes mistake. • Uses reasoning support (simple expert system rules are used) and explanations of complicated concepts while the user is trying to solve the problem presented in the package. Thus this approach can help students make decisions and complete tasks better and also provide explanation for reasoning, enabling continual performance and improvement.	• The student interacts and observes meaningful tasks i.e. such as curvilinear motion of cylindrical components in a 3-D virtual environment. • A simple algorithm that allows student to plot path / path in a 3-D virtual environment.	• The student interacts and observes meaningful tasks i.e. such as movement of the support and rotation of structure. • A simple algorithm that allows student to plot path in a 3-D virtual environment.

continued on following page

also designed to integrate the acquisition of higher mental processes such as constructing, evaluating with the acquisition of problem solving skills, and awareness-raising about learning skills appropriate to enhancing practical knowledge acquisition.

The development of TAPS packages in this study have provided insights to designing and constructing problem solving task in selected engineering problems, which represents an improvement over previous conventional computer aided learning approaches. This improvement is most obvious in the way the technologies employed, help the learner interact and solve the engineering problems. With these technologies, learners have a variety of features that could be used to do multi task simultaneously in solving the engineering problems.

Table 4. continued

General Outcomes and Benefits	• The use of tweening technique was found to be helpful where animations were used to illustrate motions such as movement of the support and rotation of structure • The use of 2-D animated graphics and colors helped students understand better in solving the problem presented in the TAPS package • Benefits noted from this package include interactive multimedia provide comprehensive coverage combining full-motion, audio, video, and 2-D animated graphics. • Interactive multimedia provides the student an integrated learning environment, which combines explanations with illustrative examples, this is in agreement as stated by Cairncross (2002)	• The TAPS package help students to manage the sequence of steps the student should perform to solve the problem and control the 2-D animated mechanisms leading from problem statement till the solution • It is generally accepted that this TAPS package can help students process better cognitive-perceptions which, can result in fast and better understanding of the problem • Benefits noted from this package include interactive multimedia provide comprehensive coverage combining full-motion, audio, video, and 2-D animated graphics.	• Helped students to understand and visualized the problem on curvilinear motion which, was otherwise difficult to explain to students • Dynamic models present challenges to students beyond what they have learnt in the traditional way, this is in agreement as stated by Liang (2002) • Benefits noted from this package include interactive multimedia provide comprehensive coverage combining full-motion, audio, video, 2-D and 3-D animated graphics. • More emphasis is given to the visualization of 3-D objects because 3-D could enhance the process of learning, increase the level of understanding enabling students to understand more effectively through interactivity. This environment can provide a rewarding learning experience that would be otherwise difficult to obtain (Liarokapis *et al.*, 2007)	• Allows student to experiment simulated problem-solving problem in 3-D environment and manage a complex task in real time • Stereoscopic images and animated video files gives a better 3-D view of the robotic arm and path as it is not the same as static images as shown in the textbooks. The stereoscopic images help students enhance depth perception that could reduce learning time as compared to the conventional. • It is generally accepted that the stereoscopic images could create great interest and enthusiasm for students in understanding the problem presented in the TAPS package, particularly, students experiencing difficulties in understanding the Engineering Mechanics Dynamics subject • Multimedia and virtual reality techniques with simulation initiates a new appearance for learning applications - real time presentations of 3-D data, this is in agreement as stated by Klett (2002) • With simulations, students are able to take a more active role in learning, this is in agreement as stated by Jesica and Tara (2005) • Dynamic representations enable more efficient communication of complex concepts; this is in agreement as stated by Hennessy *et al.*, (2007)

REFERENCES

Ambler, S. W. (1998). *User Interface Design Tips, Techniques, and Principles*. Retrieved from http://www.ambysoft.com/essays/userInterfaceDesign.html#Principles

Bowman, D., & Hodges, L. (1997). User Interface Constraints for Immersive Virtual Environment Applications. In *Proceedings of IEEE VRAIS*, (pp. 35-38).

Cairncross, S. (2002). *Interactive Multimedia and Learning: Realizing the Benefits*. PhD Thesis, Napier University, Scotland.

Cairncross, S., & Mannion, M. (1999). How multimedia functions in engineering education. *Engineering Science and Education Journal*, *8*(3), 100–107. doi:10.1049/esej:19990302

Caramazza, A. M., McCloskey, & Green, B. (1981). Naive beliefs in 'sophisticated' subjects: Misconceptions about trajectories of motion. *Cognition*, *9*, 117–123. doi:10.1016/0010-0277(81)90007-X

Carter, J. (2002). A framework for the development of multimedia systems for use in engineering education. *Computers & Education Journal*, *39*, 111–128. doi:10.1016/S0360-1315(02)00029-5

Champagne, A. L., & Klopfer, J. & Anderson. (1980). Factors influencing the learning of classical mechanics. *American Journal of Physics*, *48*, 1074–1079. doi:10.1119/1.12290

Clement, J. (1982). Students' preconceptions in Introductory Mechanics. *American Journal of Physics, 50*, 66–71. doi:10.1119/1.12989

Diaz, D. (1999). *The Pedagogies of Multimedia Presentation*. Cuesta Community College. San Luis Obispo, CA. Retrieved from http://www.ltseries.com/LTS/pdf_docs/mmpedago.pdf

Fogler, H. S., Montgomery, S. M., & Zipp, R. P. (1992). Interactive Computer Modules for Chemical Engineering Instruction. *Journal of Computer Applications in Engineering Education, 1*(1), 11–24.

Goldberg, F. M., & Anderson, J. M. (1989). Student difficulties with graphical representations of negative values of velocity. *The Physics Teacher, 27*, 254–260. doi:10.1119/1.2342748

Gramoll, K. (2001). An Internet Portal for Statics and Dynamics Engineering Courses. *Proceedings of the International Conference of Engineering,* (pp. 1 – 6).

Gunstone, R. F., & White, R. (1981). Understanding of gravity. *Science Education, 65*, 291–299. doi:10.1002/sce.3730650308

Gupta, A. (2002). Content Development for eLearning in Engineering Education. *Interactive Educational Multimedia Journal, 4*, 12–23.

Hennessy, S., Wishart, J., Whitelock, D., Deaney, R., Brawn, R., & Velle, L. (2007). Pedagogical approaches for technology-integrated science teaching. *Computers and Education Journal, 48*(1), 137–152. doi:10.1016/j.compedu.2006.02.004

Hibbeler, R. C. (2001). *Engineering Mechanics: Dynamic*", (9th Ed.). Upper Saddle River, NJ: Prentice-Hall.

Hill, F. S., Jr. (2001). *Computer Graphics Using Open GL,* (2nd Ed.). Upper Saddle River, NJ: Prentice Hall.

Iannella, R. (1992). Designing 'Safe' user interfaces. *Australian Computer-Journal, 24*(3), 92–97.

Janson, J. L. (1992, January 27). Computer-based training helps firms trim budgets. *PC Week*, pp. 90 – 93.

Jesica, M., & Tara, M. (2005). Educational Applets for Active Learning in Properties of Electronic Materials. *IEEE Transactions on Education, 48*(1).

Kahn, T. M. (1992). Multimedia literacy at Rowland: A good story, well told. *Technological Horizons in Education, 19*(7), 77–83.

Kinshuk, R. (1996). *Computer Aided Learning for Entry accountancy students.* PhD Thesis, De Montfort University, UK.

Kinshuk, R., & Patel, A. (2003). Optimizing Domain Knowledge Representation with Multimedia Objects. *Learning and Teaching with Technology - Principles and practices*. London: Kogan.

Klett, K. (2002). Improving Engineering Student Learning in a Web-Based Learning Space Due to Virtual Reality Techniques and Advanced Interactivity. In *Proceedings of the 2002 eTEE Conference Switzerland*, (pp. 195 – 101).

Liang, W. (2002). Using Multimedia in Teaching Dynamics. In *Proceedings of the 2002 Conference for Industry Education Collaboration, American Society for Engineering Education* (Session 4424).

Liarokapis, F., Mourkoussis, N., White, M., Darcy, J., Sifniotis, M., Petridis, P., et al. (2007). Augmented Reality and Web 3D to Support Engineering Education. In *Proceedings of the Engineering and Technology Education Conference, Greece,* (pp. 24 -28).

Marchisio, L., Ronco, E., & Saacco, R. (1993). Modeling the user interface. *IEEE Communications Magazine*, *31*(5), 68–74. doi:10.1109/35.212424

McDermott, L.C., & Rosenquist, M.L. &, Van, Zee. E. H. (1987). Student difficulties in connecting graphs and physics: Examples from kinematics. *American Journal of Physics*, *55*, 503–513. doi:10.1119/1.15104

Pangalos, G. J. (1993). Designing the user interface. *Computers in Industry*, *22*(2), 193–200. doi:10.1016/0166-3615(93)90066-A

Ratan, K., & Mitty, P. (1997). Using Contemporary Tools to Teach Dynamics in Engineering Technology. *International Journal of Engineering Education*, *13*(6), 407–411.

Reif, F., & Sue, A. (1992). Cognition for interpreting scientific concepts: A study of acceleration. *Cognition and Instruction*, *9*, 1–44. doi:10.1207/s1532690xci0901_1

Scott, N. W. *(1996)*. A study of the introduction of educational technology into a course in engineering dynamics: classroom environment and learning outcomes. *PhD Thesis.*

Squires, R. G., Andersen, P. K., Reklaitis, G. V., Jayakumar, S., & Carmichael, D. S. (1992). Multimedia-Based Applications of Computer Simulations of Chemical Engineering Processes. *Journal of Computer Applications in Engineering Education*, *1*(1), 25–30.

Steif, P. S. (2003). Design and evaluation of problem solving courseware modules for mechanics of materials. *Journal of Engineering Education,* (pp.1 – 30).

Vallim, B. R. (2006). Practicing Engineering in a Freshman Engineering Course. *IEEE Transactions on Education*, *49*(1), 74–79. doi:10.1109/TE.2005.856157

Wallace, M. D., & Anderson, T. J. (1993). Approaches to interface design. *Interacting with Computers*, *5*(3), 259–278. doi:10.1016/0953-5438(93)90010-Q

Chapter 8
Evaluation of Interactive Multimedia Packages

INTRODUCTION

The literature shows that many different evaluation methodologies for computer aided learning (CAL) packages have been proposed based on different philosophical views (Worthen and Sanders, 1973; Popham, 1974; Stephen and Stanley, 1985). The evaluation may be used for a variety of purposes such as refining goals, defining products or programs, and estimating costs, usability and effectiveness (Reeves, 1993). This involves the systematic review of the content, design, and instructional value and worth of computer aided learning packages. In general, any instructional software package should be evaluated before it is delivered or used in the classroom or research laboratory. This Chapter provides some general evaluation techniques used in the evaluation of such packages.

EVALUATION TECHNIQUES

The lack of controlled evaluation procedures for computer-based instruction has led to disagreement regarding their success in student learning (Baker, 1990). There is an enormous uncertainty in the edu-

DOI: 10.4018/978-1-60566-764-5.ch008

cation society regarding the effectiveness of any evaluation scheme for software (Borich and Jernelka, 1981; Bates, 1981; Tucker 1989; Micceri *et al.*, 1989). These educational technology practitioners suggest that evaluation schemes of CAL packages need further research, improvement and are highly subjective since they depend on the objectives of the designer and the context of their use.

Evaluation has generally been conceptualized as either formative or summative. The aim of the formative evaluation is the refinement and improvement of a program or learning package while the aim of the summative evaluation is to determine the impact and the outcomes of a particular program or learning package (Guba and Lincoln, 1991). A program or learning package that is acceptable for some initial evaluation should be tested for its efficiency in real environments. Reja (2003), in an effort to identify a standard evaluation procedure for CAL packages in Europe, conducted a study on 19 organizations within ten member states of the European Community and found diversity in both formative and summative evaluation activities. According to Reja's findings, formative evaluations, conducted by most organizations, identify weaknesses in a product early enough to implement design changes. The summative evaluations, which are conducted at the end of a major development to assess the various aspects of a finished product, varied from a critical appraisal of a product by an expert to extensive in-depth testing.

Shute and Regian (1993) provide a framework within which an experimental evaluation methodology for intelligent tutoring systems (ITS) could be standardized. While ITS involve computer-based tutoring that is based on artificial intelligence models, the proposed framework is also appropriate to the evaluation of interactive CAL packages (Muramatsu *et al.*, 1998). Baker and King (1993) proposed an evaluation methodology based on a checklist in which they identified "hallmarks of quality" that characterize good learning products. The authors defined twelve basic categories that embody good learning design: engagement, interactivity, tailor ability, appropriateness of multimedia mix, mode and style of interaction, quality of interaction, quality of end user interface, learning styles, monitoring and assessment techniques, built-in intelligence, adequacy of ancillary learning support tools, and suitability for single user/group/distributed use. This evaluation checklist was used to assess 43 wide ranging of computer-based learning and training products. Baker and King (1993) found that the quality of end-user design interface was very important in producing a quality product. Engagement, interactivity, as well as tailor ability were found to be the other benchmark of quality.

However, Eibeck (1996) pointed out that the shortcoming of the checklist suggested by Baker and King (1993) is that it only addresses student learning on a superficial level. Research in the area of cognitive science has indicated that certain learning models present within CAL packages have a greater potential to improve learning than others.

A number of methods could be used to determine the efficacy of a program or learning package. These include quantitative and qualitative methods and laboratory testing method. According to Legree *et al.* (1993) quantitative methods are preferred to determine overall effectiveness of a program or learning package. On the other hand Murray (1993) recommended that the qualitative methods are suitable to find the internal efficiency of the overall program or learning package and its individual components. However Wyatt and Spiegelhalter (1990) suggested laboratory testing as the most suitable method for initial evaluation on field trials. Kinshuk (1996) provided a more detailed discussion of these evaluation techniques. Pham (1998) stated a brief review of other evaluation approaches such as objective-based, decision-based, value-based and naturalistic used in evaluating quality multimedia systems. However, only two of the approaches were found to be significant namely the naturalistic and value-based. Naturalistic approach is concerned with user's views, interests and experiences. The information for naturalistic approach evaluation is generally obtained through observations and interviews of users verbally or through questionnaires. On the other hand, the value-based approach is more concerned with the overall merit and worth of the product. ISO/IEC 9126-1, (1991) defined six quality characteristics namely functionality, reliability, usability, efficiency, maintainability and portability and described a software product evaluation process model. The software quality product characteristics can be used to specify both non-functional customer and user requirements.

Qualitative Technique

Murray (1993) stated that qualitative techniques provided information as a function of personal interaction and perception. In addition the data collected is primarily in the form of words or pictures rather than numbers. The qualitative evaluation of CAL packages may give an insight to the aspects of learning and performance that are impossible to apprehend by other means. Shute and Regian (1993) stated that qualitative analysis is about finding causes and consequences.

The most commonly used technique for data gathering when conducting qualitative research is subject-based that includes observations, in depth interviews, questionnaires and focus groups. These techniques can be summarized as following:

- *Observational* techniques are methods by which an individual or individuals gather firsthand data on programs, processes, or behaviors being studied. They provide evaluators with an opportunity to collect data on a wide range of behaviors, to capture a great variety of interactions, and to openly explore the evaluation topic. By directly observing operations and activities, the evaluator can develop a holistic perspective, i.e., an understanding of the context within which the project operates.
- *Interviews* provide different data from observations: they allow the evaluation team to capture the perspectives of project participants, staff, and others associated with the project. In the hypothetical example, interviews with project staff can provide information on the early stages of the implementation and problems encountered. The use of interviews as a data collection method begins with the assumption that the participants' perspectives are meaningful and able to be made explicit, and that their perspectives affect the success of the project.
- *Questionnaires and Interviews* techniques have almost similar sense. If a series of questions are put to a subject via a written sheet, and the same questions are read aloud, then it may be said that questionnaires and interviews respectively, have been used to collect the same data (Kinshuk, 1996).
- *Focus group* session is indeed, an interview (Patton, 1990) not a discussion group, problem-solving session, or decision-making group. The technique inherently allows observation of group dynamics and firsthand insights into the respondent's behaviors, attitudes, language, etc. Therefore, focus groups combine elements of both interviewing and participant observation.

Quantitative Technique

Quantitative evaluation is primarily about identifying the characteristics of a situation or setting (Shute and Regian, 1993). Questionnaires are the most commonly used technique for data gathering when conducting quantitative research. However one disadvantage with this technique is that they limit the type of data that an evaluator can gather which could be due to lack of responses from participants. Researchers gather numerical data from small samplings on observable behavior and questionnaires and then analyze the data to make assumptions for the rest of the population (Gall *et al.*, 1996). In general most quantitative evaluation research studies are organized to determine if a technology program is meeting its goals and objectives. Other techniques used in evaluating educational materials include pre-and post-test. According to Crompton (1996), these are one of the most difficult techniques in terms of educational research.

CAL ENGINEERING PACKAGES EVALUATION TECHNIQUE

The national engineering education delivery system (NEEDS) is an electronic database developed by the Synthesis Coalition (a body funded by the national science foundation and industrial partners in USA). The NEEDS database is used for delivery and evaluating engineering education courseware and provides an access mechanism for students and faculty to a diverse range of engineering educational materials (Agogino, 1997). This evaluation criterion database is used to award the "*Premier Award*" that recognizes the contributions of engineering educators who develop outstanding non-commercialized engineering education courseware.

Since NEEDS is an electronic medium, it is highly flexible and subject to great variation. According to Pamela (1996), the quality of content and technical features of courseware on the NEEDS database is highly variable for several reasons; 1) the rapid changes in multimedia technology; 2) improved authoring environments; and 3) evolving practices of using educational courseware over the last few years. This diversity is a strength of the NEEDS database, since it encourages students, instructors and courseware developers to explore the NEEDS database, experiment, download material, and incorporate the material into lectures or reports. While this diversity of material on NEEDS is an asset, it also poses a challenge for users who are looking for reliable, tested courseware that can be incorporated into a classroom with minimal effort (Pamela, 1996).

The NEEDS database have a three-tiered structure for courseware, (1. *Non-reviewed material*, 2. *Endorsed courseware*, 3. *Premier courseware*) based on the level of review the courseware has undergone. Further information on these three-tiered structures for courseware can be found in Agoginos' (1997) works.

Review Procedure for NEEDS Courseware

As explained by Pamela (1996), when authors submit courseware for inclusion in the NEEDS Database through the NEEDS Manager, they will have the option of requesting their courseware be peer reviewed. The peer review process will determine if the courseware will receive Endorsed category on the NEEDS database.

The flow chart of courseware through the evaluation system for the NEEDS database is shown in Figure 1. All courseware submitted to the NEEDS database will be reviewed for functionality by the NEEDS Manager and then either placed on the NEEDS database, or returned to the author for modification if the courseware was not functional. If the author has requested, the courseware will also be sent to the NEEDS Editorial Board for peer review. If the courseware is not accepted by the peer evaluations for the Endorsed status, it will still remain on the NEEDS database as non-reviewed courseware.

Figure 1. Flow chart of courseware through the evaluation system for the NEEDS database

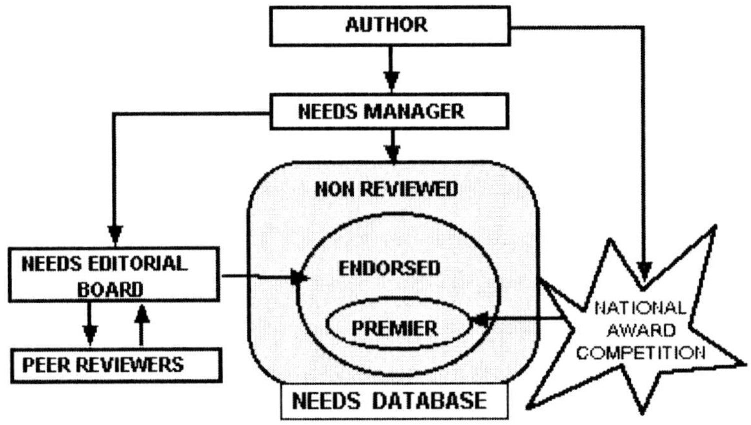

The peer-review process will be modeled only after the professional publications. A NEEDS Editorial Board will control the process. The Editorial Board consists of a NEEDS Editor and Associate Editors, all of which are engineering instructors with experience using and/or developing computer-based instruction. After receiving the courseware, the NEEDS Editor will pass it to an Associate Editor with expertise in the courseware's technical area. The Associate Editor will solicit reviews from individuals both internal and external to SYNTHESIS concerning the courseware's content and pedagogy. The reviewers will complete the questionnaire (see Appendix O) and provide a written review. The courseware will then either be accepted as it is, or recommended to be revised and re-submitted for review. Once the courseware is accepted as Endorsed, the author will be sent the written reviews and will be given the option to have this review appended to the courseware's bibliographic record.

The aforementioned criteria has been refined and streamlined over the first three competitions and are now organized under three general categories as shown in Table 1 (Muramatsu *et. al.*, 1998); (Muramatsu *et. al.*, 2000); (Muramatsu, 2002).

The initial evaluation criteria consist of nine primary areas for evaluating engineering courseware namely engineering content, engagement, impact on learning, user interface, user interaction, multimedia design, instructional use, technical performance, and accessibility from the NEEDS database. In each sub-category are a series of questions to help the reviewer judge the relative merits of CAL package in that area.

Table 1. Evaluation criteria for engineering education courseware

Engineering Content	Software Design	Instructional Design
• Accuracy of content	• Engagement	• Interactivity
• Organization of content	• Learner interface and navigation	• Cognition/conceptual change
• Consistency with learning objectives	• Technical reliability	• Content
		• Multimedia use
		• Instructional use/adaptability

Table 2. HCI Evaluation techniques (Howard and Murray, 1987)

Technique	Description
• *Expert Based*	A domain expert may employ knowledge, scientific principles and intuition to evaluate the interface.
• *Theory Based*	Consists of the model of the user and a model of the interface and occurs where mapping relationships between formal representation of the user and device are examined with a view to identify any mismatch.
• *Subject Based*	This evaluation is based on the user's judgment according to four components namely system, task, subject and metric (Reiterer and Oppermann, 1993).
• *User Based*	Consists of a user and a system and relates to personal evaluation by the user reflected in terms of patterns of system use relating to the task and overall environment.
• *Market Based*	Consists of a market and a system and relates to the final evaluation conducted by the market place reflected in terms of market performance.

Although the NEEDS evaluation criteria has been used to evaluate engineering courseware, the review methodology for the NEEDS database has not been straightforward since standard evaluation procedures have not been established in CAL packages.

Other educational researchers suggested that human computer interaction (HCI) evaluation for CAL packages could be divided into five categories namely expert, theory, subject, user and market based (Howard and Murray, 1987). These HCI evaluation techniques can be summarized as shown in Table 2.

The subject based evaluation technique is widely used for the evaluation of CAL packages as described briefly in Section 8. The Learning Technology Dissemination Initiative (LTDI) (1999) provided a useful guide for reviewing a new piece of software that could be used to evaluate the effectiveness of learning software. The aim of LTDI is to promote the use of learning technology and computer based learning materials in Scottish Higher Education.

In summary, this Chapter discussed a few evaluation techniques for educational software. Although most of the evaluation techniques were found to be suitable for the evaluation of educational software, there is no evidence that any single technique is suitable for all types of educational software. As such, further work is required to set a standard for the evaluation of educational software that could be globally accepted.

The next Chapter discusses the results of the students' background and learning styles (described in Chapter 2), the research methodologies, and evaluation of the efficiency and effectiveness of the TAPS packages.

REFERENCES

Agogino, A. M. (1997). The National Engineering Education Delivery System (NEEDS): A Multimedia Digital Library of Courseware. *International Journal of Engineering Education, 13*(5), 333–340.

Baker, E. L. (1990). Technology-assessment: Policy and methodological issues. In *Intelligent Tutoring Systems*. Hillsdale, NJ: Lawrence Erlbaum Associates.

Baker, P., & King, T. (1993). Evaluating interactive multimedia courseware – A methodology. *Computers & Education, 21*(4), 307–319. doi:10.1016/0360-1315(93)90034-G

Bates, T. (1981). Toward a better research framework for evaluating the effectiveness of educational media. *British Education Technology Journal, 12*, 215–233. doi:10.1111/j.1467-8535.1981.tb00424.x

Borich, G., & Jernelka, R. (1981). Evaluation in Computer Based Instruction. In *A state of the Art Assessment*, (pp. 161 – 209). New York: Academic.

Crompton, P. (1996). *Evaluation: A Practical Guide to Methods*. Learning Technology Initiative Dissemination Report, Heriot-Watt University.

Eibeck, P. A. (1996). Criteria for Peer-Review of Engineering Courseware on the NEEDS Database. *IEEE Transactions on Education . Special Issue on the Application of Information Technologies to Engineering and Science Education, 39*(3), 381–387.

Gall, M., Borg, W., & Gall, J. (1996). *Educational Research an Introduction.* Upper Saddle River, NJ: Longman Publishers.

Guba, E. G., & Lincoln, Y. S. (1991). *Effective evaluation.* San Francisco: Jossey-Bass.

Howard, S., & Murray, D. (1987). *A survey and classification of Evaluation Techniques for HCI.* NPL Report DITC 94/87, July 87. Retrieved July 2002 from http://www.hull.ac.uk/engprogress/Prog3Papers.htm

ISO/IEC 9126-1. (1991). *Software product evaluation standard.*

Kinshuk, R. (1996). *Computer Aided Learning for Entry accountancy students.* PhD Thesis, De Montfort University U.K.

Legree, P. J., Gillis, P. D., & Orey, M. A. (1993). The quantitative evaluation of intelligent tutoring system applications: Products and process criteria. *Journal of Artificial Intelligence in Education, 4*(2/3), 209–226.

Micceri, T., Pritchard, W. H., & Barett, A. J. (1989). Must Computer courseware evaluation be totally subjective? *British Journal of Educational Technology, 20*, 120–128. doi:10.1111/j.1467-8535.1989.tb00272.x

Muramatsu, B. (2002). *NEEDS, SMETE, MERLOT and the National STEM Education Digital Library Program.* Presentation at the Council of Engineering and Scientific Society Executives, July 25, San Jose, CA.

Muramatsu, B., Agogino, A. M., Klotz, E., McMartin, F., Robson, R., & Simutis, L. (2000). *A national digital library for science, mathematics, engineering, and technology education.* Presentation at the American Mathematics Society/American Mathematics Association Joint Mathematics Meeting, Washington, DC.

Muramatsu, B., Eibeck, P. A., Stern, J. L., & Agogino, A. M. (1998). An Evaluation Process for Engineering Courseware: The Premier Award for excellence in Engineering Education Courseware. In *NSF Engineering Innovators Conference, Washington DC.*

Murray, T. (1993). Formative qualitative evaluation for "Exploratory." . *Journal of Artificial Intelligence in Education, 4*(2/3), 179–207.

Pamela, E. (1996). Criteria for Peer-Review of Engineering Courseware on the NEEDS Database. *IEEE Transactions on Education, 39*(3).

Patton, M. Q. (1990). *Qualitative Evaluation and Research Methods,* (2nd Ed.). Newbury Park, CA: Sage.

Pham, B. (1998). Quality evaluation of educational multimedia systems. *Australian Journal of Educational Technology, 14*(2), 107–121.

Popham, W. J. (1974). *Evaluation in Education*. Berkerly, CA: McLutchan.

Reeves, T. C. (1993). Evaluating interactive multimedia. In D.M. Gayeski (*Ed.*) *Multimedia for learning, development, application, evaluation*, (pp. 92 – 112). Educational Technology Publications.

Reiterer, H., & Oppermann, R. (1993). Evaluation of user interfaces: EVADIS II – a comprehensive evaluation approach. *Behaviour & Information Technology, 12*(3), 137–148. doi:10.1080/01449299308924375

Reja, U. (2003). Open-ended vs. Close-ended Questions in Web Questionnaires. Developments in Applied Statistics. *Metodoski zvezki, 19*, 159 – 177.

Shute, V. J., & Regian, J. W. (1993). Principles for evaluating intelligent tutoring systems. *Journal of Artificial Intelligence in Education, 4*(2/3), 245–271.

Stephen, M. A., & Stanley, R. T. (1985). Computer-Based Instruction. *Methods and Development*. Englewood Cliffs, NJ: Prentice Hall, Inc.

The Learning Technology Dissemination Initiative (LTDI). (1999). Retrieved March 10, 2002 from http://www.icbl.hw.ac.uk/ltdi/index.html

Tucker, R. (1989). Software Evaluation: Who cares? In *Aspects of Educational and Technology*, (Vol. XXII: Promoting Learning, pp. 8 – 16). London: Kogan Page.

Worthen, B. R., & Sanders J. R., (1973). *Educational Evaluation: Theory and Practice*.

Wyatt, J., & Spiegelhalter, D. (1990). Evaluating medical expert systems: what to test and how? *Medical Informatics, 15*(3), 205–217.

Chapter 9
Evaluation of TAPS Packages

INTRODUCTION

This Chapter provides evaluation techniques used in the evaluation of TAPS packages. The Chapter also discusses the students learning styles and the methodology and results of statistical analysis used in the evaluation of the TAPS packages.

RESEARCH METHODOLOGY EMPLOYED FOR EXAMINING STUDENTS' LEARNING STYLES

The evaluation was carried out to examine the distribution of learning styles (discussed in Chapter 2) of the third year undergraduate engineering students and suggest effective problem solving approaches that could increase the motivation and understanding of slow learners at UNITEN. For this study, a sample target population of 60 third year undergraduate engineering students who had taken the Engineering Mechanics subject was tested. These students were selected based on their second year grade point average (GPA) of less than 2.5 as this study emphasizes on slow learners.

DOI: 10.4018/978-1-60566-764-5.ch009

Table 1. Percent frequency values

Preference	% Freq.	*Preference*	% Freq.	*Preference*	% Freq.	*Preference*	% Freq.
Strong Sensing	5.0	Strong Visual	38.3	Strong Active	1.7	Strong Sequential	3.3
Moderate Sensing	21.7	Moderate Visual	38.3	Moderate Active	18.3	Moderate Sequential	21.7
Balanced ***Sens-Int***	58.3	Balanced ***Vis-Verb***	23.3	Balanced ***Act-Refl***	70.0	Balanced ***Seq-Glob***	61.7
Moderate Intuitive	11.7	Moderate Verbal	0.0	Moderate Reflective	10.0	Moderate Global	13.3
Strong Intuitive	3.3	Strong Verbal	0.0	Strong Reflective	0.0	Strong Global	0.0
Total	100.0	Total	100.0	Total	100.0	Total	100.0

The study focused on issues of problem solving methods, user interface, and multimedia attributes of the TAPS packages. The aim was to develop quality TAPS packages that would promote learning. In this study, the survey was made anonymous and voluntary.

The students were given the index of learning styles (ILS) instrument based on Felder-Silverman model to obtain the learning styles of the students. The same students were also given a set of questionnaires that adopt the Likert-type assessment (Kinshuk 1996), i.e. based on the scale 1 (Strongly agree) to 5 (Strongly disagree) to access their perception of the TAPS packages. The results of the survey were used to determine whether the TAPS packages had potential pedagogic advantages over conventional teaching approach.

Since the objective of the evaluation of students' learning styles in this book was to design the TAPS packages tailor-made to students needs, mainly quantitative methods were employed in the evaluation.

As an adjunct to quantitative data collected, some qualitative views from students have also been sought in the evaluation study but the data obtained from various sources was not consistent and was merely in the form of observations and open-ended questions relating to their computer usage. Though this data was of limited value for statistical analysis, it provided an indication of students' keen interest towards the TAPS packages and the perceived strength and weaknesses of the packages.

Procedure Used to Evaluate the Students' Learning Styles

The ILS was administrated to the students in the form of printed questionnaire at the commencement of the first study semester in 2003. The responses to the learning style questions were then entered on-line using the Web for each respondent. The responses were processed on-line and the results of analysis were displayed as a report and printed for each respondent. Thus 60 printed reports corresponding to 60 respondents formed the basis of the data analysis and the results are presented in following Section.

Data Analysis and Results

The analysis report consists of scores on a scale of 1 to 11 (odd numbers only) for one of the dichotomy of each of four dimensions of the ILS. For example, the result for the hypothetical individual student may consist of the following scores along the four dimensions: 3 reflective, 5 sensing, 7 visual and 9 global. A score of 1 to 3 in either dichotomy of a dimension indicates a learning style preference that is fairly balanced in that dimension. A score of 5 to 7 indicates a moderate preference in the associated dichotomy of the concerned dimension (e.g. sensing dichotomy of the sensing-intuitive dimension). A score of 9 to 11 indicates a very strong preference. The result reports of the 60 respondents were consolidated into percent frequency values for the four dimensions as given in Table 1.

Data presented in Table 1 shows that 58.3% (35) of the respondents have balanced preference in the sensing-intuitive dimension. Moderate-sensing dimension shows that 21.7% (13) have moderate preference and 5.0% (3) have strong preference for sensing learners. Thus, 85% (51) of the respondents would feel comfortable with problem solving techniques geared toward sensing learners (learners who prefers facts, data, experimentation, sights and sounds in their learning).

In the visual-verbal dimension, 23.3% (14) of the respondents have balanced preference. Moderate-visual dimension shows that 38.3% (23) have moderate preference and 38.3% (23) have strong preference for visual learners. Thus, 100% (60) of the respondents would prefer problem-solving techniques geared toward visual learners (learners who prefers pictures, diagrams, charts, movies, demonstrations and exhibitions in their learning).

Table 1 shows that in the active-reflective dimension, 70.0% (42) respondents are balanced learners, 18.3% (11) are moderate active learners, and 1.7% (1) is strong active learners. Thus, 90% (54) of respondents will benefit from problem-solving techniques preferred by active learners (learners who prefers to learn by doing and participating through engagement in physical activity or discussion in their learning).

In the sequential-global dimension, 61.7% (37) of the respondents are balanced learners. Moderate-sequential dimension shows that 21.7% (13) have moderate preference and 3.3% (2) have strong preference for sequential learners. Thus, 86.7% (52) of respondents would like problem-solving techniques geared toward sequential learners (learners who take things logically step by step in their learning and will be partially effective with understanding). The data clearly shows that the majority of the engineering students surveyed in this study will benefit from problem solving methods that match the needs of sensing, visual, active and sequential learners.

The results of the student's learning styles indicate that problem-solving approaches of the engineering students geared towards sensing, visual, active and sequential learners, which formed the basis for designing the TAPS packages. To help the sensing learners, it would be desirable to model selected engineering problems that are difficult to understand from the textbooks. The solutions to the selected engineering problems should be shown in a step-by-step fashion and narrated to the students while interacting with the TAPS packages. In addition, the sensing learner will understand the concepts of engineering and problem solving steps better if the instructor gives examples of the application in a dynamic form.

Both visual and active learners will benefit from hands on activities and virtual experiments. Sequential learners will understand the material presented in the TAPS packages better if a brief summary of the selected engineering problem is given in the packages from the lecture taught by the instructor during normal classroom teaching.

The results of the data analysis indicate that a small proportion of respondents are intuitive (15%), verbal (0%), reflective (10%), and global (13%) learners. It can be suggested that certain activities and problems that are more difficult can be included in the TAPS packages to motivate this minority group. The students may be given a few relatively difficult problems whereby they can solve these problems through group effort. This opportunity will satisfy the intuitive and verbal learners who get a chance to talk about the problems in sharing of knowledge. Small case studies may be interesting to the reflective and global learners who like to work alone at their own pace, analyzing the problems from multiple views. In view of the obtained results, the author designed the TAPS packages using technologies (2-D/3-D and DVR) to make ideal problem-solving models (step-by-step approach).

RESEARCH METHODOLOGY EMPLOYED FOR EVALUATING TAPS PACKAGES

The subject based evaluation techniques were employed to evaluate the TAPS packages since these techniques have been found in the past, to be suitable for conventional CAL evaluation. These evaluation techniques were chosen because they are based directly on the user's judgment, as there is a possibility of discovering the responses that an individual gives spontaneously and thus avoiding the bias that may result from suggesting responses to individuals (Reja, 2003). In addition, data collection is possible under laboratory conditions. Two types of subject-based evaluation techniques were used in this study, namely questionnaires and observations. The students in UNITEN were also observed while they were interacting with TAPS packages in the computer laboratory and the collected information provided insights about their responses towards navigational and problem solving procedures of the TAPS packages. This method of evaluation was used to ascertain the independent navigational nature of the TAPS packages, which forms a small but vital part of the overall assessment of the effectiveness of the TAPS packages.

The same group of 60 students that participated in the ILS questionnaires study was given a set of close-ended questionnaires as shown in Appendix P in order to validate the problem solving technique adopted and the effectiveness of the TAPS packages. In addition, some open-ended questionnaires were also used to obtain feedback about the strength, weaknesses, and suggestions for improving the TAPS packages (see question 128 Appendix P).

Since better inferences can be made from the design and experimental approaches than observational studies, design approaches were used for this study. In the analysis of the data concerning the students in the design approaches, the dependent variables are the design elements which include the number of words of text per screen, the varying interactive nature of the content, the utilization of audio and video, the use of animation, as well as the use of navigational menu, color and configuration (e.g. sound volume). Problem solving skills of students constituted the independent variable.

The same close and open-ended questionnaires were used for all the TAPS packages. These questions are similar to the ones developed by Kinshuk (1996). These evaluation questionnaires were used to evaluate the TAPS packages because they were straight-forward and easy to understand. However, extra questions were added since the TAPS packages employed newer technologies such as multimedia and desktop virtual reality. The questionnaire was divided into two parts and contained a total of 128 questions. The first part contains information related to the user's background on computer usage. The second part of the questionnaire is the respondents' assessment of the TAPS packages that contains 117 close-ended questions and one open-ended question. This Section (part II of the questionnaire) provided information relevant to the evaluation of the interfaces of the TAPS packages. Each close-ended question asks whether the student is in favor or not in favor of the package on the whole. The questionnaire included fourteen Sections, with questions addressing the following areas:

- Computing knowledge background
- Comparison with other media

- Individual flexibility
- Engineering knowledge/teaching
- Error messages and documentation
- Route through TAPS packages
- Control of learning and problem solving
- Retention and learning
- Help facility
- Function of keys
- User Tools (Calculator / Notepad / Glossary / Character map table)
- Presentation of information on screen
- Holistic system
- Overall presentation

Responses to all the close-ended questions were based on a five-point Likert Scale - the possible responses being *Strongly Agree, Agree, Uncertain, Disagree* and *Strongly Disagree* that facilitates an easy and reliable analysis.

The evaluation of the TAPS packages took place in one of the computing laboratories at the College of Engineering, UNITEN. Since the number of computers was constrained with the availability of one laboratory with 30 computers, two sessions of the TAPS packages evaluation trials were done with 30 students per session. Each computer was installed with the TAPS packages. The students were brought into the computer laboratory and explained the procedure for executing and using the TAPS packages. The students were asked to go through the problem solving steps presented in the TAPS packages and then to try solving and answering the questions asked in the TAPS packages. Each session lasted approximately two hours under the observation of the author and the instructor teaching the Engineering Mechanics subject.

Upon completion of the evaluation, the students were asked to fill in the close-ended questionnaires and submit them for data analysis. The data collected from the study was used to investigate the students' views towards the TAPS packages. The feedback was used to represent measures of understanding and satisfaction, respectively. The results of the statistical analysis are presented in the next Section.

STATISTICAL ANALYSIS RESULTS OBTAINED FROM THE EVALUATION OF THE TAPS PACKAGES

The feedback obtained from 60 students who used the TAPS packages based on the fourteen Sections stated in Section 9.3 are discussed in this Section. The Likert-type assessment i.e. based on the scale 1 (Strongly agree), 2 (Agree), 3 (Uncertain), 4 (Disagree) and 5 (Strongly disagree) was adopted. Such feedback could help in improving the quality of TAPS packages particularly relating to user interface, interactivity, use of multimedia, and problem solving techniques. In this study, the most significant questions of interest for further analysis and the percentages, number of responses are shown in bold text.

The results on Computing Knowledge Background Section displayed in Table 2, clearly show that almost 50% (20+9) of the respondents felt that being a computer literate was not a prerequisite. About 27 respondents reported that after using the TAPS packages they were more confident in learning using a computer. Five respondents complained that it was difficult to tackle the selected engineering problems

Table 2. Computing knowledge background

1 = Strongly Agree, 2 = Agree, 3 = Uncertain, 4 = Disagree and 5 = Strongly Disagree	1	2	3	4	5
The packages do not require any knowledge of computing.	**9 (15%)**	**20 (33%)**	11 (19%)	12 (20%)	8 (13%)
I could not learn much because I spend time getting to know how to use the package.	0 (0%)	0 (0%)	37 (62%)	13 (22%)	10 (16%)
After using the TAPS packages I became more confident in learning from a computer.	**9 (15%)**	**18 (30%)**	33 (55%)	0 (0%)	0 (0%)
It was difficult enough to tackle the selected engineering problems without having to struggle with the computer.	0 (0%)	0 (0%)	55 (92%)	**3 (5%)**	**2 (3%)**
The system can be used sufficiently without any manuals.	10 (16%)	15 (25%)	26 (43%)	5 (9%)	4 (7%)
Normally I find these types of problems to be very simple, but I could not see how to solve the problems by using a computer.	0 (0%)	0 (0%)	10 (16%)	11 (19%)	39 (65%)

without having to struggle with the computer. However, this could be due to their unfamiliarity with the use of computer aided learning packages. The higher percentage of respondents who were confident of using and interacting with the TAPS packages indicated the importance of employing new technologies (such as multimedia) towards improving the understanding of engineering concepts.

The results about comparison with other media Section depicted in Table 3 shows that 22 respondents still prefer a human as the tutor and 19 respondents preferred the TAPS packages as the tutoring media. On the other hand 39 respondents appreciated problem solving in the subject matter with the TAPS packages, because the student's felt like being on one-to-one basis with the tutor as compared to the conventional style where one instructor tutors many respondents. Only 21 respondents were of the opinion that they were left with a lot of unanswered questions after using the TAPS packages. 25 respondents still prefer to use textbook in their learning whilst 29 respondents disagreed using a textbook. This can be clearly seen from Table 3 where 47 respondents preferred to use TAPS packages because these packages are interactive and it is easier to understand the problem solving steps. In addition, 32 respondents preferred learning and solving Engineering Mechanics problems where dynamic illustrations are employed because they could see how it is done (how the problem is solved), rather than having to ask someone to explain it. On the average, the respondents accepted TAPS packages only as an additional educational aid in comparison to human tutors.

The results about individual flexibility Section shown in Table 4 shows that 14 respondents agreed that the TAPS packages are only appropriate for advanced respondents of engineering. On the other hand, 13 respondents knew how to solve the problem without needing to go through the complete steps as required in the TAPS packages. 45 respondents showed that the TAPS packages kept adjusting its advice according to the students' needs and progress. This supports the fact that the TAPS packages have the potential in guiding the students in solving the engineering problems. TAPS packages allow individual flexibility whereby 55 respondents agreed they could skip into the more demanding material without having to go through the easier material again.

The results about engineering knowledge teaching Section shown in Table 5 concludes that 40 respondents agreed that they did not need any previous knowledge of engineering to use the TAPS packages and 44 respondents agreed that they have gained a good introduction to the concept of solving

Table 3. Comparison with other media

1 = Strongly Agree, 2 = Agree, 3 = Uncertain, 4 = Disagree and 5 = Strongly Disagree	1	2	3	4	5
I would prefer to learn from a human tutor than from these packages.	**17 (28%)**	**5 (9%)**	19 (32%)	**9 (15%)**	**10 (16%)**
It would have been a better improvement to have a tutor close by, so that I could ask questions.	21 (34%)	5 (9%)	10 (16%)	13 (22%)	11 (19%)
I liked problem solving in the subject matter with these packages, because it's like being one to one basis with the tutor.	**16 (27%)**	**23 (38%)**	13 (22%)	6 (10%)	2 (3%)
I was left with a lot of unanswered questions after using the packages.	**14 (23%)**	**7 (12%)**	7 (12%)	22 (37%)	10 (16%)
This form of problem solving and learning was really clear, because unlike an instructor, the program does not miss out a lot of steps.	27 (45%)	19 (32%)	9 (15%)	2 (3%)	3 (5%)
It is easier to get involved solving Engineering Mechanics problems using these packages than in a classroom tutorial.	18 (28%)	22 (37%)	5 (9%)	11 (19%)	4 (7%)
I prefer a textbook because I can flip it forward and backwards.	**10 (16%)**	**15 (26%)**	6 (10%)	**19 (32%)**	**10 (16%)**
It's much easier to understand the problem solving steps by using these packages than reading an Engineering Mechanics textbook because these packages are interactive and uses multimedia in it's contents.	**27 (45%)**	**20 (33%)**	6 (10%)	7 (12%)	0 (0%)
It was good to be able to discover the relationships between variables and engineering concepts, which I could not see in the textbooks.	15 (26%)	8 (13%)	14 (23%)	13 (22%)	10 (16%)
I would rather learn Engineering Mechanics by working problems using pencil and paper.	11 (19%)	7 (12%)	19 (32%)	13 (22%)	10 (16%)
I like learning and solving Engineering Mechanics problems this way because I can see how it's done, rather than asking someone to explain it.	**21 (34%)**	**11 (19%)**	5 (9%)	13 (22%)	10 (16%)

Engineering Mechanics problems. Additionally, 46 respondents agreed that the packages gave a very good background on solving the Engineering Mechanics problems and 48 respondents agreed that the TAPS packages have helped improve their knowledge by guiding them through the learning process. 47 respondents showed interest to learn solving other problems in Engineering Mechanics topics using these packages and 34 respondents agreed that the TAPS packages allowed them to learn accurately the

Table 4. Individual flexibility

1 = Strongly Agree, 2 = Agree, 3 = Uncertain, 4 = Disagree and 5 = Strongly Disagree	1	2	3	4	5
These TAPS packages are only appropriate for advanced students of engineering.	**9 (15%)**	**5 (9%)**	3 (5%)	23 (38%)	20 (33%)
The computer usage got harder just as I got better at the problems.	7 (12%)	3 (5%)	7 (12%)	21 (34%)	22 (37%)
I knew how to solve the problems, but the TAPS packages persisted in following its long and laborious steps.	**10 (16%)**	**3 (5%)**	15 (25%)	17 (28%)	15 (25%)
The TAPS packages kept adjusting its advice according to my needs and progress.	**25 (41%)**	**20 (33%)**	7 (12%)	3 (5%)	5 (9%)
I could not go into the more demanding material without having to go through the easier stuff again.	0 (0%)	3 (5%)	2 (3%)	**30 (50%)**	**25 (41%)**

Table 5. Engineering knowledge teaching

1= Strongly Agree, 2 = Agree, 3 = Uncertain, 4 = Disagree and 5 = Strongly Disagree	**1**	**2**	**3**	**4**	**5**
I did not need any previous knowledge of engineering to use the TAPS packages.	**27 (45%)**	**13 (22%)**	9 (15%)	6 (10%)	5 (9%)
I have gained a good introduction to the concept of solving Engineering Mechanics problems.	**33 (55%)**	**11 (19%)**	6 (10%)	10 (16%)	0 (0%)
I think the TAPS packages are only useful for students who are already familiar with Engineering Mechanics and want to improve.	19 (32%)	17 (28%)	11 (19%)	10 (16%)	3 (5%)
It is difficult to learn Engineering Mechanics with these TAPS packages.	5 (9%)	3 (5%)	7 (12%)	10 (16%)	35 (58%)
The packages give a very good background on solving the Engineering Mechanics problems.	**29 (48%)**	**17 (28%)**	11 (19%)	3 (5%)	0 (0%)
Most students would find it hard to solve Engineering Mechanics problems using the TAPS packages.	3 (5%)	2 (3%)	35 (58%)	11 (19%)	9 (15%)
The TAPS packages allow me to fully test my understanding of Engineering Mechanics.	0 (0%)	4 (6%)	37 (62%)	12 (20%)	7 (12%)
After using the TAPS packages I still do not understand how to solve Engineering Mechanics problems.	0 (0%)	7 (12%)	6 (10%)	9 (15%)	38 (63%)
After using the TAPS packages I can easily use my knowledge to solve any Engineering Mechanics problem better.	9 (15%)	5 (9%)	31 (51%)	9 (15%)	6 (10%)
I would find it easy to use these TAPS packages to teach someone else about engineering mechanics.	27 (45%)	13 (22%)	9 (15%)	6 (10%)	5 (9%)
The TAPS packages have helped improve my knowledge by guiding me through the learning process.	**33 (55%)**	**15 (25%)**	6 (10%)	3 (5%)	3 (5%)
I would like to learn solving other problems in Engineering Mechanics topics using these packages.	**38 (63%)**	**9 (15%)**	10 (16%)	3 (5%)	0 (0%)
The TAPS packages allowed me to learn an accurate problem solving steps of Engineering Mechanics and corrected my mistakes.	**19 (32%)**	**15 (25%)**	23 (38%)	3 (5%)	0 (0%)
Even now I do not understand the problems presented in the TAPS packages.	0 (0%)	0 (0%)	21 (34%)	23 (38%)	16 (28%)

problem solving steps of Engineering Mechanics and helped in correcting their mistakes. These results indicate that TAPS packages generally enhanced students knowledge.

The results about error messages and documentation Section shown in Table 6 found that 35 respondents agreed that the computer messages displayed on the screen were good enough to guide them through difficulties however 45 respondents disagreed that they liked the TAPS packages because it gave too many hints at frequent intervals whenever they made any mistakes. Whereas 53 respondents agreed, that they liked the way the TAPS packages gave explanation when they made mistakes. A majority of 59 respondents agreed that it was a good idea to have the TAPS packages display a small hint whenever they got the answer wrong before giving the correct answer. Infact, 54 respondents agreed that the explanations showed clearly why they obtained the wrong answer. Some 22 respondents agreed that it would be better if a clearer explanation of the aim of the problem solving exercises and the required information to complete the exercises had been given. This provides evidence of the appropriateness of the error messages and documentation integrated in the TAPS packages.

Table 6. Error messages and documentation

1 = Strongly Agree, 2 = Agree, 3 = Uncertain, 4 = Disagree and 5 = Strongly Disagree	**1**	**2**	**3**	**4**	**5**
The computer messages displayed on the screen were good enough to guide me when I got stuck.	**20 (33%)**	**15 (25%)**	20 (33%)	5 (9%)	0 (0%)
The error messages were unhelpful.	2 (3%)	8 (13%)	14 (23%)	7 (12%)	29 (49%)
I liked the TAPS packages because it guided me what to do whenever I made any mistakes.	3 (5%)	2 (3%)	10 (16%)	**15 (25%)**	**30 (50%)**
The TAPS packages allow me to make mistakes.	45 (75%)	5 (9%)	10 (16%)	0 (0%)	0 (0%)
The TAPS packages should not allow me to enter any incorrect values.	55 (91%)	5 (9%)	0 (0%)	0 (0%)	0 (0%)
The TAPS packages are so unforgiving, it does not allow me to do anything else, until I provide the correct answer.	20 (33%)	7 (12%)	10 (16%)	13 (22%)	10 (16%)
There should be an error message if I enter an incorrect negative, positive numbers or symbols (i.e. "/","[") .	10 (16%)	10 (16%)	34 (57%)	5 (9%)	1 (2%)
I like the way the TAPS packages gave explanation when I made a mistake.	**47 (78%)**	**6 (10%)**	5 (9%)	2 (3%)	0 (0%)
The TAPS packages directly give explanations without giving me another chance.	0 (0%)	3 (5%)	15 (25%)	37 (61%)	5 (9%)
It was a good idea to have the TAPS packages display a small hint whenever I got the answer wrong before giving the correct answer.	**56 (93%)**	**3 (5%)**	1 (2%)	0 (0%)	0 (0%)
The TAPS packages provide good feedback only when I have entered an answer.	31 (51%)	3 (5%)	13 (22%)	13 (22%)	0 (0%)
The TAPS packages did not give enough praise for the correct answer.	0 (0%)	0 (0%)	13 (22%)	12 (20%)	35 (58%)
The TAPS packages helped me with every step in solving the Engineering Mechanics problems.	28 (47%)	8 (13%)	14 (33%)	7 (12%)	3 (5%)
Through the feedback I immediately knew whether I was right or wrong.	55 (92%)	3 (5%)	2 (3%)	0 (0%)	0 (0%)
The explanations provided by the TAPS packages were hard to understand.	0 (0%)	9 (15%)	29 (48%)	17 (28%)	5 (9%)
The explanations showed clearly why I had got the wrong answer.	**49 (81%)**	**5 (9%)**	6 (10%)	0 (0%)	0 (0%)
The TAPS packages give the same type of feedback at all times regardless of what I have done.	52 (87%)	8 (13%)	0 (0%)	0 (0%)	0 (0%)
The TAPS packages give individual feedback to each student depending upon how far they have progressed to the solution.	9 (15%)	12 (20%)	34 (57%)	3 (5%)	2 (3%)
It would be better if clearer explanation of the aim of the problem solving exercises and the required information to complete the exercises had been given.	**13 (22%)**	**9 (15%)**	28 (46%)	7 (12%)	3 (5%)

The results shown in Table 7 revealed that 34 respondents knew what all the items on the various buttons meant and 35 respondents agreed that the instructions given on how to proceed through the problem-solving task were clear. Additionally, 37 respondents agreed that they liked learning and problem solving using this approach because they could check over things and avoid repeating the same mistakes. However, 34 respondents felt that they would have found it more helpful if given a suggested

Table 7. Route through TAPS packages

1= Strongly Agree, 2 = Agree, 3 = Uncertain, 4 = Disagree and 5 = Strongly Disagree	**1**	**2**	**3**	**4**	**5**
I never knew what all the items on the various buttons meant.	9 (15%)	6 (10%)	11 (19%)	**29 (48%)**	**5 (9%)**
The instructions given on how to proceed through the problem-solving task were unclear.	5 (9%)	2 (3%)	18 (30%)	**15 (25%)**	**20 (33%)**
I like learning and problem solving in this approach, because you can check over things and do not repeat the same mistakes again and again.	**28 (45%)**	**9 (15%)**	11 (19%)	7 (12%)	5 (9%)
Sometimes I found it hard to keep track about which bits/chunks of tutorial I have completed.	12 (20%)	5 (9%)	15 (25%)	18 (30%)	10 (16%)
The TAPS packages allowed me to solve the problem presented in multiple orders.	38 (62%)	11 (19%)	5 (9%)	2 (3%)	4 (7%)
I would have found it more helpful to be given a suggested route through the TAPS packages.	**15 (25%)**	**19 (32%)**	15 (25%)	8 (13%)	3 (5%)
The prompts, asking me to enter text/numeric/symbols, were not clear from the screen.	8 (13%)	9 (15%)	11 (19%)	23 (38%)	9 (15%)
I want to be able to stop the exercise and go back to the problem solving tutorial, and then return where I left off.	0 (0%)	0 (0%)	11 (19%)	14 (23%)	35 (58%)
On screen problem solving tutorial are difficult to follow.	13 (22%)	4 (7%)	26 (43%)	7 (12%)	10 (16%)
It provided little understanding of what you are doing or how to do it.	3 (5%)	8 (13%)	11 (19%)	15 (25%)	23 (38%)

Table 8, Control of learning and problem-solving

1= Strongly Agree, 2 = Agree, 3 = Uncertain, 4 = Disagree and 5 = Strongly Disagree	**1**	**2**	**3**	**4**	**5**
I felt the TAPS packages allowed me to work at my own pace and directions.	**25 (41%)**	**19 (31%)**	11 (19%)	5 (9%)	0 (0%)
I found the TAPS packages were too slow because I could do the exercise much more quickly with pencil and paper.	4 (7%)	12 (20%)	30 (50%)	6 (10%)	8 (13%)
It was too quick, I would like more control of how much time I was given to solve the problems.	2 (3%)	3 (5%)	30 (50%)	16 (27%)	9 (15%)
I wanted slower speed in the beginning and quicker when I got some experience, but the TAPS packages kept on at the same speed which was annoying.	0 (0%)	3 (5%)	51 (85%)	6 (10%)	0 (0%)
I learnt a lot about problem solving task in the subject matter by trying different things with the TAPS packages.	**8 (13%)**	**13 (22%)**	32 (53%)	4 (7%)	3 (5%)
I am sure I will retain most of what I have learnt.	**12 (20%)**	**12 (20%)**	15 (25%)	9 (15%)	12 (20%)
I learnt a lot from the TAPS packages, by learning how to solve various Engineering Mechanics problem.	**17 (28%)**	**13 (22%)**	14 (23%)	11 (19%)	5 (9%)
If I use the TAPS packages again, I could solve similar problems quicker.	**19 (31%)**	**18 (30%)**	13 (22%)	7 (12%)	3 (5%)
I will not be able to solve the problems if they are presented in a different format/order than these packages.	2 (3%)	6 (10%)	44 (73%)	5 (9%)	3 (5%)

route through the TAPS packages. The majority of the respondents found it easy to navigate throughout the TAPS packages.

The results about control of learning and problem-solving Section shown in Table 8 shows that 44 respondents agreed that they felt the TAPS packages allowed them to work at their own pace and directions. 21 respondents agreed that they learnt a lot about problem solving task in the subject matter by trying different things with the TAPS packages (i.e. by using the user tools). Additionally, 24 respondents agreed that they were sure that they would retain most of what they have learnt. On the other hand, 30 respondents agreed that they learnt a lot from the TAPS packages, by learning how to solve various Engineering Mechanics problem and 37 respondents agreed if they used the TAPS packages again, they could solve similar problems quicker. TAPS packages were thus found to give the students a good control on their learning and problem solving skills.

The results about help facility Section stated in Table 9 shows that 47 respondents agreed that they would prefer more help from the TAPS packages. However, it can be clearly seen that about 23 respondents agreed that the TAPS packages offered timely help and 39 respondents agreed that the help given was general instead of context sensitive most of the time. The results indicate that the help facility in the TAPS packages still needs improvement.

The results about function or keys Section charted in Table 10 shows that 43 respondents agreed that, when navigating through the packages, they were clear about which options they were allowed to select.

The results about user tools Section noted in Table 11 shows that only 9 respondents agreed that the calculator facility was very useful and 48 students agreed that the calculator should be customized to have more functions, which are needed to solve the problems. 54 respondents agreed that the notepad was very useful for them to type/make important notes and save it for reference. All the 60 respondents responded favorably that the glossary of commands table was very useful as they could look up / search for a related word of the problem matter and understand its meaning better and 45 respondents agreed that the character map table was very useful as it contained important symbols unavailable on the standard keyboard and allowed them to copy and paste into the textboxes on the screen. Hence, the user tools were found to be very helpful by the students.

The results about presentation of information on screen Section in Table 12 shows that 50 respondents agreed that the screen layout made it easy to communicate with the computer. However, 26 respondents were confused by the large amount of information on the screen. 39 respondents also found that the way the problem is presented on the screen makes it easy to work out the solution and 51 respondents agreed

Table 9. Help facility

1 = Strongly Agree, 2 = Agree, 3 = Uncertain, 4 = Disagree and 5 = Strongly Disagree	**1**	**2**	**3**	**4**	**5**
I would prefer more help from these TAPS packages.	**25 (41%)**	**22 (37%)**	5 (9%)	8 (13%)	0 (0%)
The help facility should have given a few hints, rather than the correct answer.	5 (9%)	3 (5%)	15 (25%)	14 (23%)	23 (38%)
The TAPS packages do not offer much help where it is most wanted.	9 (15%)	6 (10%)	22 (37%)	**10 (16%)**	**13 (22%)**
The TAPS packages gave a general help every time instead of context sensitive help.	2 (3%)	3 (5%)	16 (27%)	**19 (31%)**	**20 (33%)**

Table 10. Function or keys

1= Strongly Agree, 2 = Agree, 3 = Uncertain, 4 = Disagree and 5 = Strongly Disagree	**1**	**2**	**3**	**4**	**5**
I found it very difficult to know which buttons/keys corresponded to the command I wanted.	2 (3%)	7 (12%)	21 (35%)	16 (27%)	14 (23%)
When navigating through the packages, I found I was clear which options I was allowed to select.	**23 (38%)**	**20 (33%)**	17 (29%)	0 (0%)	0 (0%)
I didn't like some of the buttons/keys, I sometimes got lost because of them.	11 (19%)	9 (15%)	11 (19%)	14 (23%)	15 (25%)
When I was going through the TAPS package(s), pressing the same key would produce a different command.	0 (0%)	0 (0%)	60 (100%)	0 (0%)	0 (0%)
I got confused because same command was available through different keys in different screen.	0 (0%)	0 (0%)	60 (100%)	0 (0%)	0 (0%)
It was easy to exit from a particular route through the packages.	0 (0%)	0 (0%)	60 (100%)	0 (0%)	0 (0%)
I could not skip some parts of the problem solving tutorials.	7 (12%)	5 (9%)	29 (47%)	8 (13%)	11 (19%)
I could not skip some parts of the exercises.	0 (0%)	0 (0%)	60 (100%)	0 (0%)	0 (0%)
It was easy to delete any answer, which, I knew was wrong.	30 (50%)	22 (37%)	8 (13%)	0 (0%)	0 (0%)
I was not allowed to answer a question in the exercises provided more than one time.	0 (0%)	0 (0%)	9 (15%)	20 (33%)	31 (52%)

Table 11. User tools (Calculator / Notepad / Glossary / Character map table)

1= Strongly Agree, 2 = Agree, 3 = Uncertain, 4 = Disagree and 5 = Strongly Disagree	**1**	**2**	**3**	**4**	**5**
The calculator facility was very useful.	**2 (3%)**	**7 (12%)**	33 (54%)	11 (19%)	7 (12%)
I found it hard to use the calculator to transfer the answer to the package.	5 (9%)	10 (16%)	17 (29%)	16 (27%)	12 (20%)
Calculator should be customized to have more functions, which are needed to solve the problems.	**37 (61%)**	**11 (19%)**	8 (13%)	4 (7%)	0 (0%)
The notepad was very useful in that I could type/make important notes and save it for reference.	**45 (75%)**	**9 (15%)**	6 (10%)	0 (0%)	0 (0%)
The glossary of commands table was very useful in that I could look up / search for a related word of the problem matter and understand its meaning better.	**49 (81%)**	**11 (19%)**	0 (0%)	0 (0%)	0 (0%)
The character map table was very useful as it contains important symbols unavailable on the standard keyboard and allowed me to copy and paste into the textboxes on the screen.	**31 (52%)**	**14 (23%)**	12 (20%)	3 (5%)	0 (0%)

that the screen was attractive and clear. On the other hand 40 respondents agreed that they appreciate more font types, sizes, and colors in the packages.

The results about holistic system Section in Table 13, shows that 37 respondents agreed that they were comfortable with the TAPS packages on the whole. However, 24 respondents stated that although they enjoyed using the packages, it can be improved even more, so they would be tempted to use it again

Table 12. Presentation of information on screen

1= Strongly Agree, 2 = Agree, 3 = Uncertain, 4 = Disagree and 5 = Strongly Disagree	**1**	**2**	**3**	**4**	**5**
The screen layout presented an easy way to communicate with the computer.	**38 (63%)**	**12 (20%)**	7 (20%)	3 (5%)	0 (0%)
Sometimes I was confused by the large amount of information on the screen.	**15 (25%)**	**11 (19%)**	20 (33%)	5 (9%)	9 (15%)
The way the problem is presented on the screen makes it easy to work out the solution.	**29 (49%)**	**10 (16%)**	15 (25%)	6 (10%)	0 (0%)
There was not enough information on the screen for me to solve the exercise questions.	2 (3%)	7 (12%)	17 (29%)	9 (15%)	25 (41%)
The screen was attractive and clear	**42 (61%)**	**9 (15%)**	5 (9%)	4 (7%)	0 (0%)
The colors on the screen were confusing	5 (9%)	7 (12%)	25 (41%)	13 (22%)	10 (16%)
The TAPS packages looked dull and boring.	2 (3%)	1 (2%)	19 (32%)	19 (32%)	19 (32%)
I like to see the different font types, sizes and colors in the packages.	**15 (25%)**	**25 (41%)**	8 (13%)	7 (12%)	5 (9%)

Table 13. Holistic system

1= Strongly Agree, 2 = Agree, 3 = Uncertain, 4 = Disagree and 5 = Strongly Disagree	**1**	**2**	**3**	**4**	**5**
I did not like the TAPS packages on the whole.	2 (3%)	7 (12%)	14 (23%)	**19 (32%)**	**18 (30%)**
I really liked using the TAPS packages because it was fun to use.	15 (25%)	11 (19%)	12 (20%)	12 (20%)	10 (16%)
I enjoyed using the packages, but it could have been improved more, so I would be tempted to use it again.	**9 (15%)**	**15 (25%)**	23 (38%)	6 (10%)	7 (12%)
I simply used the TAPS packages to read what is on the screen.	2 (3%)	3 (5%)	33 (54%)	10 (16%)	12 (20%)
The learning process went on too long.	4 (7%)	6 (10%)	35 (58%)	9 (15%)	6 (10%)
It is easy to use the packages and does not require much practice to be familiar.	**22 (37%)**	**19 (31%)**	10 (16%)	4 (7%)	5 (9%)
I would prefer, if I could be given problem-solving exercises using these sorts of packages.	**29 (49%)**	**11 (19%)**	13 (22%)	4 (7%)	3 (5%)
The packages did not manage to convince me that I could learn the problem solving techniques/method better.	4 (7%)	9 (15%)	20 (33%)	11 (19%)	16 (25%)
I liked the way the packages coach in solving the problems.	17 (29%)	12 (20%)	21 (35%)	6 (10%)	4 (7%)
It is a good way to revise something that I have forgotten.	8 (13%)	11 (19%)	24 (39%)	12 (20%)	5 (9%)
The packages are very helpful especially for a student like me.	12 (20%)	13 (22%)	16 (27%)	10 (16%)	9 (15%)

Table 14. Overall perception

1= Strongly Agree, 2 = Agree, 3 = Uncertain, 4 = Disagree and 5 = Strongly Disagree	**1**	**2**	**3**	**4**	**5**
Most students may not be interested to use such TAPS packages to assist them in learning the subject matter.	4 (7%)	9 (15%)	14 (23%)	**18 (30%)**	**15 (25%)**
I solved some of the exercise questions but was not sure why the answer(s) were incorrect.	2 (3%)	1 (2%)	25 (41%)	13 (22%)	19 (31%)
I would like to have more problem solving examples.	**14 (23%)**	**9 (15%)**	17 (29%)	15 (25%)	5 (9%)

(some suggestions for improving the TAPS packages are given in Section 9). 41 respondents agreed that it is easy to use the packages and do not require much practice to get familiar and 40 respondents agreed that they would not mind using these sort of packages for problem-solving exercises.

The results about overall perception Section shown in Table 14 found that 33 respondents may be interested to use such TAPS packages to assist them in learning the subject matter and 23 respondents agreed that they would prefer to have more problem solving examples.

SUMMARY OF OPEN ENDED QUESTIONNAIRES

In general, students commented on the strengths and weaknesses of the TAPS packages and provided many useful suggestions in the open-ended questionnaires. Some comments were also received during observation and these were recorded. These suggestions will be taken into account for further improvement of the TAPS packages in future. Below is a summary of some of the selected comments:

Strengths

1. I am impressed by the fact that the concepts are shown visually, and think it saves the instructors time to cover additional information.
2. The notepad feature was very helpful in making important notes as I was solving the engineering problems.
3. The glossary of commands table was very useful because it only provides important words with it's meaning that are related to the selected engineering problems.
4. The packages allow me to repeat and see the steps in solving the problem as many times as I want and it became clear each time I repeated a step.
5. I find the packages to be very user-friendly.
6. May enhance the student's understanding but it is more suitable for self-learning activities.
7. It is very helpful in the sense that it will enable a student like me who is weak in the subject matter to understand the concept being explained.
8. It is suitable and learnable.
9. I find it easier to visualize the problems presented in the packages and doing problem solving using TAPS packages.

10. Most computer learning packages in general only provide the answers when the solution is required but these TAPS packages provide the solutions in details i.e. every step is clearly shown and explained till the solution is reached.
11. Clear and slow explanations are given with my level of English language understanding. Some of the computing packages I have used utilize difficult levels of English language that makes it difficult to understand.
12. The multiple-choice questions (MCQs) are not repeated in the same order. I like the idea of randomizing the MCQs, as I know if I have really understood the topic.
13. I liked the idea of having more computer graphics and animations as it could help maximize the computer visualization capabilities in the packages.
14. I find these sorts of multimedia packages to be very useful for engineering students as different animated colors, fonts and sketches could be used to highlight and shown in understanding the difficult engineering concepts as compared to text-books.
15. I believe if I had used these sorts of packages earlier, I could have performed better in the examination.
16. I liked navigating with the 3-D interface as it allowed me to view the objects from multiple views.
17. The 3-D robotic motion and the path shown by the TAPS package really helped me to understand the topic on curvilinear motion.
18. The desktop virtual reality images produced for the TAPS packages were excellent.
19. The use of colors and arrows on the analysis of engineering structure was good idea where only the important concepts were made to be dim, blink, rotate and move to direct my attention. It's difficult to learn these concepts from the text-book.
20. Introducing these TAPS packages is a good idea particularly to students with weaker mathematical and science background to help them learn engineering concepts faster.
21. The menu is straightforward in that it's obvious to the learners and reflects the content of the packages.
22. I believe the knowledge represented in the TAPS packages could be transferred faster to the learner as compared to the text-book.
23. I do not fear making mistakes using these TAPS packages, because if I make a mistake I can repeat the tutorials and exercises again.
24. The simulations and animations really motivate me to learn more about the selected topics as compared to the way I learnt in textbooks.
25. The 3-D models seem to be quite realistic and helps motivates learners to explore, learn and discover the problem solving strategies presented in the TAPS packages.
26. I am intrigued by the way TAPS package guides in solving the engineering problem. The package is capable of checking text and numeric inputs if they are correctly entered.

Weaknesses

1. I find it difficult to ask about clarification about something that is not clear to me.
2. These sorts of packages are only suitable for certain students.
3. I did not find the calculator as an important tool for helping me in solving the engineering problem.

4. Some of the font colors and sizes are not consistent.
5. Sometimes the system slows down, since too many large audio and video files are incorporated in the packages.
6. More examples are preferred and it should start from easy to difficult ones.
7. I took this subject a year ago and have forgotten most of the theoretical concepts, therefore it would have been better to include more theoretical information and additional references.
8. I do not have a PC hence may not be able to learn using these packages while off- campus.
9. If I wish to stop and continue my work later, there is no option to save my work in the disk.
10. I feel disoriented by using these TAPS packages because not much help is provided on using the packages.

Suggestions

1. I expect to see more functions in the calculator tool. It should work like a scientific calculator. These functions will be helpful in calculating difficult / time consuming equations and formulas.
2. The mechanism to copy and paste the character from the character map table should be improved to a drag and drop manner.
3. The packages should have more theoretical contents so that the subject could be understood better.
4. The system should have a mechanism to allow the student to store the student's work i.e. a built in database so that it can be retrieved later.
5. Try to use a minimum of three colors per screen and use bright colors for displaying important concepts. Avoid using dark background colors. Text colors should be consistent in all screens.
6. Important information such as help, tip or hint should be kept visible at all times.
7. Try asking questions later, rather than immediately after a chunk of information is presented on the screen.
8. Do not provide too much of interactivity on a single screen, it could disorient users.
9. Provide a hand-book about how to use the TAPS packages rather than directly asking users to go to the PC and work on it.
10. Try to avoid giving too many hints as I feel the learner will not put in much effort to learn.
11. The packages should be designed to provide multiple solutions rather that just one solution.

EVALUATION OF STUDENTS LEARNING STYLES AND SELECTION OF TAPS PACKAGES

This Section attempts to assess the students learning styles and their preferences towards the features offered in the TAPS packages. The purpose is to understand student's attitude towards using the TAPS packages as well as to gauge which features (i.e. DVR, 2D simulation, 3D animation or Coach-based) of the TAPS packages would be most beneficial to a given group of learners.

In this exercise, the students that participated in the Felder-Solomon's index of learning styles questionnaires were selected. These students were divided into four groups based on their learning styles as follows (*n = number of respondents*):

(1) ***Sensory***: are learners who prefer facts, data, experimentation, sights, sounds and physical sensations (*n = 30*)
(2) ***Visual***: are learners who prefer pictures, diagrams, charts, movies, demonstrations and exhibitions (*n = 59*)
(3) ***Active***: are learners who learn by doing and participating through engagement in physical activity or discussion (*n = 36*)
(4) ***Sequential***: are learners who take things logically step by step and will be partially effective with understanding. (*n = 37*)

From the open-ended questionnaires administered at the beginning of the TAPS packages as shown in Appendix P, it was found that these groups of learners had very little or no experience in using computer-based learning packages. Nevertheless, from the observations made (as mentioned in Section 10.3) during the evaluation of the TAPS packages in the computer lab by the instructor/author and also based on the following comments obtained from the open-ended questionnaires as presented in Section 9:

The 3-D models seem to be quite realistic and helps motivates learners to explore, learn and discover the problem solving strategies presented in the TAPS packages

The simulations and animations really motivate me to learn more about the selected topics as compared to the way I learnt in textbooks

It could be concluded that the students showed great interest in using the TAPS packages.
In addition, comments such as:

I am impressed by the fact that the concepts are shown visually, and think it saves the instructor's time to cover additional information

I liked the idea of having more computer graphics and animations as it could help maximize the computer visualization capabilities in the packages

I find these sorts of multimedia packages to be very useful for engineering students as different animated colors, fonts and sketches could be used to highlight and shown in understanding the difficult engineering concepts as compared to text-books

The desktop virtual reality images produced for the TAPS packages were excellent

The use of colors and arrows on the analysis of engineering structure was good idea where only the important concepts were made to be dim, blink, rotate and move to direct my attention because it is difficult to learn these concepts from the text-book

I liked navigating with the 3-D interface as it allowed me to view the objects from multiple views

Further highlight the overall perception of students towards the incorporation of multimedia and virtual reality in simulating engineering concepts that were otherwise difficult to comprehend from the text-book.

The analysis carried out based on the four groups of learners i.e. sensory, visual, active and sequential, indicate that different groups prefer different features of the TAPS packages as summarized in Table 15. The number of students with their individual learning styles is derived from the Felder-Solomon's questionnaires (Appendix A) and their feedback (Appendix P) was compiled based on their learning styles. The comments (Appendix P) with the following variables: improves, helps, understand/understood, learn/learning, interest, interesting, easy, provides, teaches and interactive were selected. The comparison was made on the basis of the comments for each of the TAPS packages. The results indicated that sensory group of learners prefer (DVR, 3D and 2D), visual group prefers (2D and 3D), active learners prefer (3D and DVR) and the sequential learners have high preference for (coach-based and 2D) TAPS packages. The main key features of each TAPS package has been explained in details in Chapter 7, Table 4.

However, when comparing the performance of the learners in the laboratory, the active and sensory learners performed almost equally well when using the 3D animation and DVR TAPS packages. This result is indicative of the relationship between laboratory performance and the active and sensory learners that prefer interactive problem solving tasks. On the other hand, sequential learners performed better when using the coach-based TAPS package. This can probably be attributed to the design philosophy of this TAPS package i.e. leading from problem statement through a series of steps and solutions. Visual learners appear to be comfortable with 2D simulation TAPS package as it matches their learning styles where the learners prefer to see 2D pictures, diagrams and charts in their learning.

In summary, the analysis showed that different groups of learners have different preferences of the features offered in the TAPS packages and this in turn would affect the effectiveness of the TAPS packages in meeting its objectives as well as the performance of the students. However, at this stage it is not possible to link the students' academic performance with the preferred features of the TAPS packages. This is because some students may have multiple learning styles thus more difficult to target, and specific activities meant to target a learning style do not always accomplish the goal due to differences in terms of individual preferences. Additionally, students' background knowledge may also affect performance depending on their exposure to modern technologies, motivation and flexibility of learning. At present,

Table 15. Students feedback for TAPS packages based on different groups of learners

	Sensory	**Visual**	**Active**	**Sequential**
DVR	• *Provides* high impact presentation. • Maintains *interest* and involvement. • Makes *learning* more intuitive.		• *Improves* comprehension and retention. • Maintains *interest* and involvement. • *Helps* communicate complex engineering problems faster.	
2D simulation	• The concept of a structure could be *understood* better as compared to the static images in the textbook. • Animation of the structure and computations could be shown simultaneously and *help* students *understand* better.	• *Provides* the ability to explore multimedia data. • Usage of colors and animation adds attraction to *learn* and makes the subject *interesting*. • The forces shown on the free body diagram of the truss in an animated form can be easily seen and *understood*.		• Very *interactive* and problem solving steps can be repeated until *understood*.

this study is focused on testing the design and functionality of the TAPS packages. More work would be required to understand how the different learning styles of students are linked to the preferred TAPS features and how this would influence their overall performance in the subject matter.

DISCUSSION

The general outcomes of the statistical data collected from the learners who used the TAPS packages to visualize and solve the selected engineering problems can be summarized as following:

- From these results, it appears that virtual problem solving aids with dynamic illustrations enhanced student learning. For example, the 2-D and 3-D animations clearly helped the students in understanding curvilinear motions better, which were otherwise difficult to understand from the textbook alone.
- The overall percentage of positive response towards the TAPS packages showed increased awareness amongst the learners (students who need additional support in applying principles presented in lectures to problems) on the importance of the course towards their future development but concurrently felt that their mathematical background did not help much in solving the engineering problems. It was also found that the TAPS packages are capable of helping them understand the selected engineering problems.
- The results further indicate that most of the students preferred learning using TAPS packages because it is on one-to-one basis with the tutor as compared to the traditional classroom learning. Currently, at UNITEN, the engineering courses are conducted with large numbers of students ranging from 50 – 60 students per tutor (human). This can be cited as one reason for the slow learners wanting the TAPS packages as an additional educational aid.

In summary, this Chapter discussed a few evaluation techniques for educational software. Although most of the evaluation techniques were found to be suitable for the evaluation of educational software, there is no evidence that any single technique is suitable for all types of educational software. As such, further work is required to set a standard for the evaluation of educational software that could be globally accepted. From the statistical results, it is expected that the knowledge gained by the students who used the TAPS packages is sufficient in helping them to understand the engineering problem solving techniques better than the traditional approach in particularly for slow learners. The aim of the TAPS packages was to help and see if students could understand the selected engineering problems better and appreciate such technology in gauging their level of understanding in the subject matter. In general, most students were of the opinion that the design elements employed to develop the TAPS packages has caused positive effects in their problem solving skills. Most notable effects include increased interest and motivation to learn from the TAPS packages.

REFERENCES

Kinshuk, R. (1996). *Computer Aided Learning for Entry accountancy students.* PhD Thesis, De Montfort University U.K.

Reja, U. (2003). Open-ended vs. Close-ended Questions in Web Questionnaires. Developments in Applied Statistics. *Metodoski zvezki, 19*, 159 – 177.

Chapter 10
Effectiveness of the TAPS Packages

INTRODUCTION

This Chapter discusses the effectiveness of TAPS packages and provides a brief account of the differences between the approach of the TAPS packages used in this study with that of commercial simulation packages accompanying the Engineering Mechanics Dynamics textbook.

EFFECTIVENESS OF THE TAPS PACKAGES

In general, there exist various ways to measure the effectiveness of learning packages. Tessmer (1995) provided a number of variables that could be used to evaluate the effectiveness of multimedia contents such as aesthetics, transparency, forgiveness, matching between the metaphors and the learning experiences, informativeness, seamlessness of contents and media as well as the achievement of the desired learning experiences and learning outcomes. While much work has been devoted to research on the impact of technology in education, there is little known about its effectiveness. Furthermore, there are certain gaps in these research efforts that need to be addressed and require further investigation, specifically the

DOI: 10.4018/978-1-60566-764-5.ch010

lack of theoretical framework (Institute of Higher Learning Policy, 1999). In most evaluation studies, the important questions concern the comparative effectiveness of various types of learning packages when measured against traditional ones rather than the innovation of the delivery model itself and the factor that contributes to its effectiveness (Institute of Higher Learning Policy, 2000).

According to Psaromilingkos (2003) the effectiveness of a learning courseware is influenced by a number of variables such as: (a) quality of the learning resources (instructional material, exercises); (b) changes of the preferred mode of study (with or without the use of computer technology); (c) computer mediated instructions with peers and instructors and means of communication (e.g. email); (d) the quality of services that the software and hardware infrastructure provide (course management tools, multimedia conferencing systems); (e) time spent on the task using the system and (f) the learner's profile (learning style, previous experience, etc).

However, the overall effectiveness of the TAPS packages in this study was examined to confirm its design and measured using quantitative techniques such as open-ended questionnaires feedback, as stated in Section 9.5, a selection of questionnaires from the close-ended questionnaires (Table 10.1), and observational results mentioned in Section 10.3. The questions from the close-ended questionnaires were selected on the basis of the majority of respondents who selected "*strongly agree*" and "*agree*". The list of these questionnaires and results (in bold text based on number of students - responses) are shown in Table 1.

The results (in bold text) shown in Table 1 complemented the conclusion that the problem solving techniques and the preferred mode of study were the most significant predicting variables for effectiveness of the TAPS packages. Moreover a number of suggestions were made for the enhancements of the TAPS packages as stated in Section 9.

The use of multimedia coupled with desktop virtual reality was found to be an added advantage for students using the TAPS packages. On the whole, the self pace independent problem solving style of learning was much appreciated by slow learners.

EFFECTIVENESS OF THE TAPS PACKAGES BASED ON THE OBSERVATIONAL RESULTS

The following are the observations made during the trial of the study by the author and the instructor:

- Students rarely called the instructor for an explanation / query related to the TAPS packages.
- The user-friendly nature of the TAPS packages seemed to keep the student engrossed in problem solving.
- Most of the students were able to finish the task allocated to them.
- Most of the students left the lab with a positive remark about the TAPS packages and also felt enlightened about the subject matter.

These observations also help establish the effectiveness of TAPS packages to a certain extent.

The impact of learning by employing animation techniques, graphics, 2-D / 3-D and desktop virtual reality environments in this study promotes the usage of multimedia presentations (problem-solving), especially in technical discipline areas since these presentations will have a direct impact on the quality of the engineering materials at every level. Multimedia may become a standard level of instruction; its

Table 1. Effectiveness of the TAPS packages

1=Strongly Agree, 2=Agree, 3=Uncertain, 4=Disagree and 5=Strongly Disagree	**1**	**2**	**3**	**4**	**5**
I liked problem solving in the subject matter with these packages, because it's like being one to one basis with the tutor.	**16**	**23**	13	6	2
This form of problem solving and learning was really clear, because unlike an instructor, the program does not miss out a lot of steps.	**27**	**19**	9	2	3
It is easier to get involved solving Engineering Mechanics problems using these packages than in a classroom tutorial.	**17**	**22**	5	11	4
It's much easier to understand the problem solving steps by using these packages than reading an Engineering Mechanics textbook because these packages are interactive and uses multimedia in its contents.	**27**	**20**	6	7	0
It was good to be able to discover the relationships between variables and engineering concepts, which I could not see in the textbooks.	**15**	**8**	14	13	10
I like learning and solving Engineering Mechanics problems this way because I can see how it's done, rather than asking someone to explain it.	**21**	**11**	5	13	10
The TAPS packages kept adjusting its advice according to my needs and progress.	**25**	**20**	7	3	5
I have gained a good introduction to the concept of solving Engineering Mechanics problems.	**33**	**11**	6	10	0
The packages give a very good background on solving the Engineering Mechanics problems.	**29**	**17**	11	3	0
The TAPS packages have helped improve my knowledge by guiding me through the learning process.	**33**	**15**	6	3	3
I would like to learn solving other problems in Engineering Mechanics topics using these packages.	**38**	**9**	10	3	0
The computer messages displayed on the screen were good enough to guide me when I got stuck.	**20**	**15**	20	5	0
I like the way the TAPS packages gave explanation when I made a mistake.	**47**	**6**	5	2	0
It was a good idea to have the TAPS packages display a small hint whenever I got the answer wrong before giving the correct answer.	**56**	**3**	1	0	0
The explanations showed clearly why I had got the wrong answer.	**49**	**5**	6	0	0
I like learning and problem solving in this approach, because you can check over things and do not repeat the same mistakes again and again.	**28**	**9**	11	7	5
I felt the TAPS packages allowed me to work at my own pace and directions.	**25**	**19**	11	5	0
I learnt a lot from the TAPS packages, by learning how to solve various Engineering Mechanics problem.	**17**	**13**	14	11	5
If I use the TAPS packages again, I could solve similar problems quicker.	**19**	**18**	13	7	3
When navigating through the packages, I found I was clear which options I was allowed to select.	**23**	**20**	17	0	0
The notepad was very useful in that I could type/make important notes and save it for reference.	**45**	**9**	6	0	0
The glossary of commands table was very useful in that I could look up / search for a related word of the problem matter and understand its meaning better.	**49**	**11**	0	0	0
The character map table was very useful as it contains important symbols unavailable on the standard keyboard and allowed me to copy and paste into the textboxes on the screen.	**31**	**14**	12	3	0
The screen layout presented an easy way to communicate with the computer.	**38**	**12**	7	3	0

Table 1. continued

1=Strongly Agree, 2=Agree, 3=Uncertain, 4=Disagree and 5=Strongly Disagree	**1**	**2**	**3**	**4**	**5**
The way the problem is presented on the screen makes it easy to work out the solution.	**29**	**10**	15	6	0
It is easy to use the packages and does not require much practice to be familiar.	**22**	**19**	10	4	5
I would prefer, if I could be given problem-solving exercises using these sorts of packages.	**29**	**11**	13	4	3

strength lies in its ability to simulate real life situations while engaging the senses. Interactive multimedia TAPS packages allow students to put themselves in real life decision making situations while providing immediate feedback that allows learners to see, feel, hear and experience the consequences of their decisions in an unprecedented way. Problem solving via using TAPS packages is specific, whereas traditional classroom lecture style learning is more abstract and cannot involve students in the consequences of their decisions in comparison to multimedia. Thus multimedia makes problem solving real (more natural). As multimedia expands, retention and instructional quality will most likely improve due to the real life nature and sensory motivation of the medium. The learners who used the TAPS packages learn and understand the problem solving steps faster as the TAPS packages enhanced their interest and engaged learners at their own level of comprehension. For example if the learner is slow in absorbing the information, the learner may use his/her own pace of time to absorb the multimedia material.

The quantitative results obtained from the evaluation provided evidence that the TAPS packages developed have a good potential as an adjunct to traditional learning aids. TAPS packages can be seen as an important educational tool in the future to be used not only by learners who need additional support in applying principles presented in lectures to problems, but also by all engineering students in their learning. TAPS packages are seen as futuristic integral part of delivering knowledge in engineering and also other fields of education where similar difficulties are faced in presenting the subject matter to the students via conventional textbooks. The present results of evaluation may not present a true reflection of the educational value due to following reasons, limited time allocated for the evaluation, limited time exposure in using computer based learning packages by engineering students and slow process of executing the TAPS packages. These limitations are explained in more details in Section 10

Limited Time Allocated in the Evaluation

In this study, more time was spent in the development of the TAPS packages as compared to the evaluation of the TAPS packages because of the difficulty of the development process and understanding the Engineering Mechanics Dynamics subject. The time allocated to test each of the TAPS packages was limited to half an hour, including trying out the exercises and quizzes at the end of the problem solving tutorials by the students. This was not sufficient to introduce each student to all the features developed in the TAPS packages, work through the entire problem solving tasks, and answer all the questions given in the TAPS packages. One of the reasons for this scenario was the limited availability of computers. In addition, since the author's background is not mainly in engineering, much time was spent in gaining the necessary information of the selected engineering problems from the Engineering Mechanics instructor and the development of the TAPS packages. Ideally, a separate evaluation session should be allocated to allow students to gain familiarization with the computer-based problem solving environments. TAPS packages were developed to provide a standardized fashion of problem solving steps to numerous engineering problem prototypes of the subject matter. Therefore, by introducing the student to the TAPS packages problem solving steps once, the student will be able to use any problem solving prototype without requiring any further assistance.

Limited Exposure of Using Computer Based Learning Packages

The TAPS packages evaluation was, in many cases, the student's first exposure to using computer based learning packages. Although most students participating in the TAPS packages evaluation were computer-literate, some students may still require a human tutor to be around while engaging with the TAPS packages.

Slow Process of Executing the Taps Packages

Since the computers in the laboratory where the TAPS packages were evaluated were not equipped with high processing memory chips, it was found unsuitable to execute multimedia-based packages. In general, multimedia-based packages demand high processing memory and disk space to execute multimedia files such as audio and video. In addition, the information presented in the form of multimedia is usually synchronized and this caused the computer system to slow down. In many cases, students had to wait for more time than expected to move on the next screen in the TAPS packages. As such, these sorts of packages should be executed on computers with high memory and disk space.

There are many computer-based learning packages in the domain of engineering, which have been developed for higher learning institutions and also used commercially in western countries. However, the development of TAPS packages has made one of the pioneering efforts targeted for learners who need additional support in applying principles presented in lectures to problems in Mechanical Engineering at University Tenaga Nasional.

COMPARISON OF TAPS PACKAGES WITH OTHER SELECTED ENGINEERING PACKAGES

In the present study, a comparison was made between the available computer simulation packages that accompany the Engineering Mechanics Dynamics textbooks with the TAPS packages developed. For this purpose, simulation packages accompanying four textbooks were selected for this study:

- Engineering Mechanics Dynamics, by R. C. Hibbler, Prentice Hall, Inc., 2004
- Engineering Mechanics Dynamics by A. Bedford & W. Fowler, Prentice Hall, Inc., 2005
- Vector Mechanics For Engineers: Dynamics, by F.P. Beer, E.R. Johnston & W. E. Clausen, McGraw Hill Inc., 2004
- J. L. Meriam & L. G. Kraige, Engineering Mechanics Dynamics, John Wiley & Sons, Inc., 1997

In general, the textbooks by Hibbler and by Bedford & Fowler are accompanied by a student study pack which comprise of a workbook of selected questions and simulation packages in the form of a CD based on working model software. These simulation packages typically contained questions and were accompanied by basic 2-dimensional simulation models. Both these simulation packages were developed by the same developer as shown in the opening screens of the CDs and therefore have similar working environment and platforms. As for the textbooks by Beer & Johnson and by Meriam & Kraige, such CDs were not provided with the textbook. However 2-D tutorial simulations of selected problems are available via a secured publisher's website.

A comparison was also made of the features and attributes of the simulation provided in the CDs. In general, the simulation packages in the accompanying workbook contain pre-set simulations of text examples that include questions for further exploration. These simulations however are not very useful in general (because of the fast pace and lack of interactivity) unless the students are specifically assigned to look at them. The analysis carried out on the content of these CDs indicated that the best types of these simulations have been in the demonstration mode. This is in agreement with Cornwell (2004), who further added that such simulations in the demo mode do not strengthen student's problem solving abilities or their ability to apply fundamental mechanics principles in solving exercise.

In this comparison, the simulation packages that accompanied the textbooks were analyzed based on seven aspects, i.e. presentation and clarity, learning approach, level of interactivity, support and environment, assessment of performance, feedback, and 3-D effects. The major differences between these commercial simulation packages with the TAPS packages approach in teaching and learning of Engineering Mechanics Dynamics are summarized as follows:

1) In normal practice students are usually informed of the objectives of a problem that they are attempting to solve (this is further explained in Chapter 7). However, the objectives of these problem-solving activities are not specified in the simulation packages as compared to TAPS packages. For example, Figure 1 shows a snapshot of an exercise taken from the simulation package accompanying the Hibbler textbook. As it can be noted from this Figure, the student is given instruction to run the simulation at a specified speed and then at different speeds. To understand the purpose of this simulation, it is essential for the student to first read the question in the workbook prior to engaging with the simulation. This was the general case with all other simulations accompanying the Engineering Dynamics textbooks by other authors and with those made available on the publisher's website.

This is different from the approach taken in the development of the TAPS packages where the objectives of the problem are presented at the beginning of each exercise. The approach taken in the TAPS packages is to fully integrate the system to be independent of the textbook, such that student's attention is not deviated during learning. The questions, objectives, formulas, diagrams and charts are presented in sequential order within the TAPS packages. In addition, students are also provided with essential built-in tools such as a calculator, notepad and nomenclature table to help them in solving the engineering problems directly without needing external devices or tools such as calculator, pen, notepads or reference textbook, as is the case with most commercial simulation packages. All these options are made available within the TAPS environment and student can get direct access by a click of a mouse.

2) In general, the problem statement presented in the simulation packages accompanying the textbook does not present itself in a step-by-step approach but rather simultaneously presenting the questions and activities on the same screen at the same time. This can be clearly seen in Figures 1 and 2. For example, in Figure. 2, there are five questions posed to the student at one time as shown.

Figure 1. Snapshot of a problem taken from the simulation packages accompanying the textbook by Hibbler (2004)

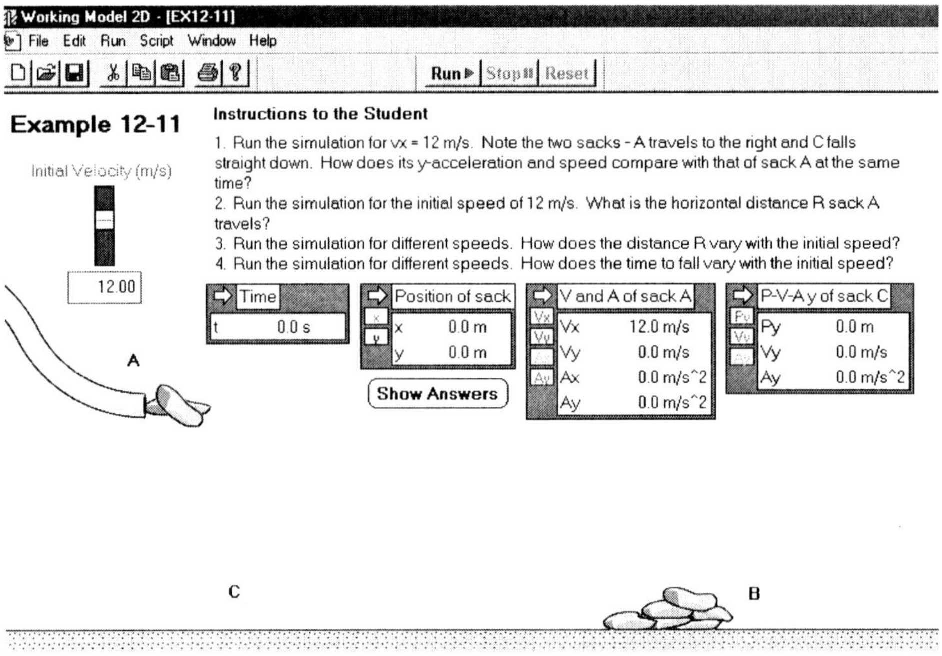

Figure 2. Snapshot of a problem taken from the simulation packages accompanying the textbook by Bedford & Fowler (2005)

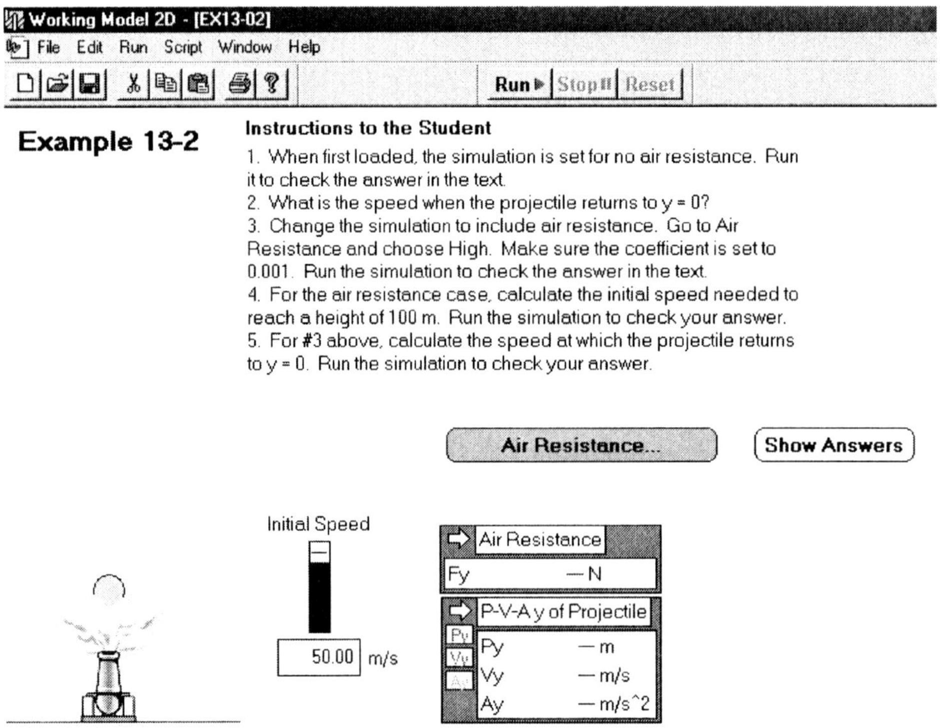

Too many questions being addressed at one time would only increase the cognitive workload (complexity of the task) of the student in solving the problem. Students having difficulty in understanding normal lectures would be affected to a greater extend by this approach of computer tutoring. Moreover, in the typical simulation packages shown in Figures 1 and 2, the student is not coached to understand the problem in the question before engaging with the simulation provided on the screen.

In comparison, in the TAPS package, each problem or a sub-section of a problem is addressed one at a time or after a student has solved each sub section of the problem as explained in Chapter 7. Therefore the student is not burdened with solving multiple questions in a single problem. TAPS packages are developed to include multimedia features and simple intelligent functions such as alerting a student by displaying messages (hints) on screen if a wrong formula is applied or a wrong answer given in solving the selected engineering problem. However, if the user still cannot solve the problem, the student could approach the TAPS package by clicking on the "solve" button to aid the student in solving the problem. The solution is given in a step-by-step manner showing how the answer is obtained. Additionally, desktop virtual reality features were incorporated to encourage students to interact and engage with the TAPS package. These efforts have focused on conveying technical knowledge to the student to solve the engineering problem in such a way so as to support the acquisition of theoretical knowledge.

3) From the point of view of the level of interactivity and visualization, the simulation study packages were found to be moderate and in some instances, rather poor. For example, when a student input any value into the textbox provided and click the "Show Answers" button, no effort was made by the commercial computer simulation packages to coach the students in understanding how the answer was concluded and the accompanying animation (by solving the problem in details or step-by-step approach like the one supported in TAPS packages). Instead, when the "Show Answers" button is activated, one would only see the answers as typically shown in Figure 3 for the problem given in Figure 2.

Since interactivity is the greatest advantage that multimedia contributes to teaching (Lieu, 1999), this key element was considered to be missing in the simulation packages accompanying the Mechanics Dynamics textbooks. Both, interactivity and visualization have been the central part of the development approach taken in the TAPS packages. As an example, the approach in the TAPS packages would be to present the case into manageable steps and the student is prompted for what to do next. Data input is by means of textboxes and symbolic or numeric text (this is further explained in Chapter 7). Context-sensitive help is implemented through message boxes. In addition, if the student had given the wrong answers, the TAPS packages would prompt the student for "Hints" or narrates and explains to the student why the answer is wrong. Therefore, in contrast to the simulation packages available in the market place, the TAPS packages have the ability to reinforce learning.

4) The problems contained in the CD-ROM simulation packages are presented with simple colors and 2-D graphics with limited animation. In addition, there was no sound effect, in particular the questions were not narrated to explain to the students the nature of the question or the steps involved in solving the problem in question. The advantage of the TAPS packages is that every problem is narrated and explained to the student with high quality graphics, video and images to help students visualize the engineering concepts (example of this can be seen in Chapter 7, Figure.5). The use of 3-D images and animation in the TAPS packages provided added benefit in terms of enhancing visualization of motion which otherwise would be difficult to visualize even in 2-D simulation environment. This is explained further in Chapter 7.

Figure 3. Snapshot of answers provided for the problem in question (Bedford & Fowler 2005)

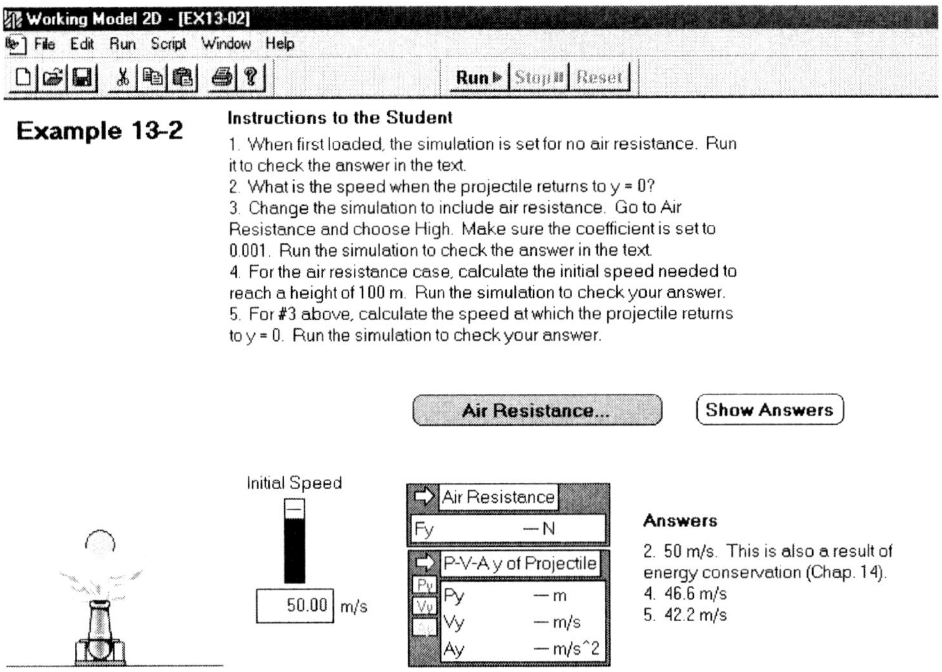

5) As the simulation packages accompanying the textbooks are not autonomous and require supplementary information be drawn time to time from the workbook or textbook, this would eventually discourage students from engaging with the simulation packages.

One other feature that the simulation packages do not offer as compared to the TAPS packages is the flexibility to control the delivery of information and the rate of delivery. In the TAPS packages approach, students have the option to read the text or be narrated with explanations by clicking the appropriate buttons. This form of interactivity allows the students to control visualization and adapt the material to their learning styles.

6) In the TAPS packages, students are also assessed while engaging with the tutorial simulation. Assessment of the student in attempting to solve a given exercise is being monitored by the system and timely feedback related to the student's interactions while solving the problem is provided at the end of the session. The total score or marks gained by the student in solving the problem could be stored in a database and this is explained further in Chapter 7 of this book. Additionally, the authoring tools used in implementing the TAPS packages allow for the option to include a built-in database where the student activities could be saved and retrieved whenever required. These additional features are not implemented in the commercial simulation packages.

7) Another key difference between the simulation packages accompanying the textbook and TAPS packages is the flexibility to incorporate 3-D geometric models as well as stereoscopic views of models and images. In the TAPS packages, the stereoscopic images help students enhance depth perception that could reduce learning time as compared to the conventional method via textbook. This feature is further discussed in Chapter 7 of the book.

In summary, based on the results of this study, it can be concluded that the TAPS packages are instructionally effective and that students' subjective impressions of the packages were, on the whole, positive.

In addition this Chapter has provided a brief account of the differences between the TAPS packages approach used in this study with that of commercial simulation packages accompanying the Engineering Mechanics Dynamics textbook. The differences found were indicative of better presentation and clarity, step-by-step approach to solve engineering problems, user-friendly environment, unbiased assessment of performance and flexibility to incorporate 3-D geometric models in the TAPS packages. As stated in Chapter 1 of this book, the approach was to address the shortcomings of the commercial simulation packages by developing a simulation package to incorporate the various multimedia attributes and simple artificial intelligence concepts so as to create a virtual learning environment that has the potential to encourage learning and understanding of Engineering Mechanics principle. This may be particularly effective for reinforcing understanding of a difficult concept where slow learners find it difficult to absorb during normal classroom lectures. The TAPS packages were designed to provide coaching rather than just merely taking the form of an electronic page containing questions and answers as observed in the commercial simulation packages.

Although desktop virtual reality technology was also employed in developing the TAPS packages, it is envisaged that immersive virtual reality development tools such as the use of high-end software and hardware (head mounted helmet, wired data gloves, and 3-D goggles) could further accelerate the visualization and learning process of students. Since these hardware and software tools are not available at University Tenaga Nasional at present, it is clearly seen that by securing them in the future, there is a possibility to employ and develop immersive virtual reality tools to develop better TAPS packages that will enhance its popularity with the students.

REFERENCES

Cornwell, P. J. (2004). Interactive Web Examples for Dynamics Developed using DazzlerMax. In *Proceedings of the 2004 American Society for Engineers Education Annual Conference & Exposition* (Session 3268).

Institute for Higher learning Policy, (1999). *What's the difference: A review of contemporary research on effectiveness on distance learning in Higher education*? Retrieved February 20, 2004 from http://www.ihep.com/Pubs/PDF/Differece

Institute for Higher learning Policy, (2000). *Quality on the line: Benchmarks in success in Internet-based distance education.* Retrieved February 20, 2004 from http://www.ihep.com/Pubs/PDF/Differece

Lieu, D. (1999). Using interactive multimedia tutorials for engineering graphics education. *Journal for Geometry and Graphics*, *3*(1), 85–91.

Psaromilingkos, Y. (2003). Re-Evaluating the Effectiveness of a Web Based Learning System: A Comparative Case Study. *Journal of Educational Multimedia and Hypermedia*, *12*(1), 5–20.

Tessmer, S. (1995). Formative multimedia evaluation. *Training research Journal*, 1, 127 – 149.

Chapter 11
Challenges and Trends of TAPS Packages in Enhancing Engineering Education

INTRODUCTION

It can be envisaged that the use of multimedia computer technology as replacement, or supplement to, human educators in engineering education would become widespread in the future. Such technology can be employed to demonstrate and correlate real life application and theory thereby promoting deep learning. Interactive courseware for higher learning institutions may be extremely useful where trained human resources in the engineering education sector are limited. This Chapter discusses the current trends of incorporating new technologies with TAPS packages in the teaching of engineering subjects.

EMPLOYMENT OF TAPS PACKAGES

A number of TAPS packages are being implemented and enhanced at University Tenaga Nasional (UNITEN) for the use in the teaching and learning of Engineering Mechanics subject. A snapshot of students interacting with TAPS packages is shown in Figure 1.

In the aim to promote and enhance the existing TAPS packages, additional hardware and software are being used and tested for its effectiveness. These are further discussed in the next sections.

DOI: 10.4018/978-1-60566-764-5.ch011

Figure 1. Students interacting with TAPS packages in one of the computer labs in UNITEN

Graphics Tablet

A graphics tablet is an input device used by artists which allows one to draw a picture onto a computer display without having to utilize a mouse or keyboard. A graphics tablet consists of a flat tablet and some sort of drawing device, usually either a pen or stylus. A graphics tablet may also be referred to as a drawing tablet or drawing pad. While the graphics tablet is most suited for artists and those who want the natural feel of a pen-like object to manipulate the cursor on their screen, non-artists such as engineering instructors may find it (Graphics Tablet) useful in their teaching as well. A graphics tablet may come in a range of sizes, from smaller 3" by 4" (7.6 by 10.2 cm) models to larger 7" by 9" (17.8 by 22.9 cm) ones. Even larger graphics tablets exist, up to enormous 14" by 14" (35.6 by 35.6 cm) tablets targeted towards professional designers and architects. Some well known graphics tablets include Wacom™, Aiptek™, and KB Gear™.

For the usage with TAPS packages, the high pressure sensitivity of the graphics tablet, allows the instructor/student to control a number of aspects of their drawing, including color and line thickness, simply by pressing the stylus/graphics pen more or less heavily, mimicking drawing with an actual pen. Most graphics tablets also have function buttons on the side, so that the user can perform common actions, such as switching a tool in a drawing program from paint to erase, without having to use the mouse or keyboard. This is useful for an engineering student for example when the student needs to switch to a virtual calculator or notepad to perform calculations or make notes. Figure 2 shows a demonstration usage of the graphics pen where important engineering concepts are being highlighted directly onto an image in the TAPS package.

Figure 2. A TAPS package being used with a Graphics Tablet

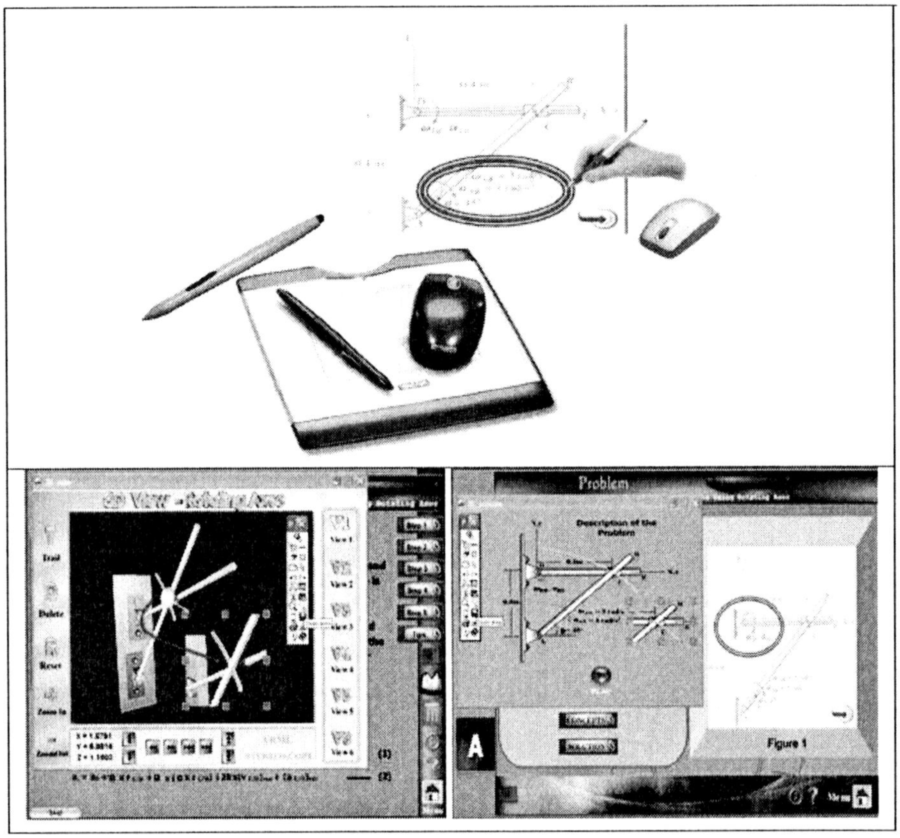

Interactive Whiteboards

An interactive whiteboard (IWB) is a large interactive display that connects to a computer and projector. A projector projects the computer's desktop onto the board's surface, where users control the computer using a touch sensitive pen, finger or other device. The board is typically mounted to a wall or on a floor stand as shown in Figure 3.

In general IWBs are used in a variety of settings such as in classrooms at all levels of education, in corporate board rooms and work groups, in training rooms for professional sports coaching, broadcasting studios and more.

The use of IWBs has had a profound effect on schools, teachers, and learners (Cuthell, 2005). A considerable body of evidence from schools across the United Kingdom has shown the transformational effect of the technology on teaching and learning (Cuthell, 2002 & 2004). What the boards enable teachers to do is to support the whole range of learning styles of the learners in the class. The learners themselves feel empowered: The ability to visualize and recall the lesson supports learning; the range of resources that can be embedded within the IWB lesson software and the interactivity itself has engaged almost all of the learners and enhanced their progress (Cuthell, 2005). When using an IWB the learners themselves see themselves as engaged with the process of learning, rather than simply progressing through the scheme of work.

Figure 3. Interactive white board with a stand (a) and mounted on the wall (b)

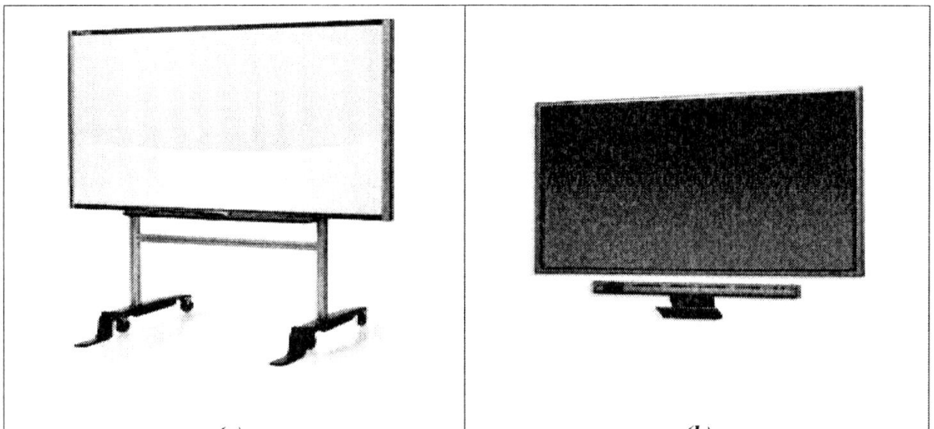

(a) (b)

In UNITEN, progress is being made to employ IWBs to teach engineering students. At present the IWB at UNITEN is being used as a research tool for teaching and learning only. Figure 4 shows a demonstration usage of the IWB where a projector is being used to project the computer display i.e. TAPS packages onto the IWB. The animated engineering problems can be paused and important engineering concepts can be highlighted to help students visualize better.

The IWB (shown in Figure 4) which can be used interactively by learners during classroom teaching may offer new opportunities for engineering students to solve engineering problems, by using graphical and other representations. Hence they can more easily articulate engineering knowledge and receive instructor (and peer) feedback. The IWB provides collaborative opportunities for reasoning, hypothesis testing and interpretation that go well beyond those afforded by more established classroom devices.

According to Hennessy (2008), the ever-present concern to maintain lesson pace means that ironically IWB use may afford even less thinking time and opportunity for student/user input than other forms of educational technology. Indeed instructor-only operation of the IWB avoids reducing pace through committing time for turn taking (Moss et al., 2007), and it simultaneously retains instructor control.

Augmented Reality

In general multimedia environments have offered new ways for learners to interact with various educational resources. Since, printed learning materials have been favored and used particularly for systematic study, they have been dealt with as totally different media that yield distinct learning environments and learners can only get its alternative merits at each environment. Augmented reality (AR) is an emerging technology that overlays virtual objects onto real scenes that has potential to provide learners with a new type of learning material (Asai, 2005). The architecture of AR technology can be found in (Liarokapis, 2002 & 2004).

AR has the ability to enhance real scenes viewed by the user, overlaying virtual objects over the real world, and works to improve the user's performance and perception of the world. Some advantages of using AR technology in teaching and learning include (1) the user can get three dimensional information based on a real scene, (2) the user can see objects from his/her own viewpoint, and (3) the user

Figure 4. TAPS packages being projected onto interactive white board

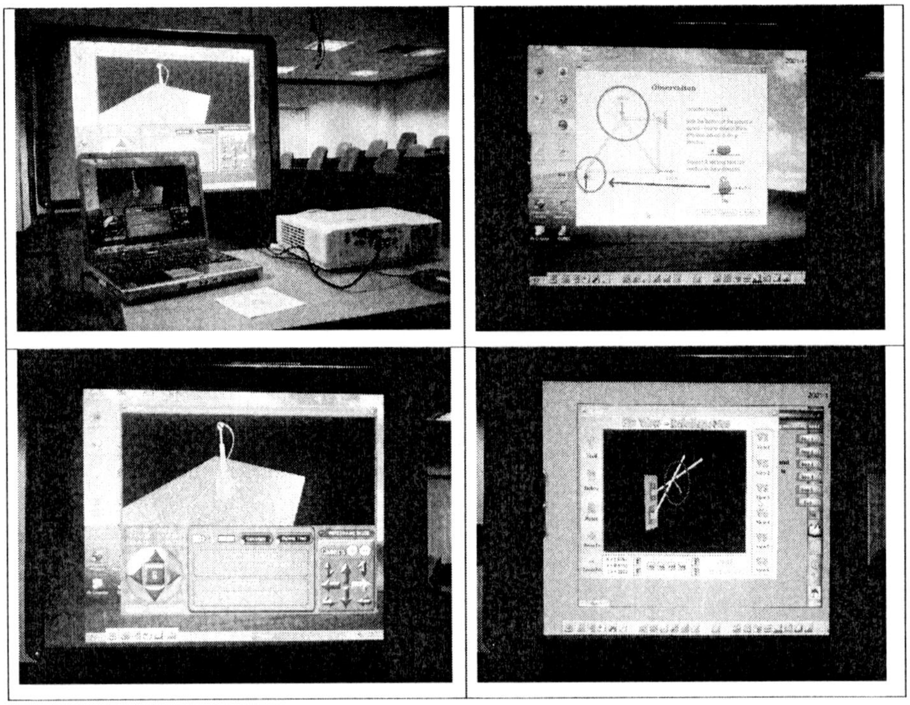

can interact with both virtual and real objects wirelessly. With these advantages, some researchers have employed AR technology to develop their educational software for teaching and demonstration purposes (White 2001, Lang 2004, Asai 2005, and Liarokapis 2005).

White (2001) presented a new approach to the teaching of top down design of hardware design language (VHDL) using a novel virtual interactive teaching environment. The environment enabled the students to learn more efficiently using virtual multimedia content, while exploiting extended markup language (XML) and augmented reality. According to White, the environment can be adapted for teaching other subject areas. Asai (2005) conducted an experiment to investigate on the characteristics of AR instructions and its appropriate way of human computer interaction. He dealt with chemical properties of caffeine as a topic for augmented instructions and prepared a two-page document for the experiment. A handheld personal computer (PC) was compared to a head mounted device (HMD) as a presentation system for augmented instructions. The result of the experiment suggested that a handheld PC was more suitable than a HMD as a presentation system for augmented instructions in terms of long time use.

AR is now being used in engineering education, for example Liarokapis (2004), presents an educational application that allows users to interact with Web3D content using virtual and augmented reality. The study enables an exploration of the potential benefits of Web3D and AR technologies in engineering education and learning. Preliminary study found that by employing AR technology, students could understand more effectively through interactivity and multimedia content. An example of the study by Liarokapis is illustrated in Figure 5 where the user can interact with a 3D model of the object (i.e. Piston) and can compare it to real objects in a natural way.

Another project (Handheld Augmented Reality Project or HARP) by Harvard University with the collaboration of the Massachusetts Institute of Technology (MIT) and the University of Wisconsin at

Figure 5. AR visualization of a piston [Liarokapis, 2004]

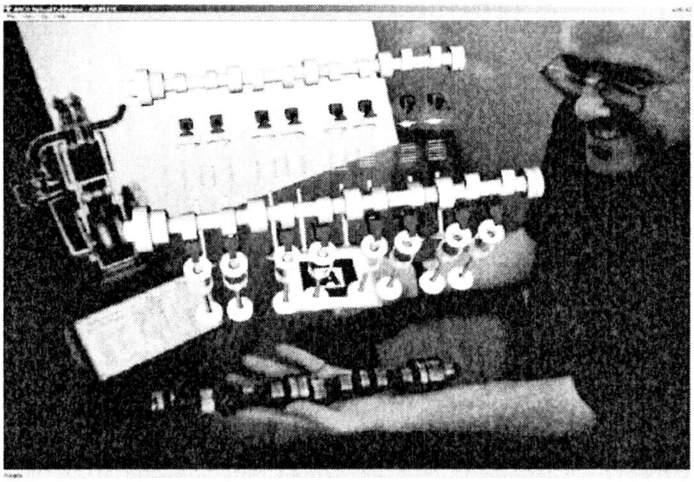

Madison, uses handled computers to enhance teaching and learning through a series of activities that draw on the attributes of students' surroundings.

Since AR has the potential to enhance engineering education, future TAPS packages will employ AR technology in its upcoming project called Development of a Computer Augmented Reality Engineering Mechanics Learning System (CAREMeLS) at University Tenaga Nasional.

TRENDS AND ROLES OF INFORMATION COMMUNICATIONS TECHNOLOGY IN ENHANCING ENGINEERING EDUCATION

Recent developments in engineering education and training suggest the need for closer cooperation between the industry and universities in the planning, design and implementation of engineering curricula to ensure engineering graduates receive the requisite education and training which prepares them for a successful career in industry. In the knowledge-based society of the 21st century, the critical and only sustainable competitive advantage will be to have the ability to learn faster than the competitors (Melsa, 2009). Melsa also points out that in order to meet the 21st century, universities must become learning organizations which are skilled at creating, acquiring and transferring knowledge, and modifying their behaviour to reflect new knowledge and insights. To cater these additional needs, Melsa (2009) proposed that a model for engineering education must be generated where the:

- The programmes must be *learning-based*,
- The experience must be *practice-oriented* and,
- The programmes must demand *active involvement of the student*.

Other engineering practitioners suggest the use of ICT to enhance the quality and accessibility of higher education (Hattangdi & Ghosh, 2009). According to these practitioners the benefits that ICT integration in education can provide, right from breaking time and distance barriers to facilitating collaboration and knowledge sharing among geographically distributed students.

ICT has changed the dynamics of various industries as well as influenced the way people interact and work in the society (UNESCO, 2002; Bhattacharya and Sharma, 2007; Chandra and Patkar, 2007). Internet usage in home and work place has grown exponentially (McGorry, 2002). ICT has the potential to remove the barriers that are causing the problems of low rate of education in any country. It can be used as a tool to overcome the issues of cost, less number of teachers/instructors, and poor quality of education as well as to overcome time and distance barriers (McGorry, 2002).

The integration of ICT increases the flexibility of delivery of education so that learners can access knowledge anytime and from anywhere. It can influence the way students are taught and how they learn as now the processes are learner driven and not by instructors. This in turn would better prepare the learners for lifelong learning as well as to contribute to the industry. It can improve the quality of learning and thus contribute to the economy. In addition wider availability of best practices and best course material in education, which can be shared by means of ICT, can foster better teaching. ICT also allows the academic institutions to reach disadvantaged groups and new international educational markets. Thus, ICT enabled education will ultimately lead to the democratization of education. Especially in developing countries like Malaysia, effective use of ICT for the purpose of education has the potential to bridge the digital divide.

In the present Information society, there is an emergence of lifelong learners as the shelf life of knowledge and information decreases (Hattangdi & Ghosh, 2009). People have to access knowledge via ICT to keep pace with the latest developments (Plomp *et al.*, 2007). In such a scenario, education, which always plays a critical role in any economic and social growth of a country, becomes even more significant. Education not only increases the productive skills of the individual but also his/her earning power. It gives him/her a sense of well being as well as capacity to absorb new ideas, increase his/her social interaction, gives access to improved heath and provides several more intangible benefits (Kozma, 2005). The various kinds of ICT products available and having relevance to education, such as teleconferencing, email, audio conferencing, television lessons, radio broadcast, interactive voice response system, audiocassettes and CD & DVD ROMs etc have been used in education for various purposes (Sharma, 2003; Sanyal, 2001; Bhattacharya & Sharma, 2007).

Cross and Adam (2007) listed four main rationales for introducing ICT in education as shown Table 1. The significance of these rationales can also be extended to engineering education. Other powerful ICTs include laptops wirelessly connected to the Internet, personal digital assistants, low cost video cameras, and cell phones have become affordable, accessible and integrated in large sections of the society throughout the world. It can restructure the learning process, promote collaboration, make education more widely available, foster cultural creativity and enhance the development in social integration (Kozma, 2005).

In summary many technologies are being implemented and researched for its benefits to enhance the learning process such as graphics tablet, interactive white boards (IWBs) and augmented reality as

Table 1. Rationales for introducing ICT in education (Cross and Adam, 2007)

Rationale	Basis
Social	Perceived role that technology now plays in society and the need for familiarizing students with technology.
Vocational	Preparing students with jobs that require skills in technology.
Catalytic	Utility of technology to improve performance and effectiveness in teaching, management and many other social activities.
Pedagogical	To utilize technology in enhancing learning, flexibility and efficiently in curriculum delivery.

discussed in section 11. On the other hand innovative use of Information and Communication Technology (ICT) can potentially further contribute in supporting the learning process, particularly in engineering education. ICT can enable wider access of information and knowledge, participation and interaction. Although such fundamental changes in the curriculum could take place and help transform the way students learn in many ways, this transformation may require new skills, capabilities and attitudes i.e. are students and instructors prepared to use these technologies in their teaching and learning, are higher learning institutions committed to employ these new teaching aids.

REFERENCES

Asai, K. (2005). Augmented Instructions – A Fusion of Augmented Reality and Printed Learning Materials. In *Proceedings of the 5th IEEE International Conference of Advanced Learning Technologies*, (pp. 1 – 3).

Bhattacharya, I. & Sharma, K. (2007). 'India in the knowledge economy – an elect

Chandra, S., & Patkar, V. (2007). ICTS: A catalyst for enriching the learning process and library services in India . *The International Information & Library Review*, *39*(1), 1–11. doi:10.1016/j.iilr.2006.11.001

Cuthell, J. P. (2002). *How do interactive whiteboards enhance learning?* Retrieved February 24, 2004, from http://www.virtuallearning.org.uk/iwb/Learning_theories.pdf

Cuthell, J. P. (2004). *Interactive whiteboards survey.* Retrieved May 22, 2005, from http://www.virtual-learning.org.uk/2003research/whiteboards_survey.doc

Cuthell, J. P. (2005). Seeing the meaning. The impact of interactive whiteboards on teaching and learning. In *Proceedings of WCCE 05 Stellenbosch South Africa.*

Hattangdi, H., & Ghosh, A. (2009). Enhancing the quality and accessibility of higher education through the use of Information Communication Technology. *International Journal of Educational Management*, *21*(6), 1–12.

Hennessy, S., Deaney, R., Ruthven, K., & Winterbottom, M. *(2008).* Pedagogical strategies for using the interactive whiteboard to foster learner participation in school science. *Learning, Media & Technology, University of Cambridge.*

Kozma, R. (2005). National Policies That Connect ICT-Based Education Reform To Economic And Social Development. *Human Technology*, *1*(2), 117–156.

Lang, U. (2004). Virtual and Augmented Reality Developments for Engineering Applications. *European Congress on Computational Methods in Applied Sciences and Engineering,* (pp. 1 – 16).

Liarokapis, F. (2002). Multimedia Augmented Reality Interface for E-Learning (MARIE). *World Transactions on Engineering and Technology Education*, *1*(2), 173–176.

Liarokapis, F. (2004). Web3D Augmented Reality to Support Engineering Education. *World Transactions on Engineering and Technology Education*, *3*(1), 11–14.

McGorry, S. Y. (2002). Online, but on target? Internet-based MBA courses: A case study. *The Internet and Higher Education*, *5*(2), 167–175. doi:10.1016/S1096-7516(02)00089-1

Melsa, J. L. (2009). Trends in engineering education in the USA. *Engineering Science and Education Journal*, 239–244.

Moss, G., Jewitt, C., Levacic, R., Armsrong, V., Cardini, A., & Castle, F. (2007). The Interactive Whiteboards. *Pedagogy and Pupil Performance Evaluation: an Evaluation of the Schools Whiteboard Expansion (SWE) Project: London Challenge*, London, DfES.

Plomp, T., Pelgrum, W. J., & Law, N. (2007). SITES2006 International comparative survey of pedagogical practices and ICT in education. *Education and Information Technologies*, *12*(2), 83–92. doi:10.1007/s10639-007-9029-5

Sanyal, B. C. (2001). New functions of higher education and ICT to achieve education for all, Expert Roundtable on University and Technology-for- Literacy and Education Partnership in Developing Countries, International Institute for Educational Planning, UNESCO, September 10 to 12, Paris. *Open and Distance Learning Trends, Policy and Strategy Considerations*, 14 *UNESCO*.

Sharma, R. (2003). Barriers in Using Technology for Education in Developing Countries. Washington, DC: IEEE.

White, M. (2001). A virtual Interactive Teaching Environment (VITE) Using XML and Augmented Reality. *The International Journal of Electrical Engineering Education*, *38*(4), 316–329.

Chapter 12
Conclusion and Further Work

CONCLUSION

Mechanical engineering course subjects such as Mechanics Dynamics, combine a mix use of mathematics, schematic diagrams, and text descriptions. Frequently, students are unclear of basic principles of Engineering Mechanics Dynamics, and as such they do not know which mathematical relationships are to be applied in solving a particular problem. Additionally, as the name "dynamics" implies, the very nature of this subject is not "static" and thus requires learners to visualize motion; for example, in a given time period, a particle may be moving in a straight line and after some seconds the particle may experience a curvilinear motion. If the learner fails to see this, the learner will not be able to employ the right equations to solve the problem.

As such, an effort was made to evaluate the feasibility and effectiveness of employing technologies such as multimedia and desktop virtual reality to enhance the problem solving skills and learning of students.

In this book, the development of computer-aided learning software termed as technology assisted problem solving (TAPS) packages is demonstrated in Chapter 7. The book provided an overview of developing TAPS packages using multi design approaches. The work is one of the pioneering efforts to

DOI: 10.4018/978-1-60566-764-5.ch012

address the need for computer based problem solving software packages for the domain of engineering. The development processes of TAPS packages are shown in (Figure 1).

More specifically, the conclusions of the study are as follows.

Technologies

The use of multimedia and desktop virtual reality in the development of TAPS packages has helped to address the potential benefits of employing technologies which provide a combination of multimedia and dynamic illustrations in engineering problem solving tasks. The ability of mixing different formats of media for the development of TAPS packages has greatly enhanced the ability to convey engineering concepts and descriptions in a better and simple manner (as stated in Chapter 6). The TAPS packages

Figure 1. Development processes of TAPS packages

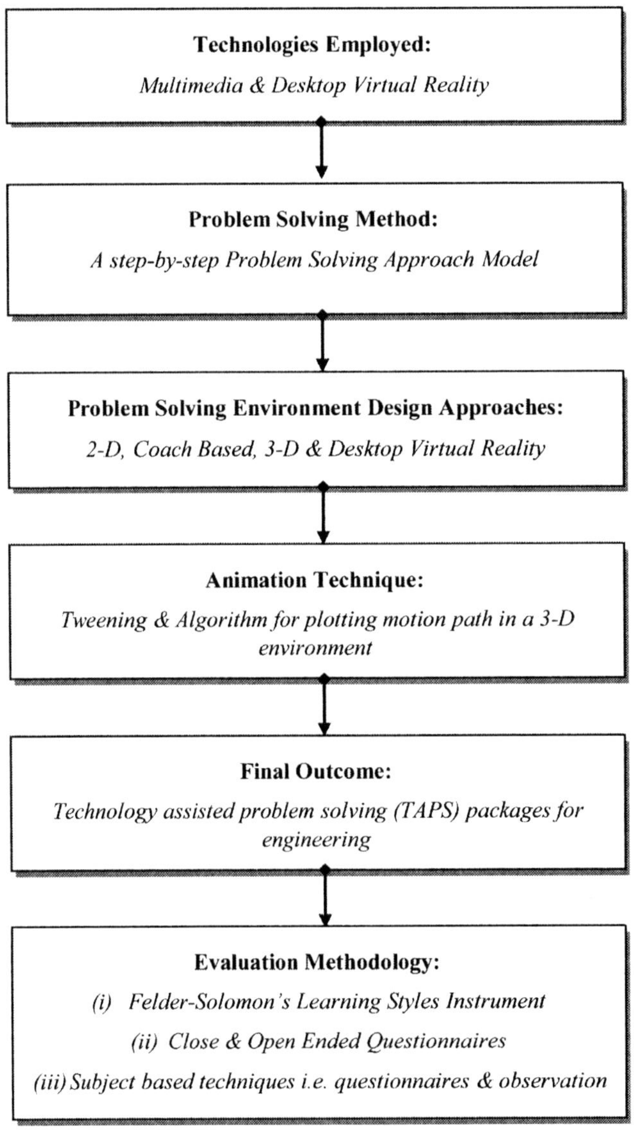

helped students appreciate the laws of motion i.e. where they apply and where they do not apply and learn problem solving better when the available knowledge is well structured at appropriate levels of detail to meet the needs of learners. Also, complex engineering concepts can be supported by audio to narrate information and videos to explain engineering concepts by the immediate availability of interconnected dynamic presentations. TAPS packages helped in engaging the students to learn problem solving and support conceptualization of the material being presented as compared to the conventional textbook. Multimedia and desktop virtual reality thus has the potential to create high-quality TAPS packages to support and to enhance the learning and problem solving experience as compared to existing engineering packages (Chapter 7 & Chapter 9).

For the learners, enormous interactivity is available at the click of a mouse. This supports the user to navigate freely while interacting with the TAPS package. For the department, while up-front design and development time is usually increased, the technology-enhanced package is easier to update and easier for users to participate in. For both users and faculty, feedback and evaluation can be instant and transparent.

The learning material is more consistent. The use of multimedia shifts the balance in favor of capturing better practice in each area of learning. The best instructor lecture, or the best explanation of how to solve a problem can be recorded, and made available to all present and future users.

The multimedia-based material is richer than that provided through the combination of lectures and textbooks. The multimedia system makes it easy to provide high quality images (rather than slides at lectures), audio (rather than tape-based language laboratories), and video (rather than classroom television). Furthermore, materials can be interacted with repeatedly, rather than the single opportunity of a lecture.

Problem Solving Method

Teaching engineering subject can be tedious, difficult, time consuming and requires the instructor to repeat the entire exercise several times until the student understand. A problem given in engineering can lead to a series of steps, from a problem statement to the solution. As such, it is better to group the problem solving steps under appropriate titles such as "*1. Problem*", "*2. Observation*", "*n…step*" and show the problem solving process in a step-by-step fashion. For this reason, a problem-solving model was designed and employed as described in this book (Chapter 6). Although a good student may know several ways to solve a given problem in engineering, the problem-solving model helped slow learners (learner experiencing difficulties with Engineering Mechanical Dynamics subject) to understand which step is to be applied first, followed by the other subsequent steps in a structured approach. Therefore, the problem-solving model was found to be suitable for constructing linear engineering multimedia based TAPS packages and generally can be designed in an easy and visually appealing format.

Analysis of Problem Solving Environments

2-D (Design approach 1) – The 2-D TAPS package revealed that 2-D graphics behavior could appear more natural and predictable than the mono medium style in virtual environments. In addition, the animation that has been incorporated in the package provided a consistent 'look and feel' to a 2-D direct manipulation user interface. The study showed that multimedia technology could be employed to aid learning of various topics pertaining to engineering such as the analysis of engineering structures.

The developed prototype has been found to be useful in reinforcing understanding of concepts such as equations of equilibrium through animation. In addition, users of the package found that it was easier as compared to the traditional approach to investigate all possible loading conditions of an engineering structure (a truss) to determine the most severe loading experienced by a truss member.

Coach based (Design approach 2) – The coach based TAPS package developed in this study uses multimedia and dialogue to explain the concepts. The TAPS package assists students to solve Engineering Dynamics problem, namely rectilinear kinetics, erratic motion. During the problem solving process, the information presented using multimedia features such as interactivity, delivery control, and simple expert systems rules helped students to enforce important concepts and provided user-friendly interaction in solving the engineering problem presented. The interactivity provided the virtual environment the capability to adopt tutoring and feedback that could significantly accelerate the learning curve of students. These capabilities were found to be essential for positive learning. In the TAPS package, interactivity allowed the students to work at their own pace, in the order desired, and repeat sequences at will. The interactive learning materials offered students essential feedback and the computer could display consequences of students' actions. For example, if a mistake was made while solving the problem and interacting with the TAPS package, the package could provide immediate feedback in the form of hints or the student could approach the package to solve the problem by clicking the "solve" button. The strength of this coach based TAPS package is that it can provide analysis to compute the unknown reactions in a step-by-step approach. As the text and figures are written and displayed on the screen, the TAPS package would prompt the students at various checkpoints to see if the student understands the step that has been executed. At this point, if for some reason the student is unclear, the student can move back to the previous step. This TAPS package also enhanced students' knowledge on applying the correct equations for effective computation. Additionally, the user tools provided in the coach based TAPS package allowed students to perform multi task such as using a calculator, glossary of commands, and character map table while interacting. Generally, the study found that weaker students benefited and appreciated the most from such TAPS package, in that they found that the subject becomes more interesting, thrilling, simplified, and understandable when compared to the traditional method of classroom learning.

3-D (Design approach 3) – The 3-D TAPS package helped engineering students' to visualize and understand curvilinear motion of particles in a 3-D environment. This TAPS package allowed visualization of 3-D presentation of learning contents. The 3-D environment promoted learning by discovery as it allowed students' to generate the motion path of a particle experiencing a curvilinear motion. One of the problems in understanding curvilinear motion from the textbook is that the examples are given on a 2-D static plane or surface schematic diagrams that makes the visualization difficult. The static nature of the object, which is used to elaborate the practical relevance of the subject, is not good enough for the purpose of analysis. For the purpose of visualization, analysis and problem solving, a greater impact was obtained when the 2-D static rod and collar was modeled and shown in a 3-D environment. Additionally, the 3-D environment could facilitate a faster and more complete analysis of design alternatives by providing ways to rapidly create models or prototypes of proposed designs and to then simulate them in a more realistic way. The access to various activities features such as interactivity, metaphors, navigation, manipulation, pop-up windows, activation of 3-D objects, and dynamic presentation of contents enabled the student to select the situation relevant information and to execute the autonomous movement through the learning process. The students could also delete and generate the motion path trail of the rod and collar repeatedly until the student is clear about the particle's path of motion.

Desktop virtual reality (Design approach 4) – The desktop virtual reality (DVR) TAPS package allowed 3-D time based presentation of learning contents, which has been visualized. The DVR environment promoted explorative learning by allowing students' to generate the motion path of a particle experiencing a curvilinear motion and at the same time applying the correct formulas and performing the necessary computations to show the acceleration and velocity of the particle with regards to the time (in seconds) input by user. This sort of complex problem is difficult to be shown and taught by using still images. However, this DVR based explorative learning offered the integration of possible visualizations using the DVR environment, which supported the free investigation of complex contents, especially for particles that experiences a curvilinear motion. The realistic approach due to advanced modeling, animation and texturing techniques as well as the interactive access to varied contents increase the acceptance of 3-D visualizations. This DVR TAPS package was found to be significant in facilitating the students to understand the curvilinear motion better. This highly interactive TAPS package has encouraged students to actively participate and explore complex relationships of engineering concepts and increased the development of problem solving skills and activities through self motivated-exploration and discovery. The use of stereoscopic images in this TAPS package benefited the students to gain a better view of the trail (path) and motion of the 3-D robotic arm.

Animation Techniques

Tweening technique - The tweening technique employed enhanced and smoothened real-time motion and was found to be very useful as it facilitated the understanding of engineering structures on the screen. This technique combined with multimedia features allows the demonstration of boundary conditions, which in turn reinforces Mechanics concepts.

Algorithm for plotting curve motion path in a 3-D environment - The mechanism to generate the curve motion path in the 3-D environment helped enhanced students' understanding of curvilinear motion and could be used for other similar engineering problems to enhance learning.

The development of the above mentioned four TAPS packages have been described in details in Chapter 6.

Evaluation Methodology

The evaluation of this study reveals that generally there are significant differences in students' characteristics and improvements in understanding the subject matter, in particularly for slow learners, when using additional aids such as TAPS packages in their learning. The differences in students' characteristics may partly occur because the preferred learning styles of the students and the author coincide. The majority of students, who participated in using the TAPS packages agreed that they could understand the engineering problem solving techniques better, hence improved their understanding in the problem solving techniques as compared to learning in the traditional way. Individual differences among learners, variations among tasks, contexts, and the nature of learner interaction with the packages all contributed significantly to differential learner success in learning engineering problem solving mediated by computers.

The index of learning styles (ILS) instrument was found to be a suitable tool for designing, evaluating, and improving the contents of the TAPS packages. Additionally, the ILS instrument helped in analyzing and understanding the teaching methods that were mostly preferred by students. The most noted preferred teaching method by students found in this study was sequential learning (86.7%), where learning occurs logically in a step-by-step manner. This method clearly matched the problem-solving model designed and shown in Chapter 6, which was employed to implement the TAPS packages. Therefore it can be concluded that the Felder-Solomon's Learning Style Questionnaire is a suitable instrument for the purpose of designing, evaluating, and improving TAPS packages.

Contribution of Technology Assisted Problem Solving (TAPS) Packages

Among the four TAPS packages developed, significant contribution was shown in the desktop virtual reality TAPS package that demonstrated the motion of a robotic arm in 3-D space. Better visualization technique was introduced i.e. the design of an algorithm to show the motion path taken by the robotic arm from one point to another in a 3-D space. This technique contributed significant level of visualization and understanding among engineering students to understand the motion of curvilinear motion. In addition, the algorithm could be used to show similar motion paths for other engineering problems. Another important contribution that could be seen in the TAPS package is the way it provided feedback. The TAPS package is capable of informing the learner the questions that were incorrectly answered, and providing a brief explanation to clarify the student's misunderstanding. The TAPS package could also suggest the learner to revise a topic thereby increasing the students' motivation to learn.

In general all the TAPS packages were found to be effective in promoting learning and the outcome of this study revealed that technologies such as multimedia and desktop virtual reality approach enhanced user understanding of the underlying theory of Engineering Mechanics Dynamics, promote interactivity as well as visualization and users are able to solve engineering problems such as engineering structures, rectilinear kinematics, and curvilinear motion quickly and efficiently (based on the feedback from the closed and open-ended questionnaires).

Engineering education is an area that holds great interest and potential for developing TAPS packages. One reason for this is the ability of providing experimental learning. The TAPS packages presented in this study has achieved its objectives, as students were able to describe the position, velocity, and acceleration as two-dimensional vectors, recognize two-dimensional structures motion, and visualize particles experiencing curvilinear motion. Students were able to apply the relevant Mechanics Dynamics theories and kinematics equations to solve the problems presented.

The study has shown that multimedia and desktop virtual reality are powerful learning technologies that could help slow learners to understand the underlying Engineering Mechanics Dynamics principles, visualize curvilinear motion in a dynamic manner and more importantly, to promote deep learning. The interactivity that was incorporated in the multimedia package allowed users to manage and control the delivery of the material and act as a guide / coach in problem solving. The initial step of incorporating multimedia and desktop virtual reality technologies in virtual learning environment has enabled to understand the process and challenges involved in developing TAPS packages.

Designing and implementing a TAPS package is not an easy task. In general, although most of the guidelines stated in this book could be useful in facilitating in the development of good user interfaces for interactive multimedia TAPS packages, a standard user interface that could be accepted globally is still awaited.

The objectives of the book as stated in Chapter 1 were met at various stages in the study and are summarized as follows:

1. To evaluate whether slow learners could better visualize and solve engineering problems with the aid of technology assisted problem-solving (TAPS) packages. The information in the Chapter 10, Sections 10.1, 10.2 and Table. 10.1 provide evidence for the same.
2. To explore how best to assist students to understand engineering concepts via interactive and virtual environment. The Chapter 7explores the various approaches used in developing the TAPS packages. The feedback obtained from the Felder-Solomon's ILS questionnaires was employed for the same. This is depicted in Chapter 7, Table 4.
3. To determine the best approach in integrating the various elements in the TAPS packages with the aim of promoting learning and understanding of engineering concepts suitable for slow learners. The information in Chapter 9, Table 15 shows that different students have different preferences of the TAPS packages however the step-by-step approach which is the core of each of the TAPS packages developed, was preferred by all the students.

From the aforementioned it can be concluded that, with the aid of TAPS packages slow learners could better visualize and solve engineering problems. Engineering concepts via interactive and virtual environment were simplified for the comprehension of the students. The best approach in integrating the various elements in the TAPS packages with the aim of promoting learning and understanding of engineering concepts suitable for slow learners was found to be "step-by-step approach".

RECOMMENDATIONS FOR FURTHER WORK

In general, it is difficult to access the success of a newly investigated, developed, and evaluated learning package and to say that it has met the requirements of every student or instructor. Although it can be concluded from the results of this study that the majority of students were of the opinion that the TAPS packages has helped them in understanding the problem solving techniques better, these results could be used in refining the packages in future. Therefore, there are a number of research issues of interest, which could be the focus of further investigation. These issues can be summarized as follows:

The first recommendation that comes out of this study is the need to collect more qualitative data. This could provide more detailed analysis of changes in students' behavior due to application of TAPS packages and long-term impact of such packages in engineering education. The evaluation method used in this study certainly was adequate for measuring areas of concern for improvement of the TAPS packages. However, more information was gathered after the evaluation was administrated, as the respondents felt more convenient to discuss their experience of using the TAPS packages. As larger respondent samples are collected, more qualitative feedback can be gathered that will help establish a set of construction criteria that will make the TAPS packages even better engineering problem solving aids.

Future versions of TAPS packages should employ a standard user interface to enhance the problem-solving environment. The improvements should be based on the suggestions given in the evaluation of the TAPS packages carried out in this study. In addition, students from other local higher learning institutions should be enrolled for evaluating the TAPS packages to see if they have different perceptions on the use of TAPS packages.

Although the TAPS packages are only in its initial phase of development, this study provides sufficient evidence to continue its development especially to aid teaching and learning of mechanical engineering at UNITEN. In this book, four selected engineering problems were developed to investigate its' pedagogical effectiveness. The effectiveness of the TAPS packages was examined with particular, to confirm its design (see Chapter 10). The results of the effectiveness of the TAPS packages (laboratory testing) have provided an initial validation of the design approaches adopted in this study. The future work could include the development of more engineering problems that could be used by students in their learning e.g. within an entire semester.

The aspect of TAPS packages requiring the most improvement is the student assessment model. In future versions of TAPS packages, the student assessment model should not be influenced by the results from the multiple choice questions quiz only. Factors such as the time spent on each of the problem-solving task, the use of supporting tools, and the detail level of help requested by the student should be used to influence the way information should be presented and recorded in the database.

User interface guidelines provide a good initiative to develop TAPS packages. However, the following are some points that could be pondered upon in the existing guidelines so as to make them a globally accepted standard:

- The number of students necessary to be tested when evaluating a computer based learning package.
- Definite ways of testing the students using the computer based learning package i.e. independently, laboratory, based on single/multiple institutions.
- The guidelines should be specific in nature (i.e. defined more precisely) and at the same time be accepted world wide by higher learning institutions.
- Is basic computer knowledge of the student a compulsory requisite for evaluating in such trials?
- Time allocated for each trial should be pre-defined so as to standardize the results (e.g. a semester / an annual basis).

While the ideas in this book were all worthy of design and exploration, this study provided insights into the development and usage of TAPS packages as an adjunct to traditional teaching and an alternative to resource-intense problem solving tutorials for engineering students especially slow learners. This book presented an approach to the provision of interactive multimedia problem solving tutoring of technical skills. Although numerous examples of interactive multimedia packages can be found in the literature, very few could guide students in problem solving in the domain of engineering. The findings of this book may provide useful problem solving tutoring systems in other domains as well since the approach taken in this book has shown successful development of TAPS packages in four environments namely 2-D, coach based, 3-D, and desktop virtual reality.

Chapter 13
Educational Technoethics Applied to Career Guidance

Pilar Alejandra Cortés Pascual
University of Zaragoza, Spain

ABSTRACT

Educational orientation should be set within a specific socio-historical context, which is nowadays characterized by the Society of Information. From this starting point, we think that the understanding of both an ethical analysis of technology as well as of the means of communication, which individuals will have to deal with in their professional development, must be considered as content linked to professional orientation. This idea becomes more definite in the concept of educational technoethics and it is studied from two parameters: the intrinsic values that technology and the means of communication include (the aim of technoethics) and their use as mediators of ethical values (means of technoethics). Therefore, the proposal that is currently being implemented in the project "Observation Laboratory on Technoethics for Adults" (LOTA) as well as its implications for professional orientation are concisely presented from both points of view. The present text is a review and update of a previously published article (Cortés, 2006).[1]

To Pedro, my brother and partner of athletics and life

INTRODUCTION

The information society entails lifelong training in general professional competencies and, in certain cases, in those specific to information and communication technologies (ICTs). Of course, it is true that due to the resources they provide technologies are being employed to search for employment and training, especially via web pages and some *online* and computer programs. As part of its aims and contents careers guidance therefore includes finding out about (knowledge guidance) and knowing how to use (skills guidance) technological resources and means of communication for work-related choices and adaptation (Cogoi, Sobrado, Hawthorm, R. and Korte, 2005; Hartley and Almuhaidib, 2007). In particular, with regard to the IAEVG's international competencies (2003) for educational and vocational guidance practitioners with regard to *career development* and *placement,* both categories are linked to careers guidance, and although they suggest use of computer and networked resources for said field (*skills*), our understanding is that the *attitudinal* and *capacity* component of ICTs that we are adding here could also feature.

These relationships between ICTs and careers guidance are necessary, but the inter-relationship of a third component is proposed: ethical values. In view of the socio-contextual factors framing the current educative panorama, such as post-modern thinking and the knowledge and information society, it is necessary to study the *triangle* formed by careers guidance, education in values and technology. In other words, if we talk about the space formed by this trio of variables it is because society itself demands that we do so, and, in educational terms, we will need to come up with a response. In this respect, in our opinion the relationship between careers guidance, education in values, and technology involves two lines of study: the first involving the reinforcement of career values demanded by the present knowledge and technology society (Cortés, 2006), and the second dealing with *technoethics* as a component of careers guidance (advice on *attitudes and capacities*). That is, career guidance has to intervene, assess, advise, programme or provide a response to a consultation in three directions: knowing about ICTs, knowing how to use ICTs and having the right attitude to ICTs.

TECHNOETHICS VERSUS EDUCATIONAL AND CAREER GUIDANCE

In this section we will consider the last direction, that is to say academic and professional guidance on the ethical contents entailed by use of technologies, in other words, *guidance on technoethics.* We shall commence with *educational technoethics,* a concept we developed in previous works (Cortés, 2005a; 2006) and which here we also integrate within the careers guidance field. A significant part of the research undertaken with respect to educational technology and means of social communication focuses on the 'what' and 'how' of their existence and use, but there is a lack of works that include an axiological dimension. Nevertheless, Grill (1997) argues that the first thing a professional should do is look for the 'why' of things from attitudinal perspectives, and states that technology in itself is not a problem, but rather technopolism understood as the ethical changes that become the cause of problems such as, for example, addictive behaviour at work vis a vis technology, or excessive pressure from use of technology in work environments.

The need to axiologically analyse educational technologies in careers guidance is stressed in order to meet full training and educational needs in society both at present and in the future. As Cortina (2001) states, there is a need for an ethic of co-responsibility to guide the current social process and one of IT globalization so that this technical progress serves human beings, without foregoing an ethics of minimum values, which for Cortina (1998) is represented by freedom, solidarity, equality, responsibility and honesty. And it is true that technology and the means of communication for social communication require an ethical analysis in order that they can be employed suitably and coherently, as emphasised by others including Hawkrinde (1991), Nichols (1994), Postman (1995), Sunstein (2003), and Ortega and García (2007). This should leave its mark on careers guidance processes, both those of educational centres, from lower to higher levels, and those of the family and other exo and macrosystemic environments (Bronfenbrenner, 1979), in accordance with constructivist principles of careers guidance (Watson and McMahon, 2006).

Our argument is linked to the *Science, Technology and Society*[2] (STS) line of research which arose in opposition to the unidirectional technological model (+ science = + technology = + wealth = + well-being), because the latter does not really correspond to a true conception of science and technology in view of the fact that social factors including moral values, professional interests, political pressures and economic determinants are inherently combined in the aforementioned process, with a considerable influence on the scientific-technological (Bloor and Henry, 1996). In this sense, López (2003) suggests that, on the one hand, humanistic information be provided to natural science and science students in the form of critical sensibility and, on the other hand, that knowledge on science and technology be offered to humanities and social science students.

In this respect, our proposal is dual and inclusive: understanding the intrinsic values that technologies entail (*end*) and using them as mediators to transmit values (*means*). From this point onwards, different studies in both senses will be presented, which are complemented with research experience developed at two Adult Education centres by means of the "Observational Laboratory on *Technoethics* for Adults" (OLTA) project. A number of reflections referring to the issue dealt with will be included in the final part of the chapter.

By *end* we mean that the technologies and *mass-media* include within themselves a value-oriented connotation. However, the aim is not to employ an exclusively negative discourse on said end, akin to that which has normally featured (Nichols, 1987; Ward, 2003), and the fact that the person himself or herself is ethically in charge of the technological also has to be taken into consideration (Bunge, 1974; Medrano and Cortés, at press). The main issue is that schools should address these *ethos* questions, as Katz denotes them (1992), by means of the educational curriculum or in accordance with guidelines "to help us understand and diffuse the inevitable conflicts in our practice of educational technology".

Likewise, Braun (1992), Pruzan and Thyssen (1994), Postman (1995) and Bilbeny (1997) reflect on actions for working in school contexts with the aim of analysing the social revolution that computers are producing in human competencies. Pruzan and Thyssen (1994) propose that a code of values be agreed at each company, and that said idea is extrapolated and adapted to other contexts, for instance teaching, and more particularly careers guidance. We are interested in their work, and believe that an educational mediator, possibly the careers officer, would be an appropriate figure to consider conflicts between the values demanded by the Knowledge Society and those included within the curriculum for student development.

Bilbeny (1997) also proposes a "revolution of the etemas", that is, a common moral minimum for the technological impact from a cognitivist and constructivist perspective, based on three principles: thinking of oneself, or the initiation of moral autonomy (*moral point of view*); imagining oneself in the other's place, or the commencement of reciprocity (*ideal role-taking*); and thinking in a way that is consequent with oneself, or the beginning of reflexibility (*moral insight*). In our opinion, the values that should continue to be upheld are those identified by Cortina (2001), quoted above, but via the three procedures suggested by Bilbeny, in other words, moral autonomy, empathy, and reflection, and we would add a further one, which to a certain extent is implicit: commitment, regardless of our social and professional role, and even more so for educators (teachers, parents, monitors, careers advisors...).

Bilbeny (1997) also suggests re-learning sensibility, that is, recuperating the emotional, empathy, sensibility, reflection and education in ICTs. Thus, "with the growth of the digital society our understanding of the moral has to radically change" (Bilbeny, 1997: 188). Bilbeny also considers the question of the spatial. If with the communication technologies and means of communication there is there is an evolution from a close, presence-based and sensitive relationship, to one which is distant, virtual and subjective, the latter would have different relationship and moral connotations. Consequently, an interaction at a distance may offer people the chance to get to know different people, but the danger indicated by Sunstein (2003) is that in the end these individuals, by means of the aforementioned virtual media, end up seeking common spaces, thereby isolating themselves from other experiences and ideas that are more real. The author (Sunstein, 2003) stresses that some internet sites end up being very selective and very exclusive, and that this process could create a barrier, that is to say a *digital divide* or, as Castells puts it (1997), a *digital dividing line*, which also exerts an influence on professional differences. For example, *online* guidance may reduce these differences, but we must remain aware of the ethical limitations of failure to establish a *close and direct* relationship between advisor and advisee (Mallen, Vogel and Rochlen, 2005).

Up to this point, the conclusion of our analysis has been that *technoethics,* which in itself entails an ethical purpose, must be taken into account in current society and in educational environments by means of a commitment to and proposals for educational and professional guidance. But, on the other hand, guidance in *technoethics as a means* also implies that the technological and information resources may themselves be transmitters and mediators of contents and activities of an axiological nature. Said posture accords with certain of the ideas of Ryan, Bednar and Sweeder (1999) on the need to cultivate the moral via educational technology, because of its motivational capacities, with the aim of safeguarding against vanity, a typical trait of American culture according to the authors, and a morality exclusively sustained by reasoning. The authors present the project known as *Social Projector Virtual Gatherings,* in which they contend that morality currently involves a marriage between feelings and moral behaviour, and, therefore, sympathy, duty, impartiality or justice, and self-control. The project has the goal of ensuring that justice operates in the lives of the students, and also that feelings of duty and sympathy interact in the practice of personal and social equity; therefore becoming general competencies (*attitudinal and capacity*) for all professional fields.

In particular, for the practice of educational guidance, Ryan et al. (1999) propose four types of strategy: virtual assemblies for working on ethical topics, in which different people would participate by means of virtual contact; social action via the internet, as a medium for searching for texts linked to humanity issues, community work or actions of solidarity; creation of IT simulations to address matters in which action is required and decisions taken as a professional, for example, in the face of environmental problems; and lastly, use of video productions featuring "real" stories with an ethical basis to be used in role-playing.

In addition, the Utah State Educational Technology Department has created a technological application program (Jensen, 1993) for middle and higher level students with the aim of providing academic and professional guidance in areas such as: industrial technology and agriculture; business and marketing; and economics and health occupations. It includes 18 practical sessions, with an average duration of 40 minutes, with titles such as "What am I like?", "Personal assessment", "Making a decision", "Decision and emotion" and "Real occupational case histories". A further module, more in tune with our theme in the present chapter, is "The scale of values", in which the student is offered information on how a person's ethical development is constructed. There is a questionnaire on which social and ethical values (pleasure, power, recognition, morality, creativity, work etc.) are the most relevant for their professional development, students are asked to order their values in terms of personal preference, and finally there are self-assessment questions on the session. The *IQ Team project* (Nevgi, Virtanem and Niemi, 2006), developed by the Finnish Virtual University, offers a further proposal for student guidance and tutorials in which it was noted that that students' collaborative skills improved. And this is because virtual community environments, if they used well (Allan and Lewis, 2006), foment lifelong *capacity* competences, as promoted by the *Center on Educational and Training for Employment*[3] (Ohio State University).

MAIN FOCUS

In our opinion, the proposals for incorporating the axiological in technological career guidance programmes are really interesting, but it is just as important to ensure guidance in the training of a professional who has to understand the whys and wherefores of technological factors in relation to his or her practices and attitudes (Grill, 1997). There is agreement with the idea of Repetto and Malik (1998) concerning the importance of learning to use the ICTs in an affective way, that is to say, via the development of ethical and quality standards. In this respect, the Association for Educational Communication and Technology (AECT) discusses professional ethics in relation to research in collaborative electronic environments (Ravitz, 1997), considering issues related to the proliferation of antisocial information (racist groups, child pornography...), as well as ethical practices on the web in three areas: respect for copyright, level of privacy (Cottone and Tarvydas, 2003; McCrickard and Butler, 2005) and level of accessibility of information (Lin and Kolb, 2006). These issues, all highly topical, might form a theme of ethical educational analysis in technical and humanities disciplines in the line indicated above (López, 2003).

Guidance in the knowledge society must be considered in two ways: ICT literacy and "literacy" in non-discrimination and equality of access to the ICTs. In this last sense, there is increasing advocacy (Pantoja, 2004; Touriñan, 2004; Ortega, 2004; Rumbo, 2006) for inclusion of learning to learn and learning to live together within the contents of ICT education for the training of citizens able to seek information in a context of plurality and the democratization of said information. Thus, different electronic addresses can be found on the internet[4] which, in our opinion, link the two concepts (technoethics as end and as means), given that they provide web-based analysis of the ethical in academic and work environments (cyber-constructivist perspective) (Luppicini, 2003).

It is necessary for students to progress towards technological literacy, but without neglecting other types of learning, which are *more human and more social* (Flecha and Rotger, 2004; Ortega and Chacón, 2007). A number of writers link the latter with professional responsibility or professional codes of conduct. As Martínez states (2003), the education of a future professional must not only be focused on problem-solving, since he or she must also acquire a moral occupation in view of the fact that many of the decisions taken at work involve a conflict of values. For example, a conflict of interest involving standards, such as that in which a worker has to decide between respecting the confidentiality of work information or disseminating and/or reporting it. For his part, Pantoja (2004) mentions the dilemma between the modernization of professional systems and non-discrimination and equality of access to training. The isolation of individuals as opposed to fomenting interpersonal relationships can also be added (as occurs in teleworking or networking).

In summary, the inclusion of technology as both the objective and contents of career guidance is defended from three standpoints (see Figure 1): knowing about the technologies, knowing how to use

Figure 1. Relationship between knowing, skills, attitudes and capacities in technology and career guidance.

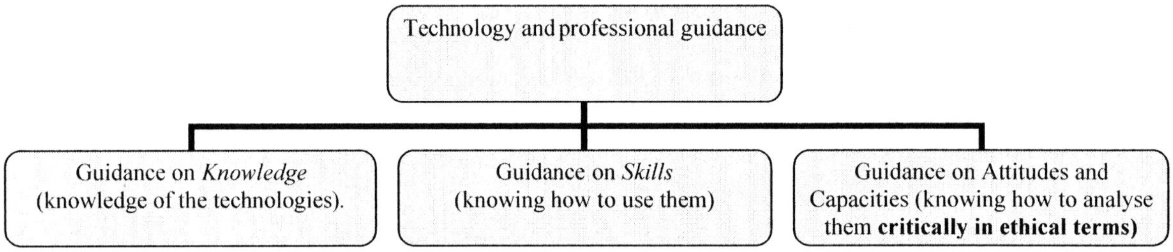

them and knowing how to analyse them critically, the aspect on which the final part of our foregoing discourse was based.

OBSERVATIONAL LABORATORY ON TECHNOETHICS FOR ADULTS (OLTA): FOR FUTURE TRENDS

All of this epistemological contextualization is part of the genesis of the "Observational Laboratory on Technoethics for Adults" (OLTA)[5] project, developed at two lifelong education centres from 2003 to 2005, under the coordination of the author of the present chapter. Most of the proposals in the field of Adult Education have been focussed on distance training, technologies as an extension of the memory or the internet for inter-generational and generational relations. However, in educational and professional guidance programmes for adults there is a lack of the type of approach defended here, which is more reflexive on resources and means of communication with social repercussions. The OLTA project was started as contribution to a possible a solution. Its aim is to teach skills for analysis, criticizing, choosing and reflecting on the new information and communication technologies by means of an axiological interpretation.

During 2003-2004, the project developed a series of modules: "Gathering of prior knowledge and project presentation" (Cortés, 2005b), "Critical and value-oriented analysis of television and radio media instruments", "Debate on the positive and negative aspects of the technologies", "ICT-related dilemmas", "ICT-based dramatization", "The influence of marketing in professional success", and "Internet in the world of work". During the 2004-2005 academic year, a multimedia platform was created; including a web page (http://usuarios.lycos.es/tecnoeticazgz), CD and printed material featuring all project information and results, and study days on "media credibility" were organized, that is to say on the ethical code of communication and information professionals. In the future, we would like to extend it to more professions. We will now outline very general conclusions of some of these modules. We would like to indicate in advance that this OLTA proposal received an average evaluation of 4.2 on a scale of 1 (very negative) al 5 (very positive) from a total of 72% of the adults consulted at the end of the project.

In the first module we investigated the preconceptions of 150 adults (Cortés, 2005b), (110 women and 40 men) aged between 18 and 63. The following is an example of one of the questions that they answered: *Do you think that the internet has changed the world of work and society? How?* In accordance with the methodological proposal of Pascual-Leone(1978), the analysis identifies three types of preconceptions in responses to that question, approximating to a greater or lesser degree to the most widely-held definition depending on the concept of the internet (Castells, 1997), that is whether it is detailed or globalised (the most ideal); imprecise or one-off (responses describing an ideal idea, not in a complete way, but rather in terms of a partial understanding); and anecdotic or occasional (those describing the concept, with only one item of ideal information and/or an example).

Thus, responses were close to offering an anecdotic definition including the idea that the internet "unites, although it divides people because it creates addiction" (40.8%), with protocols such as that the web facilitates making friends easily, aids communication, creates addiction and divides people. In second place, 50 (32.9%) participants offered no response, and in third place, 36 (23.7%) adults answered with contents close to the imprecise level, with only two responses at the detailed level. One man of 53 commented that "the internet brings people together at a distance while separating them where there is proximity. It helps you to find out about far away things, while we look into the distance for things we are ignorant of close to hand". From this, we might infer that axiological reflection on the part of the respondents with regard to the influence of the internet was lacking or minimal. In our opinion, this analysis or diagnostic should occur in proposals for professional guidance and intervention, given that the internet is practically an essential working medium, increasingly known and used, and that therefore, given this data, more critical and axiological analysis of all its labour and personal repercussions should be promoted.

We found another example in third module that could lead to a debate in the classroom as follows: "Let us imagine that we are members of an assessment board deciding whether or not it should give economic support to a research project on *in vitro* fertilization. Would we need to consider ethical factors or would we base our decision exclusively on scientific principles? Why or why not? And what would said ethical principles be if we were to take them into account?" 79 students participated in this activity, 46 women and 33 men, between the ages of 18 and 56. The majority, 85%, was in favour of considering the aforementioned ethical aspects, but there was disagreement as to whether the economic aid had to be public or not; given that if it came from public funds, a first principle would be that the results should be of benefit for the population as a whole. Another principle on which there was relative agreement (65%) was that no human lives should be endangered. There was general agreement (70%) that science degrees at different universities should feature the study of subjects related to ethics in science. 75% of the adults awarded this debate with a 5 (very positive), on a scale of 1 to 5, with it being the module that received the highest score.

Finally, in the last module, three professionals from the world of radio, television and journalism brought to bear their viewpoints with regard to how the *mass-media* exercises its credibility function. We shall only mention two conclusions: the objectivity of the media is impossible, and it very much depends on the ideology of the publishing or communication group; but there should be certain standards within the professional code of conduct agreed by everyone and which should feature the participation of government spokespersons, viewers, listeners and/or readers. There was also a discussion and a consensus reached that the principles, set out by the government and the public and private television stations in the *Agreement to promote self-regulation on television contents and children* (2005),[6] must be included in career guidance and training of a future professional, in this case, of the media; in particular, television.

CONCLUSION: TOWARDS AN AXIOLOGICAL ANALYSIS OF CAREER GUIDANCE

Having reached the end and based on the foregoing, we would like to conclude with some ideas that it will be necessary to take account of in formal educational fields in future. *E-learning* and *e-guidance* initiatives require this perspective (*attitudinal and capacity* guidance in relation to ICTs) in employment training, given that the world of work is evolving in accordance with socio-economic factors and technological advances. Careers guidance must be alive to these movements not only to ensure technically efficient professionals, but also to guarantee responsible citizens who know how to live together. Thus, we will be able to seek a balance between the existing socio-economic order and the democratic society can be built by lifelong education (Pantoja, 2004; Rumbo, 2006).

In addition, it is our understanding that the relationship between official responsibility and professional responsibility is crucial in understanding many of the latter's features and limitations, and that the link with educational commitment is therefore crucial. We have presented a real adult education initiative (OLTA), which, as described above, obtained a very satisfactory result from its recipients. Participating teachers also evaluated the project positively.

But on this point there are also other proposals for lifelong *techno-ethical* guidance at different levels of education, that is to say primary, secondary, post-secondary etc. We agree with Olcott's (2002) suggestion that in all technological training programmes there must be a section on ethics and technology. Thus, in primary and secondary education, from our perspective, we propose that these should be included as cross-cutting themes in curriculum subjects, for example by means of activities such as "*the internet: a divide and a bridge*" to analyse the digital web frontier between those who do and do not have access to the world of the web, as well as the possibility to create links and disseminate information. Other examples might be its employment repercussions in the form of inclusion and exclusion, or "*ethics and nanotechnology*" with the aim of analyzing the limits of technologies in the study of life sciences, or "*the computer*", in order to consider the computer's potential and limitations (computer games, IT programs,...) in social relations, academic guidance and insertion at work. We believe that a good way to approach these themes is via the use of intervention strategies in values including ethical dilemmas, group work, techniques for clarification of values, debates and role-playing. In addition to making use of material resources, both printed and technological and audiovisual (videos, slides, educational forums, the internet...).

Each profession has a code of conduct on which standards and principles should dictate practice. Therefore, in branches or modules of professional training and university courses the contents of said code has to be studied alongside other more technical and conceptual questions and procures. In part, this is included within the new European Higher Education Space, which refers to the fact that students not only learn contents, but also attitudes via the participatory and personal competencies. And it would be necessary to analyse it regardless of the specific training itinerary and occupational field, but in our case we are more specifically referring to people who use technological instruments and are immersed in telecommunications, for example engineers, scientists, administrators, computer programmers or the director of a television news programme. Although of course also in the training of future educators and advisors, who need to know it in order to be able to provide guidance on it. In this respect, a university experience, presented by us in Cortés' book (2004), uses an educational seminar on the subject of the new technologies to ensure that the students learn about *technoethics* activities for their future development as teachers. And Martín (2006) also presents ten simulated cases on science, technology and society within the framework of ethical values.

In this direction, it is noticeable that the influence of the world of technology and communication and the life sciences is currently leading to many debates on professional codes of conduct. Camps (2003) points out that committees of experts are becoming increasingly necessary for debating and, eventually, adopting decisions on ethical factors related to protection, especially protection of individuals and the environment, as mentioned in the EULABOR (2005) report, for example. In any case, these themes cannot just be dealt with in initial training, since they also transverse lifelong training, as in the case of adult education, which was the focus of the project mentioned in the present chapter, OLTA. In said project, using different modules, the idea was for the students to evolve towards reflecting in different ways about the ethical repercussions related to the use of ICTs and means of communication in their daily and working lives. It ought to be pointed out that OLTA was favourably evaluated by the students. And in this way two (individual and institutional benefit) of the three levels (community benefit is missing) of the ethical incorporation of ICTs are reached (Riley, 2004). Certain of the didactic activities proposed here could be included in a digital portfolio of professional development (Milman and Kilbane, 2005), serving the person receiving guidance as a catalyst in his or her own process of autonomous assessment and acquisition of skills in use of the ICTs.

Thus, an essential study topic at present (Hargreaves, 1999; Metros and Woolsey, 2006) is the development of personal competencies in new technologies in a critical and reflective way. We would therefore like to propose the line of investigation discussed in the present chapter and invite research on it,[7] that is to say on *professional guidance in attitudes and capacities in relation to the ICTs.*

REFERENCES

Acuerdo para el fomento de la autorregulación sobre contenidos televisivos e infancia. (2005). Gobierno de España y Televisiones españolas. Retrieved April 15, 2005, from http://www.cnice.mecd.es/tv_mav/n/f6_normativa.htm

Allan, B. & Lewis, D. (2006). The impact of membership of a virtual learning community on individual learning careers and professional identity. *British Journal of Educational Technology, 37(6)*, 841–852

Bilbeny, N. (1997). *La revolución en la ética. Hábitos y creencias en la sociedad digital.* Barcelona: Anagrama.

Bloor, D. & Henry, J. (1996). *Scientific knowledge: A sociological analysis.* London: Athlone.

Braun, J. (1992). Caring, citizensship, and conscience. The cornestones of a values education curriculum for elementary schools. *International Jorunal of Social Education*, 7(2), 47-56.

Bronfenbrenner, U. (1979). *The ecology of human development: Experiment by nature and design.* Cambridge: Harvard University Press.

Bunge, M. (1974). *Por una tecnoética, en etica, ciencia y técnica.* Buenos Aires: Editorial Sudamericana.

Cabero, J. (2001). *Tecnología educativa. Diseño y utilización de medios en la enseñanza.* Barcelona: Paidós.

Camps, V. (2003). Ética para las ciencias y técnicas de la vida. In A. Ibarra & L. Olivé (Eds.), *Cuestiones éticas en ciencia y tecnología en el siglo XXI* (pp. 225-244). Madrid: Biblioteca Nueva.

Castells, M. (1997). *La era de la información. Economía, sociedad y cultura.* (3 vols). Madrid: Alianza.

Cogoi, C., Sobrado, L., Hawthorm, R. & Korte, A. (2005). ICT skills for guidance practitioners: Final results of the research. Work group, CD-rom en *AIOSP Conferencia Internacional.* Lisboa.

Cortés, P. A. (2004). *Una mirada psicoeducativa de los valores. Seminario aplicado a las nuevas tecnologías de la educación.* Prensas Universitarias de la Universidad de Zaragoza: Zaragoza.

Cortés, P. A. (2005a). Educational technology: as a means to an end. *Educational Technology Review, 13*(1), 73-90.

Cortés, P.A. (2005b). Las preconcepciones sobre la tecnoética en los adultos. *Revista Mexicana de Psicología, 22*(2), 541-552.

Cortés, P.A. (2006). Valores y orientación profesional: líneas de investigación e intervención actuales. *Contextos Educativos, 8-9*, 223-238.

Cortina, A. (1998). ¿Qué son los valores y para qué sirven? *Temas para el debate, 42*, 20-22.

Cortina, A. (2001). *Alianza y Contrato. Política, ética y religión.* Madrid: Editorial Trotta.

Cottone, R. R. & Tarvydas, V. M. (2003). *Ethical issues in counseling* (2nd ed.). Upper Saddle River, NJ: Merrill Prentice Hall.

EULABOR (2005). *European and Latin American systems of ethics regulation of biomedical research: Comparative analysis of their pertinence and application for human subjects protection.* Madrid: EPSON Entreprise.

Flecha, R. & Rotger, J.M. (2004). Innovación, democratización y mejora de la docencia universitaria en el marco de la Sociedad de la Información. *Contextos Educativos, 6-7*, 159-166.

Gill, D. W. (1997). Educating for meaning and morality: The contribution of technology. *Bulletin of Science, Technology & Society, 17*, 5-6, 249-260.

Hartley, R & Almuhaidib, S.M.Y. (2007). User oriented techniques to support interaction and decision making with large educational databases. *Computers & Education, 48*(2), 268-284.

Hargreaves, A. (1999). *Profesorado, cultura y postmodernidad.* Madrid: Ediciones Morata.

Hawkridge, D. (1991). Challenging educational technology. *ETTI, 28*(2), 102-110.

International Association for Educational and Vocational Guidance (IAEVG, 2003). *International competencies for educational and vocational guidance practitioners.* Retrieved April 14, 2004, from http://www.iaevg.org/iaevg/nav.cfm?lang=4&menu=1&submenu=5

Katz, SB. (1992). The ethic of expediency: Classical rhetoric, tecnology, and the holocaust. *College English*, *54*(3), 255-275.

Lin, H. & Kolb, J. (2006). *Ethical issues experienced by learning technology practitioners in design and training situations*. Paper presented at the Academy of Human Resource Development International Conference (AHRD) (Columbus, OH, Feb 22-26) pp. 1168-1175.

López, J.A. (2003). Ciencia, técnica y sociedad. In A. Ibarra & L. Olivé (Eds.), *Cuestiones éticas en ciencia y tecnología en el siglo XXI* (pp. 113-158). Madrid: Biblioteca Nueva.

Jensen, R. (1993). *TLC career guidance*. Currículo. Utah: Salt Lake City.

Luppicini, R. (2003). Towards a cyber-constructivist perspective (CCP) of educational design. *Canadian Journal of Learning and Technology*, *29*(1). Retrieved April 26, 2006, from http://www.cjlt.ca/content/vol29.1/01_luppicini.html

Mallen, M., Vogel, D. & Rochlen, A. (2005). The practical aspects of online counseling: Ethics, training, technology, & competency. *Counseling Psychologist*, *33*(6), 776-818.

Manosevitch, E. (2006). Democratic values, empowerment and giving voice: Children's media discourse in the aftermath of the assassination of Yitzhak Rabin. *Learning, Media & Technology*, *31*(2), 163-179.

Martín Gordillo, M. (2006). *Controversias tecnocientíficas*. Barcelona: Octaedro OEI.

Martínez, S. (2003). Ética de científicos y tecnólogos. In A. Ibarra & L. Olivé (Eds.), *Cuestiones éticas en ciencia y tecnología en el siglo XXI* (pp. 277-300). Madrid: Biblioteca Nueva.

Martínez, F. & Área, M. (2003). *El ámbito docente e investigador de la tecnología educativa en España. Algunos apuntes para el debate*. Paper presented at the meeting of the Meeting of the area of Didactics and Scholastic Organization, Valencia, University of Valencia. Spain.

Mccrickard, M.P. & Butler, L.T. (2005). Cybercounseling: A new modality for counselor training and practice. *International Journal for the Advancement of Counselling*, *27*(1), 101-110.

Medrano, C. & Cortés, P.A. (2007). Teaching and learning of values through television. *International Review of Education*, *53*(1), 5-21.

Metros, S. & Woolsey, K. (2006). Visual literacy: An institutional imperative. *EDUCAUSE Review*, *41*(3), 80-81.

Milman, N.B. & Kilbane, C.R. (2005). Digital teaching portfolios: Catalysts for fostering authentic professional development. *Canadian Journal of Learning and Technology*, *31*(3). Retrieved April 26, 2006, from http://www.cjlt.ca/content/vol31.3/milman.html

Nevgi, A., Virtanem, P. & Niemi, H. (2006). Supporting students to develop collaborative learning skills in technology-based environments. *British Journal of Educational Technology*, *37*(6), 937-947.

Nichols, R. G. (1987). Toward a conscience: Negative aspect of educational technology. *Journal of visual/Verbal Languaging*, *7*(1), 121-137.

Nichols, R. G. (1994). Searching for moral guidance about educational tecnology. *Educational technology*, *34*(2), 40-48.

Olcott, D. (2002). Ética y tecnología: desafíos y elecciones inteligentes en una sociedad tecnoética. In En Hanna, D.E. (ed.), *La enseñanza universitaria en la era digital* (pp. 225-243). Barcelona. EUB.

Ortega Carrillo, J.A. & García Martínez, F.A. (2007). Ética en los medios de comunicación e Internet: promoviendo la cultura de paz. In J.A., Ortega Carrillo & A., Chacón Medina (Coords.), *Nuevas tecnologías para la educación en la era digital* (pp. 331-353). Madrid: Ediciones Pirámide.

Ortega Ruíz, P. (2004). La educación moral como pedagogía de la alteridad. *Revista Española de Pedagogía, 227*, 5-30.

Ortega Carrillo, J.A. & Chacón Medina, A. (Coords.) (2007). *Nuevas tecnologías para la educación en la era digital*. Madrid: Ediciones Pirámide.

Pascual-Leone, J. (1978). La teoría de los operadores constructivos. In J. Delval (Comp.), *Lecturas de psicología del niño* (pp. 25-36). Madrid: Alianza.

Postman, N. (1993). *Technopoly: The surrender of cultura to technology.* New York: Vintage.

Pantoja, A. (2004). *La intervención psicopedagógica en la sociedad de la información. Educar y orientar con nuevas tecnologías*. Madrid: Editorial EOS.

Postman, N. (1995). Making a living. Making a life: Technology reconsidered. *College Board Review, 176-177*, 8-13.

Pruzan, P & Thyssen, O. (1994). The renaissance of ethics and the ethical accounting statement. *Educational Technology, 34*(1), 23-28.

Ravitz, J. (1997). Ethics in scholarly communications: intellectual property and new technologies. In *National Convention of the Association for Educational Communications & Technology*, 19th, Albuquerque.

Repetto, E., Rus, V. & Puig, J. (1994). *Orientación educativa e intervención psicopedagógica.* Madrid: UNED.

Repetto, E. & Malik, B. (1998). Nuevas tecnologías aplicadas a la orientación. In R., Bisquerra, *Modelos de orientación e intervención psicopedagógica.* Barcelona: Praxis.

Riley, J. (2004). *Ethical drivers of the implementation of instructional technology.* Paper presented at the Teaching Online in Higher Education Conference, Fort Wayne.

Ryan. F., Vendar, M. y Sweeder J. (1999). Tecnology, narcissism, and the moral sense: Implication for instruction. *British Journal of Educational Technology, 30*, 115-128.

Rumbo, B. (2006). La educación de las personas adultas: un ámbito de estudio y de investigación. *Revista de Educación, 339*, 625-635.

Sunstein, C.R. (2003). *República.com. Internet, democracia y libertad.* Barcelona: Paidós.

Touriñan, J.M. (2004). La educación electrónica: un reto de la sociedad digital en la escuela. *Revista Española de Pedagogía, 227*, 5-30.

Ward, L. M. (2003). Understanding the role of the entertainment media in the sexual socialization of American youth: A review of empirical research. *Developmental Review, 23*, 347-388.

Watson, M. & McMahon, M. (2006). My system of career influences: responding to challenges facing career education. *International Journal Educucational Vocational Guidance*, *6*, 159–166.

KEY TERMS

Digital Divide: Social, economic and political differences between communities that have information and communication technologies (whit alphabetization technological, capacity and quality) and those that not.

Observational Laboratory on Technoethics for Adults (OLTA): Project, developed at two lifelong education centres from 2003 to 2005, for to teach skills for analysis, criticizing, choosing and reflecting on the new information and communication technologies by means of an axiological interpretation.

Science, Technology and Society (STS): Line of research which arose in opposition to the unidirectional technological model (+ science = + technology = + wealth = + well-being). STS analyse science ant technology in context and social aspects (values, policy, economy, etc.).

Technoethics as an End: That technologies and mass-media also include an assessing ethics connotation.

Technoethics as a Means: The implication is that technological and informational means can be transmitters of contents and activities of an axiological kind.

Technoethics as an End and Means: A reply is required from a doubly interweaved aspect, that is to say, as a refl ection and performance about its axiological purpose and as an instrument to deal with attitudinal and ethical knowledge, its medium.

Technology and Professional Guidance: Triple analysis: guidance on knowledge of the technologies, guidance on skills and guidance on attitudes and capacities.

ENDNOTES

1 Cortés Pascual, P.A. (2006). An analysis of careers guidance from the standpoint of educational *techno-ethics. Revista Española de Orientación y Psicopedagogía*, 17,2, 181-193. With the express authorisation of said publication.

2 On its web page, the Organization of Ibero-American States: http://www.oei.es/cts.htm, proposes different publications and projects linked with the STS line.

3 Collage of Education. The Ohio State University 1900 Kenny Road Columbus OH 43210-1090. www.cete.org

4 The following are examples: *Center for Accounting Ethics* (University of Waterloo, Ontario) (http://accounting.uwaterloo.ca/ethics/index2.html), *Centre for Applied* Ethic (University of British Co-

lumbia, Vancouver) (http://www.ethics.ubc.ca) and *Wharton Ethics Program* (Wharton School, University of Pennsylvania, Philadelphia). Consulted in February 2007.

5 A project approved by the Aragón regional government (Spain) in the calls of 2003 and 2004, directed by Carlos Sanz, director of the Concepción Arenal centre (Zaragoza), and coordinated by the present writer. Isabel Segura (Teruel) also collaborated.

6 See http://www.cnice.mecd.es/tv_mav/n/f6_normativa.htm. Undertaken in Spain by the government.

7 The author of this chapter may be contacted directly at alcortes@unizar.es

This work was previously published in the Handbook of Research on Technoethics, edited by R. Luppicini and R. Adell, pp. 426-438, copyright 2009 by Information Science Reference (an imprint of IGI Global).

Chapter 14
Digital Simulation in Teaching and Learning

Youngkyun Baek
Korea National University of Education, Republic of Korea

ABSTRACT

This chapter expands upon the definition of a simulation with two categories: experiential and symbolic. It discusses the interactive, experiential trend in digital teaching and learning, and the educational merits of simulations. This chapter tries to locate digital simulation's position in these trends. In doing so, it explores the educational merits of digital simulation, discusses the learning mode of digital simulation, and outlines what digital simulation conveys to deliver educational contents. In addition, it will look at the characteristics and functions of digital simulation. Mainly this chapter focuses on how simulation is used for teaching and learning. It highlights simulation's features to be effective for teaching and learning. It also introduces challenges to simulation to overcome its disadvantages. Several examples of digital simulation in teaching and learning are explored: They are "Max Trax, Strategy CoPilot, Virtual School, simSchool, simClass, Krucible", and "Starry Night". Lastly, this chapter seeks to forecast the future of teaching and learning with a focus on information technology and simulation by finding simulation's role and contribution in learning context.

INTRODUCTION

This section summarizes interactive and experiential trends in teaching and learning. It tries to locate digital simulation's position in these trends. In doing so, it explores the educational merits of digital simulation, discusses the learning mode of digital simulation, and outlines what digital simulation conveys to deliver educational contents.

The Interactive, Experiential Trend in Digital Teaching and Learning

New ways of teaching and learning are arising, made possible by a variety of new technologies, online resources, and educational delivery methods. These new approaches induce teachers and students to perform different academic roles, share workloads in new ways, and acquire and utilize new skills and knowledge (Hovenga & Bricknell, 2004). The changes began to appear with the enormous increase in and effectiveness of information technology and computer use, but there is an even more significant change emerging now: digital games and simulations, artificial intelligence and virtual reality with immersive interactive technology. These new tools require teachers to devise new teaching methods for today's students, who have grown up with such technologies. Students are comfortable being ubiquitously connected, wired "24/7" and multi-tasking, and accustomed to using technology as and when they wish for their daily lives, including learning. This trend toward self-directed, highly interactive, rich-media experiential environments, evident in sites like YouTube, online news media sites, and iTunes, challenges educators to examine the power of interactive digital environments. MacDonald (2008) notes that today's students want, need, and expect the flexibility, convenience, interactivity, and animation afforded by the use of technology in their courses and programs.

The trend toward increased interactivity and personal experience is expected to continue into the future and to become more embedded in work and everyday life. For example, Dwerryhouse (2001) asserts that future learning is work-related learning, which involves learning embedded within the workday to promote higher levels of productivity. Self-directed learning, which is the most personalized kind of learning, is more prevalent nowadays in informal learning settings (e.g. museums and exhibits) than in the formal educational system, but with technology, could become embedded in the student's "workday" to help them achieve higher levels of productivity. Experiential learning, a hallmark of the kind of learning that is embedded in games and simulations, is assumed to be the ideal learning method for self-directed learning. Besides the advantages gained when learners take responsibility for their own improvement and advancement, experiential learning is expected to increase and deepen understanding of a subject, and to increase self-efficacy and motivation. Experiential learning fosters in-depth information processing and elaboration, as it builds up learning skills and leads to higher motivation for learning initiated by a learner's direct involvement.

According to Kolb (1984), experiential learning consists of four elements: concrete experience, observation and reflection, the formation of abstract concepts, and testing in new situations. It is suggested that the learning process begin with carrying out a particular action and then seeing the effect of the action in this situation. The second step is to understand these effects in the particular instance enough to understand what follows if the same action is taken in similar situations. The third step is to understand the general principle under which the particular instance falls. The last step is to transfer what is obtained into real life. Because experiential learning is often equated with high levels of learner activity, simulation-based learning is thought to be ideal especially for those who may be less motivated to learn with traditional materials. The contextual content of simulations allows the learner to "learn by doing" (Kluge, 2007).

With the advent of the computer age, digital simulation provides effective virtual learning experiences for learners in many fields, such as medicine, police training, engineering, physics, the military and aviation. Prensky (2001) notes that learning by doing is central to game and simulation based learning, because it turns out that "doing" is something that computer simulations are especially good at; they allow us to interact with them. Of course, there are many ways of learning by doing; drill and practice is one form of doing; exploring, discovery, and problem solving are other forms. What is essential is active participation by the learner. We thus expect to observe, research, measure and report on the impacts of active participation and learning by doing—interaction and experience—in simulation-based learning.

Digital Simulations' Popularity

The popularity of virtual environments in education is related to how they form and nurture learning community through new kinds of social interactions (Sanders & McKeown, 2007). Multi-user virtual environments, such as Second Life, provide a space in which social interactions are mediated by avatars—virtual stand-ins for each participant. Through their avatar's interactions, both planned and serendipitous, students can begin to create knowledge together (Kim, 2008; Park, Jung, & Chris, 2008). In such a world, they talk about the work they are doing in class, they share ideas, processes, and resources with one another and contribute to the base of knowledge that exists in their field. Throughout this process, they proceed from novice to expert, both in terms of knowledge and skills, but also in terms of their abilities to work collaboratively. A virtual world can be designed to meet the needs of learners engaged in self-directed meaningful activity within a community of practice of novices as well as experts.

Learning in an abstracted world, rather than the real world, has become increasingly popular, due to the highly interactive immersive environments, which provide motivation (Franklin, 2008). The use of virtual worlds, simulations and digital games may motivate learners to explore and "play," thereby encouraging a more constructivist classroom. Digital simulations are also popular because they can be used in situations where physical safety is an issue. Digital simulations can substitute for costly or hazardous situations in a real world setting. In addition, digital simulations are appealing because they promote decision-making activities by learners. These are a few of the reasons that digital simulations have always been useful for teaching and learning.

Situational learning are taking place in a simulation. Ormrod (2004) argued that in a simulation the students are dealing with tasks in an authentic learning environment that is identical or similar to those that they will eventually encounter in the real world. Such a simulation may have as much transferability as other learning environments, but with the added benefits of scalability, safety and wider possibilities for learning situations almost without boundaries. Galarneau (2005) notes that simulations afford the unique possibility of designing an authentic learning experience when it is impossible or impractical to foster such an experience in the physical world.

Even without the richness of a fully immersive virtual environment for learning, and when the measured learning results are on par with traditional methods, simulations have several advantages. Klein and Doran (1999) investigated students' performance in individual and cooperative learning structures with a computer simulation. The results indicated that there was no significant difference in the performance of students across the learning structures. However, the computer simulation was effective for promoting student learning of the technical and procedural aspects of accounting. Results for time on task also indicated that the computer simulation was an efficient tool for promoting learning. Turning to attitude, responses to their survey and interview questions suggested that a majority of students had positive feelings toward using the computer simulation. Students in the extensive small group conditions exhibited significantly more discussion about the content and tasks than students in the occasional small group conditions.

These explanations of the rising popularity of digital simulations for learning indicate some of the directions researchers need to look to explore the educational merits, including the cognitive and affective impacts. They also point to the need for better understanding of how a simulation creates its effects on learners, and the need for teacher and leader preparation programs in education to participate and contribute to the research so that education can be improved for all students.

Educational Merits

The merits of digital simulation in education can be summarized into cognitive benefits such as 'knowledge' and 'skills', and also into pedagogical benefits such as 'individualized learning.' One of the main educational merits from simulation is gaining knowledge; Swaak, van Joolingen and de Jong (1998) asserted that learners can gain several kinds of knowledge from simulation, including intuitive knowledge, implicit knowledge, and functional knowledge. Hubal, Helms and Triplett (1997) point out that "procedural knowledge is best gained ... by doing" (p. 1), which simulations provide. Moreover, because of the hidden structure of relationship among variables in simulations, they challenge learners to actively infer knowledge.

The educational benefit of using digital simulation, when combined with traditional lecture and discussion, is multi-faceted. Students who are visual learners are allowed to 'see' a process, rather than only hear about how it is done or how it works. Waller (2007) notes "the use of digital simulations for learning emerged from the need to provide hands-on practice. Besides, digital simulations promote higher-order thinking skills, such as decision-making, analytical reasoning and problem-solving" (p. 36). Simulation provides visual stimulation and feedback of an actual object, even when that object is a virtual representation of an object in the real world. Other benefits include teamwork and listening skills that must occur in order for the students to succeed in many simulations (Speelman & Gore, 2008). As Menn (1993) asserted "The goal of multimedia education is to provide a stimulating, tailored, non-judgmental environment which children can explore their creativity and develop individual learning strategies" (p. 52), simulations provide an excellent environment for students to play out their ideas.

Individualized learning is another merit of simulation (Sanders & McKeown, 2007). Simulations in asynchronous settings offer more opportunities for students to take their own paths through resources and activities together, in groups and at times that make more sense to them. Structure and guidance can still be provided as appropriate, for example, a class in Second Life might be as linear as a class in any other setting. However, teachers in virtual environments can also provide choices for the students within the simulation and they have the ability to help students construct individual paths through the virtual world, which implies that educators in training need opportunities to develop the knowledge and skill of individualizing within virtual environments.

Several researchers assert that simulations have other advantages. Magee (2006) argues that simulations offer a risk free environment for experimentation, problem solving skills assessment, and social interaction. Hopewell (2008) notes that instructors choose simulation as a teaching tool because simulation can reduce the cost of conducting real world activities, the risk to a student's or participant's life, and potentially deleterious effects. He also reports that simulations offer the teacher the possibility of placing their students into authentic situations in order to better learn the material. Baek (2008) asserts that manipulating and observing are additional merits coming from simulation. He notes that when the learning objective is to explore the relationship between variables and their impact on a whole system, a simulation is recommended.

Learning Modes of Digital Simulation

Computer simulations can broadly be divided into two types (de Jong & van Joolingen, 1998): *conceptual* models and *operational* models. Conceptual models hold principles, concepts, and facts related to the system being simulated and cover a wide range including qualitative and quantitative, continuous and discrete, and static and dynamic models (van Joolingen & de Jong, 1991). Operational models include sequences of cognitive and non-cognitive operations that can be applied to the simulated system. Examples of conceptual models can be found in economics and in physics and examples of operational models can be found, for example, in radar control tasks. Operational models are generally used for experiential learning, while conceptual simulations are generally found in a discovery-learning context.

In scientific discovery learning with simulation, the main task of the learner is to infer the characteristics of the model underlying the simulation. The learners' basic actions cause changing values of input variables and allow observations of the resulting changes in values of output variables (de Jong, 1991; Reigeluth & Schwartz, 1989). Originally simulation environments were rather limited in their means of receiving input and giving output, but according to de Jong and van Joolingen (1998), increasingly sophisticated interfaces using direct manipulation for input, and graphics and animations as outputs, are emerging, such as in the latest developments in virtual reality.

The next four sections will expand on the definition of digital simulations. In section two, "What is a digital simulation?" will be examined, and "Digital simulation as a tool for teaching and learning" will follow. Then, "Examples of digital simulation in teaching and learning" will be presented and summarized, and the final section will deal with "Digital simulation's role in the future of teaching and learning".

WHAT IS A DIGITAL SIMULATION?

This section will define digital simulation and its types. In addition, it will look at the characteristics and functions of digital simulation. In short, this section will give readers a basic idea of what digital simulation is.

Definition of Digital Simulation

Simulation is, according to Ralston and Reilly (1983), the representation of certain features of the behavior of a physical or abstract system by the behavior of another system. The representation in a computational model includes a set of rules that define some specific model that reflects or imitates reality (Shirts, 1975). Wikipedia (2007) defines simulation as "an imitation of some real thing, state of affairs, or process. The act of simulating something generally entails representing certain key characteristics or behaviors of a selected physical or abstract system". The WordNet online dictionary (http://www.wordnet-online.com/simulation.shtml), defines simulation as, "the act of imitating the behavior of some situation or some process by means of something suitably analogous especially for the purpose of study or personnel training." These definitions imply that simulation is an act of imitating the behavior of a physical or abstract system, such as an event, situation or process that does or could exist.

Reduced to it's essence, a simulation consists of placing an individual in a realistic setting where he or she is confronted by a problematic situation that requires active participation in initiating and carrying through a sequence of inquiries, decisions and actions. A simulation compares favorably with its counterpart problem in a real situation, in that they can both present an individual with an ill-defined problem with several parameters and possible courses of action.

A *computer (digital) simulation* can be defined as a program that models a system or a process, which can be natural or artificial. After critically examining several alternative definitions of a computer simulation, Humphreys (1991) suggests the following working-definition:

A computer simulation is any computer-implemented method for exploring the properties of mathematical models where analytic methods are unavailable. (p. 501)

Alessi & Trollip (2001) define an educational simulation as a model of some phenomenon or activity that users learn about through interaction. Since it is not easy to provide students with many of the experiences they need to truly understand many of the complex, hard-to-grasp materials they are studying, simulation offers a way to allow students to work on tasks or projects that would otherwise be impractical, dangerous, or prohibitively expensive (Schmucker, 1999).

Digital simulations differ from computer games, even though "games re-create or simulate in detail a real-life, first-person activity" (Schmucker, 1999). A *computer game* is defined as such by the author or inferred by the user because the activity has *goals*, is *interactive*, and is *rewarding* (Vogel et al., 2006). Unlike games, simulations are evolving case studies of a particular social or physical reality. The goal, instead of winning as in games, is to take a leading role, address the issues, threats, or problems arising in the simulation, and experience the effects of one's decisions (Gredler, 2004). Interactive simulation activities must *interact* with the user by offering the options to *choose or define parameters* of the simulation then observe the newly created sequence rather than simply selecting a prerecorded simulation (Vogel et al., 2006). In other words, a digital simulation can take any of several directions, depending on the actions and reactions of the participants and natural complications that arise in the exercise. Digital simulations differ from face-to-face or virtual role-plays, which are brief, single incidents that require participants to improvise their roles. An example of a role-playing exercise is a student who acts as a scientific specialist, assigned to a group either supporting or opposing the cloning of dinosaurs. There might be 4-6 specialists (or groups of specialists): geneticists, ecologists, etc., on each side. Each side researches for a couple of weeks and presents its argument in class, and other students can ask the specialists questions. In such an exercise, there can be no repeated experimental interactions by the participants, which contrast with that capability of a digital simulation.

Virtual environments create particular settings and attempt to draw the participant into the setting (Gredler, 2004). The concern here is whether virtual environments use simulations for learning or not. The answer depends on the nature of the problem or situation the learner is addressing and the capabilities required of the learner. That is, is it a complex, evolving reality? And what are the capabilities executed by the learner?

Kinds of Digital Simulation

There are several ways to categorize simulations. Researchers who focus on the nature of participant roles and interface with the modeled situation divide simulations into two distinct categories: experiential and symbolic (Psotka, 1995; Vanlehn, Ohlssen, & Nason, 1994).

In an experiential simulation, the participants are meant to view themselves as components within a larger, changeable situation (Brown, 1999). An experiential simulation is one in which the participant is placed in a situation that attempts to offer a degree of verisimilitude, a sense of reality. Experiential simulations can be based upon case studies or scenarios, and include role-play and activity, often collaborative, in an authentic environment that in some way or other reconstructs aspects of real life tasks (Maharg, 2006). Participants must react to situations as they emerge with minimal deviations from the restrictions placed on them by their defined roles. "Experiential simulations, in other words, are dynamic case studies with the participants on the inside" (Gredler, 1996, p. 523). Experiential simulations are social microcosms, because learners interact with real-world scenarios and experience the feelings, questions, and concerns associated with their particular role (Gredler, 2004). Usually, there are three types of experiential simulation social process, diagnostic and data management. In social process, the contingencies for different actions are imbedded in the scenario and role descriptions. In diagnostic, the contingencies are based on the optimal, near-optimal, and dangerous decisions that may be made. And in data management, the contingencies are imbedded in the quantitative relationships among the variables expressed in equations. All of them can be used in a group exercise, and the diagnostic also may be used in an individual exercise.

Symbolic simulations are different from experiential simulations. They are more abstracted representations of a system or set of processes. Symbolic simulations "depict the characteristics of a particular population, system or process through symbols...The user performs experiments with variables that are a part of the program's population" (Barles et al., 2006, p. 3). The participant remains "outside" the simulated environment while exerting control by adding, removing, or altering variables (Gredler, 1996). An example of this is the use of simulation models in economics or in science discovery learning (Windschitl & Andre, 1998). In symbolic simulations, the student functions as a researcher or investigator and tests his or her conceptual model of the relationships among the variables in the system (Gredler, 2004). Symbolic simulations comprise two kinds: laboratory-research simulations and system simulations. In laboratory-research simulations, users investigate a complex, evolving situation to make predictions or solve problems. In system simulations, they interact with indicators of system components to analyze, diagnose, and correct operational faults in the system. The advent of powerful, easily accessible, and usually networked computers has led to a particular interest in the development of symbolic simulations, as the computer can process the interaction and display the outcome between the multiple variables. However, as Barton and Maharg (2006) demonstrate, the distinctions between symbolic and experiential learning are breaking down, not only because new educational models are being used that employ both approaches, but also because technology has advanced to the point where the two categories can be integrated in the same application.

In an alternative categorization, researchers like Nurmi (n.d.) and de Jong and Joolingen (1998) divide simulation into two broad categories: operational and conceptual. Conceptual models hold principles, concepts, and facts related to the system being simulated while operational models include sequences of cognitive and non-cognitive operations in the simulated system. In operational simulation, the learning is enacted within a specific evolving situation, and in conceptual simulation, learning the content occurs by inferring and making experiments, which can take place either by using or building simulations.

In another categorization schema, Hartmann (1996) distinguishes between continuous and discrete simulations. In a continuous simulation the underlying space-time structure as well as the set of possible states of the system is assumed to be continuous. Discrete simulations are based on a discrete space-time structure right from the beginning (Wolfram, 1994).

Focused on education, Alessi & Trollip (2001) categorize simulations into two groups according to whether their main educational objective is to teach *about something* or to teach *how to do something*. The "about something" group can be subdivided into two subcategories, *physical* and *iterative* simulations, and the "how to do something" group into two subcategories, *procedural* and *situational* simulations. In *physical* simulations, a physical object or phenomenon is represented on the screen, giving the user an opportunity to learn about its underlying principles. We learn from physical simulations by manipulating the various objects or variables and observing how the overall system changes as a result. *Iterative* simulations are quite similar to physical simulations in that they teach about something. The primary difference is the manner in which learners interact with the simulation. Time is generally not included as a variable in iterative simulations. That is, whether the real phenomenon occurs very quickly or very slowly, in iterative simulations the learner runs the simulation over and over, selecting values for various parameters at the beginning of each run, observing the phenomena occur without interventions, interpreting the results, and then running it all over again with new parameter values. The purpose of *procedural* simulations is to teach a sequence of actions to accomplish some goal. In all procedural simulations, whenever the user acts, the computer program reacts, providing information or feedback about the effects the action would have in the real world. *Situational* simulations deal with the behaviors and attitudes of people or organizations in different situation, rather than with skilled performance. Situational simulations are the least common type of educational simulation, perhaps because they are more difficult and expensive to develop, given the great complexity of human and organizational behavior.

In still another approach to categorization, Aldrich (2005) undertook one of the most extensive evaluations of current products and companies. His work resulted in the observation of four distinct genres of simulations that are currently being used in adult education contexts today. These include branching stories, interactive spreadsheets, game-based models and virtual labs or products. These genres represent the stable products employed by organizations that use simulation and games in their training programs.

Characteristics and Functions of Digital Simulation

There are four types of elements or presentation methods that are usually present to varying degrees in every simulation (Alessi & Trollip, 2001): Choices to be made; Objects to be manipulated; Events to react to; Systems to investigate. Naturally, the complexity of any simulation's functions is proportional to the complexity of the problem, representation, and the methodology of simulation. This methodology includes, among its components (Winsberg, 1999, p. 19):

- A computational structure for the theory.
- Techniques of mathematical transformation.
- A choice of parameters, initial conditions, and boundary conditions.
- Reduction of degrees of freedom.
- Ad hoc models.
- A computer and a computer algorithm.
- A graphics system.
- An interpretation of numerical and graphical output coupled with an assessment of their reliability.

A thorough epistemology of simulation requires a detailed analysis of the role of each of these components and of how a skilled simulation developer can manage each of their potential contributions as sources of system behavior, support for learning objectives, and model error.

There are at least six elements involved in simulation development (Table 1. adapted from Aldrich, 2004). The first one is appropriately used linear, cyclical, and systems content; the second is used of simulation genres, which includes branching stories, virtual products/virtual labs, interactive spreadsheets, flight simulator, and 3D maps, as well as new genres to be introduced. The third element is appropriately used genre elements, including modeling, AI, graphics, and interface. The fourth is creating an atmosphere similar to the atmosphere in which the content will be used. The fifth element is presenting behavior to be modeled or recognized. Although focusing primarily on linear content, most narratives, instructions, and case studies have a non-interactive simulation aspect. The last element is feedback that shows the natural consequences of the behavior. Specifically, we can rigorously but selectively represent objects or situations, and can rigorously but selectively represent user interaction. Different simulation elements enable discovery, experimentation, concrete examples, practice, and active construction of systems, cyclical, and linear content.

While Aldrich's six elements are manifested in the simulation development, four characteristics have been identified as a contributing to the success of computer simulations for instruction (de Jong, 1991). These are: (a) a computational model underlying the simulation: (b) the presence of clearly stated instructional goals: (c) the ability of the simulation to evoke exploratory learning: and (d) the opportunity or possibility for learner activity. These four characteristics appear self-evident to those knowledgeable about instructional design. However, there has been considerable research outside of education on these factors because simulations are incorporated into industrial applications and fields outside of education (Towne, 1995). D'Augustine and Charks (1973) summarizes that games and simulation promote high interest, that they may result in greater depth of understanding of a concept or better mastery of a skill even though they may be more time-consuming, and that knowledge is likely to be imparted at a higher rate when they are used.

There are two characteristics of simulations that Winsberg (2003) argues meaningfully distinguish them from mere brute-force computation, in ways that connect them to experimental practice in an interesting fashion. Successful simulation studies do more than compute numbers. They make use of a variety of techniques to draw inferences, such as imaging, from these numbers. Simulations also make creative use of computational techniques that can only be motivated extra-mathematically and extra-theoretically. As such, unlike simple computations that can be carried out on a computer; the results of

Table 1. Simulation Elements, adapted from Aldrich (2004, p. 6)

Simulation Elements
• *Appropriately used linear, cyclical, and systems content* • Use of simulation genres, including branching stories, virtual products/ virtual labs, interactive spreadsheets, flight simulator; and 3D maps, as well as *new genres* to be introduced • *The appropriate use genre elements, including modeling, AI, graphics, and interface* • *Creating an atmosphere similar to the atmosphere in which the content will be used* • *Presenting behavior to be modeled or recognized* (Most narratives, instructions, and case studies have a non-interactive simulation aspect, although focusing primarily on linear content) • *Feedback from a decision (or series of decisions) that shows the natural consequences of the behavior*

simulations are not automatically reliable. Much effort and expertise goes into deciding which simulation results are reliable and which are not.

The factor of fidelity is an overarching issue that affects all aspects of a simulation, such as the underlying model, presentations, and interactions. Fidelity refers to how closely a simulation imitates reality. Fidelity of a simulation affects learning of its users. For example, for beginners high fidelity may hinder basic mechanism of variables which a simulation tries to transcend to users. Fidelity affects both initial learning and transfer (Alessi & Trollip, 2001).

A simulation differs from other kinds of models by its dynamic nature. It *represents an operating model* of a system (Magee, 2006). It allows an observer to view not only a single point in time in the model but also how it changes under different parameters (Greenblat, 1975). The model is not a complete representation of an event, but abstraction that focuses on a specific aspect of that event. The interrelationships represent the focus for either an issue that a researcher is investigating or a concept that an educator is teaching to their students. Simulation models are explicitly defined, as it is important to define the biases this particular view of reality introduces to both research and education. "Once they are set up they provide a venue where new ideas can be explored and complex interrelationships examined" (Greenblat, 1975, p. 10). It is through experimenting with the numerous variables that control the underlying models that a simulation can provide a variety of outcomes. A pure simulation does not provide an evaluation of the outcomes of that experimentation. These kinds of assessments of the behaviors of the participant are often external to the actual simulation (Magee, 2006).

Simulations have the capacity to mimic the chaotic and ambiguous environment of the real world (Magee, 2006). Simulations are more than just an interactive model or a collection of facts with which the learner interacts. It provides the framework for learners to build on their existing knowledge and augment existing cases they already have in their memory. They are an experience where learning is both interactive and dynamic. No one in the field of simulation or Aritificial Intelligence (AI) research is naïve enough to believe that they can ever completely model the real world in enough detail to replicate reality. But they do believe that the technology is becoming good enough that in a specific context they can make learners believe that they have encountered an accurate representation of reality. This belief is enough to begin thinking about how we can use simulations to learn in an authentic way. Allowing us to act virtually in a way that is similar to how we would act in the real world (Shank & Cleary, 1995).

While having a mimetic quality is not in itself what gives a simulation interesting methodological and epistemological features, it is an important sign of other features that do. The extensive use of realistic images in simulation is a stepping stone that simulation developers use in order to reveal the inferences from their data. It is also a tool they use in order to draw comparisons between simulation results and real systems; a move that is part of the process of sanctioning their results. It is the drawing of inferences and sanctioning of results that give rise to interesting philosophical connections between simulation and experimental practice.

There are various functions of simulations in science. The following are some of the main motives to run simulations in a science-teaching context (Hartmann, 1996. p. 6):

- Simulations as a technique: Investigate the detailed dynamics of a system
- Simulations as a heuristic tool: Develop hypotheses, models and theories
- Simulations as a substitute for an experiment: Perform numerical experiments
- Simulations as a tool for experimentalists: Support experiments
- Simulations as a pedagogical tool: Gain understanding of a process

Scientific simulations can be used as a technique to investigate the detailed dynamics of a system; or as a heuristic tool to develop hypotheses, models, and theories. Simulations can also be used as a tool for an experiment to perform numerical experiments or support experiments; and finally, simulations can be used as a pedagogical tool to gain understanding of a process.

In sum, these are characteristics of a digital simulation in teaching and learning: it has an adequate *model* of a complex real-world problem or situation with which the student interacts, a defined *role* with a set of available actions, a *data-rich environment* that permits a range of strategies from a variety of perspectives, *feedback* in the form of changes in the problem or situation, *embedded instructional goals*, and *mechanisms* for active participation and the promotion interest, which elicits deeper, more expedient, and better retention of understanding of a concept, mastery of a skill or strategy, or acquisition of knowledge.

DIGITAL SIMULATION AS A TOOL FOR TEACHING AND LEARNING

This section focuses on how simulation is used for teaching and learning. It highlights simulation's features to be effective for teaching and learning. It also introduces challenges to simulation to overcome its disadvantages.

Uses of Digital Simulation

Simulations have the potential to be used in several approaches to teaching and learning. For example, they can be used as a didactical tool, as models and conveyance for complex concepts, for discovery learning, and experiential learning. As a result of implementing properly designed simulation activities, the role of the teacher changes from a mere transmitter of information to a facilitator of higher-order thinking skills (Woolf & Hall, 1995).

Didactical Tool

Simulation can be used as a tool to give students concrete experience and background for abstract concepts and specific methods in stochastic systems (Rade, 1994). As Law and Kelton (1991) suggest, some of the areas which simulation has been found to be a useful and powerful tool include 'Designing and analyzing manufacturing systems', 'Determining ordering policies for an inventory system', 'Designing and operating transportation facilities such as freeways, airports, subways or ports', 'Evaluating designs for service organizations such as hospitals, post offices, or fast-food restaurants' and 'Analyzing financial and economics systems'. The main reason why simulation is used in these and other practical areas of applications are that it can be can be used to estimate unknown probabilities, distributions, and expectations.

Models and Conveyance for Complex Scientific Concepts

Technology enhanced constructivist learning currently focuses on how representations and tools can be used to mediate interactions among learners and natural or social phenomena (Dede et al., 1999). In simulations, learners experience being part of the phenomenon, and participate in a shared virtual context within which the meaning of the experience is socially constructed. By becoming a part of a phenomenon, learners gain direct experiential intuitions about how the natural world operates.

Discovery Learning

Simulation is well suited for discovery learning (de Jong & van Joolingen, 1998) in which the main task is to infer, through experimentation, characteristics of the model underlying the simulation. By "playing" with a simulation model and visualizing the results on a screen, students increase their understanding of the underlying processes and develop an intuition for what might happen in similar circumstances. Learning things this way is cheaper, safer, and faster than performing real experiments (Hartmann, 1996).

Extending Formal Instruction

On more fine-grained analysis, simulations are highly useful for experiencing something before formal instruction about it, as well as for applying learned contents after formal instruction. However, simulations by themselves are not as useful for delivering and reinforcing functions, because they are not designed to transmit knowledge to learners, nor directly reinforce their correct behavior. To fill these functions, simulations need pedagogical and perhaps game-based elements added. Simulation-based learning is said to lead to knowledge that is qualitatively different from that acquired from more traditional instruction (Nurmi, n.d.). Learning with simulations has been characterized as highly intuitive and heavily rooted in students' subjective knowledge base. Thus, it is found to be useful in formal instruction as an advance organizer and a stimulus for experimentation and reflection.

Design Features for Effectiveness

Magee (2006) argues that simulations have the following features that are useful for education: risk-free environment, experimentation, problem-solving skills, assessment, social interaction, and they appeal to the gamer culture.

Risk-Free Environment

Computer simulations offer the opportunity for "risk-free" exercise experience without the undue consequence of property damage or personal harm. Further, it is anticipated that the opportunity for practice with simulations is greater than with real systems, allowing learned skills to be more quickly reinforced (Brandenburg, 2006). A risk-free environment allows the learner to fail and then provides them with a chance to go back and modify their strategy until they have achieved a successful result. "Failure is seen as a necessary experience for learning in a simulated environment" (Aldrich, 2005, p. 136). There are two advantages to removing risk during teaching. One is the ability to improve the skills of learners in a way that does not affect actual outcomes (Walker, 1995). In situations where the student is learning skills that may affect human health and welfare the cost of failure in real life is often extremely high. Allowing them to fail occasionally in the course of a simulation will not affect the real world. The other positive effect is that learners do not come to fear failure. Failure becomes part of the learning process that will lead to their improvement in that area of knowledge, not the end of their involvement in it (Carstens & Beck, 2005).

Experimentation and Guided Discovery

Simulations can allow a learner to modify both their own behavior and the model parameters in order to observe how the simulated system changes. Most simulations are designed with a flexible architecture that allows their variables to be altered. By directly modifying a model, students can experiment with the behavior of the models in a number of different scenarios. They can then experiment with how their own behavior might change given the modified variables. The learner centric nature of the simulation makes the outcomes completely dependent on the player's actions (Aldrich, 2005).

Guided discovery learning is a learner-centered approach that combines didactic instruction with more student-centered and task-based approaches. According to Mayes (1992), computer simulation exercises based on the guided discovery learning can be designed to provide motivation, expose misconceptions and areas of knowledge deficiency, integrate information, and enhance transfer of learning. Guided discovery learning serves to focus on real problems and adds relevance and motivation to mastery of related basic information. In comparison to traditional lectures, it has the potential of greater involvement of the student in exploring the topic through self-directed learning. Thus, as Mayes (1992) identifies, computer simulation provides motivation, expose misconceptions and areas of knowledge deficiency, integrate information, and enhance transfer of learning.

Problem Solving Skills

The ultimate test of an individual's knowledge is not simply being able to repeat what they know but rather their capacity to convert or apply that knowledge into an appropriate pattern of behavior (Ruben, 1999). In conventional simulation, the goal is to focus the learner onto a specific set of problems that test their understanding about previously learned concepts (Greenblat, 1975). These are often scenario-based problems that reflect a situation the learner might encounter in the real world. The simulation presents the environment, authentic information sources, and the tools to let the student solve the problem and test their knowledge. The simulation then provides them with the feedback to modify their existing ideas and patterns of behavior they will need to navigate that environment. There is evidence that simulations enhance students' problem solving skills by giving them an opportunity to practice and refine their higher-order thinking strategies (Quinn, 1993).

Assessment and Transfer of Learning

Any simulation incorporates a model of some phenomenon or procedure, and the primary objective is for the learner to internalize that model. Therefore, how such internal models form and work is crucial to our understanding of learning from simulation. A good simulation can start the learner out with very helpful, immediate, and corrective feedback, which over time may be reduced and replaced with teacher and peer feedback. Perhaps one of the strongest aspects of simulation is its ability to evaluate if the theoretical knowledge of students can actually be enacted in a practical application. Often a passing grade in a theoretical subject is assumed to be the basis for correct decisions and problem solving in real-world situations (Jones, 1988). If an educator is interested in examining the type of knowledge reflected in the marks they could present a simulation to evaluate the competence of the student. They might then find that the student has become adept at passing standardized exams rather than gaining the ability to solve real problems with the knowledge they possess.

Another aspect of simulation is that it enhances transfer of learning. Transfer of learning could be near or far. Near transfer is such that transfer of skills and knowledge are applied the same way every time they are used. However, far transfer tasks involve skills and knowledge being applied in situations that change. An understanding of what facilitates transfer is also crucial to our understanding (Alessi & Trollip, 2001). Following the simulation experience debriefing sessions can be conducted with all participants to provide another opportunity for fast transfer of acuired knowledge. These sessions provide participants with individualized feedback on their performance and an opportunity to discuss the experience while encouraging participants to furnish feedback on how the simulation can be improved. Small group learning activities are used to promote classroom discussion about what they did, what they learned, and teaching strategies they experienced.

Social Interaction

Traditionally, learning theorists thought that knowledge was transferred from an expert to an individual. This is part of the traditional rationale for having many students taught by one teacher. But we now know that individual learners construct knowledge within a social community, which may include peers as experts. Classrooms are not the only place where these social groups and expert relationships can be formed and maintained. Simulations can allow learning to occur outside of the traditional classroom setting by interacting with students at any time in an online learning community. "It is also possible to gain social and collaborative skills by solving problems together within the simulation environment" (Ruben, 1999, p. 502).

Gamer Culture

One of the reasons games and simulations may have a much more significant impact on adult learners is that many of them are part of a new generation that has grown up on computer games. They, as digital natives, are multimedia oriented, thrive with redefined structure, and are so impatient that they are not willing to wait for right answers, rather they just apply a particular strategy and observe the result. Learning experiences that use a familiar form of media provide this group with a recognizable paradigm (Ruben, 1999). It is the same paradigm where they have already developed many of their problem solving skills through years of interaction (Magee, 2006) with games.

When used for instruction, these design features also illustrate de Jong's (1991) four factors for success: Presence of formalized, manipulable underlying models; Presence of learning goals; Elicitation of specific learning processes; Presence of learner activity, via a computational model underlying the simulation to evoke exploratory learning, and the opportunity or possibility for learner activity. These four characteristics describe the procedures of exploratory learning. In exploratory learning, planning, verifying, and monitoring are learner's main activities. These activities are supported directly and non-directly in simulation based learning.

Challenges

Even though clear advantages to using digital simulation for teaching and learning, there are many challenges including:

- The physical interface might be cumbersome. Certain digital simulations need head-mounted displays, cables, 3-D mice, and computerized clothing all can interfere with interaction, motivation, and learning.
- Digital simulations may have limited tracking ability with delayed responses.
- Feedback during play may be hard to be included in the immersive experience.
- Digital simulations may require users to switch their attention among the different senses for various tasks. In particular, multisensory inputs can result in unintended sensations and unanticipated perceptions.
- When learning in simulation, users may often feel lost. Accurately perceiving one's location in simulation is essential to both usability and learning.
- Digital simulation environments and tasks are often overwhelming for some students. Digital simulations particularly make demands on students' meta-cognitive skills, and in some cases, place students in complex environments.
- If students are not getting enough guidance or they use a simulation just for simple practicing of their skills, the simulation-based learning does not necessarily lead to a positive attitude towards the learning environment.
- Creating a digital simulation for education is costly and difficult. Simulation creation does not yet have a reliable, affordable set of software tools that can assist the teacher in creating tailor-made simulation environments. Digital simulation systems are difficult to create and maintain, and the skills needed to do so are so far outside a single teacher's usual domain of knowledge. As a result, simulations often have little adaptability; they tend to get used once and then laid aside.

As digital simulation technology evolves, some of the challenges to educational design will recede. At present, however, achieving the potential of immersive, synthetic worlds to enhance learning requires educators' understanding of its effectiveness as well as its challenges and their participation in creating simulations.

EXAMPLES OF DIGITAL SIMULATION IN TEACHING AND LEARNING

This section introduces examples of digital simulation in teaching and learning. They are 'Max Trax', 'Strategy CoPilot', 'Virtual School', 'simSchool', and 'simClass'. simClass is an experiential simulation. 'Krucible', and 'Starry Night' are examples of symbolic simulations.

Max Trax

Max Trax is a sports simulation game that focuses on numeracy skills and is delivered on CD-ROM. It is a full-on driving game, with all the speed, action and competition of a racing simulation - but with number skills built in. This simulation was designed to motivate people who are reluctant learners. The

Figure 1. Screen shot of Max Trax

Figure 3. Screen shot of Strategy CoPilot

initial impetus for the development of Max Trax was to create a product that would attract new learners (16-25) into learning, where the initial brief was to make a game with educational content or values that were recognizable as a commercial/entertainment game.

The majority of learners described Max Trax as both a game and a course. On Max Trax, most learners are able quickly to put together numeracy and driving for optimal game-playing. From the initial focus group testing of Max Trax, learners commented that the game was challenging and fun; better than normal lessons. This simulation can be reached at http://www.desq.co.uk/

Strategy CoPilot

The Strategy CoPilot was designed as a training resource for business, strategy management, and planning professionals and students. Strategy CoPilot is ideal for self-study and can be successfully introduced into facilitated workshops which combine simulation experience with reflection on the real business issues faced by participants. Imparta developed this program in consultation with the London Business School and other corporate sponsors, which provided funding. The main objective of the simulation is to steer the company back into the black through uncovering the main failings and restructuring accordingly. The learner is initially cast as a consultant brought in to advise on how the company can reinvigorate its growth and market share. The Strategy CoPilot software has a wide appeal, and is currently being used in a number of different adult learning contexts.

Users of the Strategy CoPilot system reported that the simulation was "extremely well put together and really helped me gain a better understanding of the thought process, the simulation exercise was fantastic, simulation made the theory easier to understand and more relevant, simulation enabled learning to be immediately applied," (Feedback to Imparta; J Heaford, e-mail correspondence 4 July 2006). This simulation can be reached at http://www.imparta.com/leadership/strategy_sim_video

Krucible

Krucible is an educational simulation game for learners aged 14-19 studying physics, launched in 2003. The software is aimed at subjects that are traditionally difficult to teach, and is a support for face-to-face

Figure 4. Screen shot of Krucible

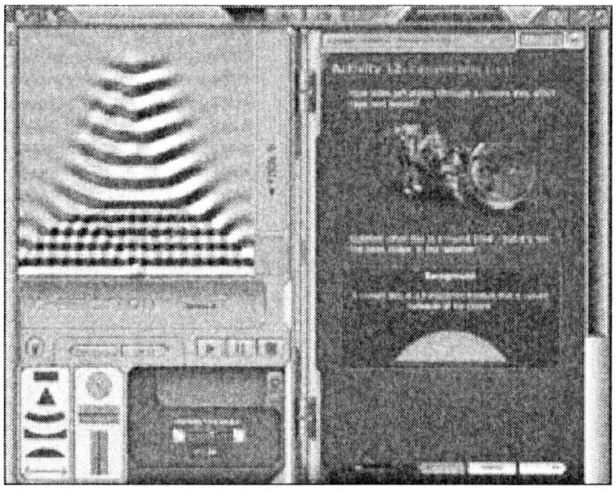

Figure 5. Screen shot of Virtual School

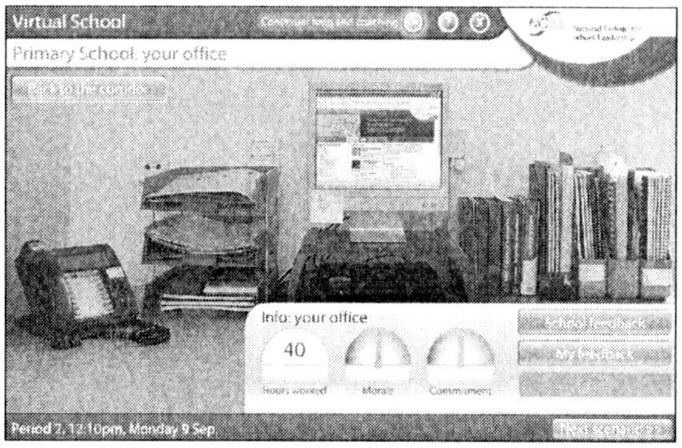

teaching, using a blended learning approach. Krucible enables difficult concepts such as waves, gravity, and terminal velocity to be explored using simulations. Users are able to control the environment and observe the effects on the object in the simulation.

One user thought that 'learning is fun, experiments can be simulated that can't be done [and] can be cheaper and quicker to simulate' (J Wright, survey correspondence 19 March 2004). He also felt that games and simulations would merge, and that they supported collaborative learning, learning with male students and problem-based learning. A science teacher from The Ockenham School in Essex, was also positive about the role that Krucible plays in his classroom activities. Information on this simulation can be found at http://education.guardian.co.uk/ evaluate/story/ 0,,1255262,00.html

Virtual School

Virtual School is an online simulated learning tool designed by the National College for School Leadership in the United Kingdom, and a game where you play as a guy named Fred who has an outstanding imagination. He writes comics and lives a normal life until he moves to Texas.

The module focuses on the professional lives of middle-level leaders, such as heads of subject or heads of year. Virtual School is a scenario-based simulation for leaders in schools and is designed to support leadership development opportunities. It was specially developed to allow learners to practice what they have learned in a non-threatening and failure-safe environment. This simulation can be reached at http://www.ncsl.org.uk/

Starry Night

Starry Night gives the user a realistic photo of the sky, not a drawing or computer simulation (Boe, 2004). For many, astronomy is more than just a matter of popping out and taking a peek at the stars - it's a serious hobby, and the Starry Night range is the software to support that legion of amateur astronomers (Clegg, n.d.). Starry Night lets users manipulate the sky. Users can discover future events and watch them before they actually happen, or replay famous events from the past. In the long run, users will gain a far better understanding of how the sky works than they can as a passive observer (Mosley, 2000).

Figure 7. Screen shot of Starry Night ("Courtesy of Starry Night")

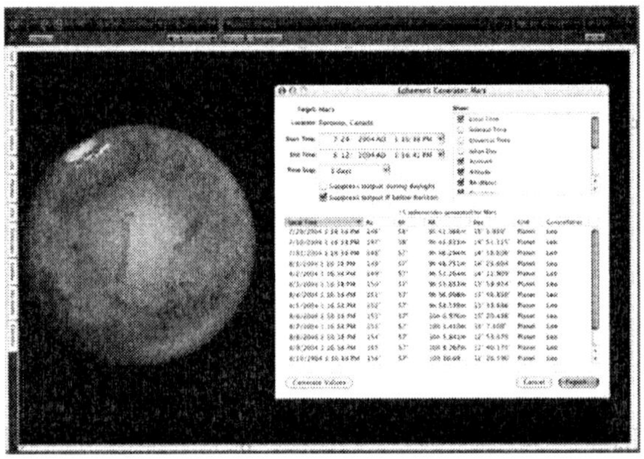

Figure 8. Screen shot of simSchool

Starry Night gives students engaging simulations and easy-to-follow lesson plans that teach the critical space science concepts. Written by teachers, for teachers, each unit includes interactive and hands-on activities that will spark your students' curiosity. This simulation can be reached at http://www.starrynight.com/

simSchool

simSchool is a classroom simulation program initially funded by the Preparing Tomorrow's Teachers to Teach with Technology (PT3) program of the U.S. Department of Education. Just as a flight-simulator immerses a player in the complexities of flying a plane, simSchool immerses novice teachers in some of

Figure 9. Screen shot of simClass

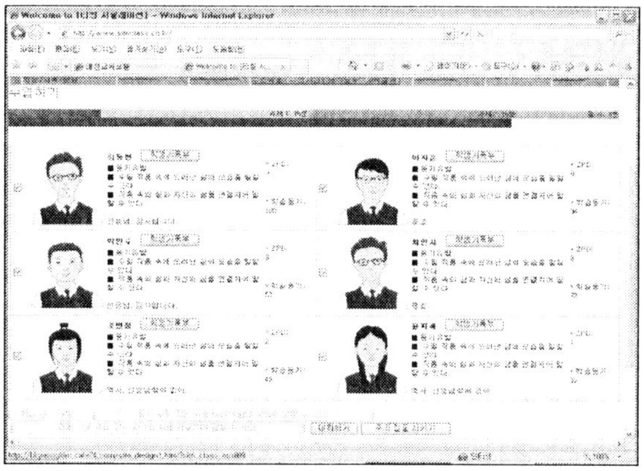

the complexities of teaching 7th-12th grade students who possess a variety of different learning characteristics and personalities. The simulation is designed to serve as a "virtual practicum" that augments teacher preparation programs by supporting the development of teaching skills prior to field experience in real classrooms (Zibit & Gibson, 2005).

simSchool offers a potentially powerful way to connect learning about teaching with practice in representative environments and situations. simSchool allows teachers and teacher educators to test out pedagogical ideas to see what combination of strategies helps all students learn. The complexities of the students and the large number of variables in the game provide a realistically complex and dynamic solution set. This simulation can be reached at http://simschool.org/

simClass

simClass is a web-based learning environment that teachers and pre-services teachers can use to practice their decision making (Kim & Kim, 2008). There are six virtual students in simClass and teachers who play this simulation have a goal to raise their achievement. Users have to design, teach, and evaluate their class considering the students' traits such as types of intelligence, motivations, and personality. Users are made aware of the effectiveness of their decisions by the responses of the virtual students, who represent boredom, challenge, anxiety, positiveness, and negativeness in verbal and nonverbal ways. At the end of the simulated class, simClass provides feedback to users about their teaching. The simulation thus provides an environment for simulating many important steps of teaching.

Teachers and pre-service teachers can improve their decision-making teaching skills by practicing teaching in a real classroom context. But this field experience can be harmful to students if novice teachers use trial and error in the classroom. Practice teaching in a simulation can be a safe way of training teachers' teaching skills, and improve their ability to make good instructional decisions. simClass provides users with opportunities for developing skills such as lesson planning, differentiating instruction, and adapting teaching to the traits of students. This simulation can be reached at: http://www.simclass.co.kr

DIGITAL SIMULATION'S ROLE IN THE FUTURE OF TEACHING AND LEARNING

This section seeks to forecast the future of teaching and learning with a focus on information technology and simulation by finding simulation's role and contribution in learning context.

Future of Teaching and Learning

Students learn in various ways; by seeing and hearing; reflecting and acting; reasoning logically and intuitively; memorizing and visualizing and drawing analogies and building. While the purpose of a teacher's activities is to achieve educational goals for each learner, teaching methods also vary considerably (Felder & Silverman, 1988). Some instructors prefer to lecture, others demonstrate or discuss; some focus on principles and others on applications; some emphasize memory and others understanding. Among the various tools of teaching, the use of technologies has been rapidly adopted for the enhancement of interactions and activities and to create a more abundant learning environment (Choi & Johnson, 2005). There can be no doubt that professional learning in all disciplines is changing fast, not only in answer to market pressures and regulatory concerns, but also in response to new technologies and the pedagogies that are being constructed around them (Barnett, 2000; Eraut, 1994; Maharg, 2006; Shaffer, 2004; Wenger, 1998).

"Simulations can be used to improve how and what children learn in the classroom with their active engagement, participation in groups, frequent interaction and feedback, and connections to real-world contexts" (Roschelle et al., 2000, p. 76). In this context, digital simulation functions as an important tool facilitating student's learning. Simulations are perceived as more interesting and motivating than many other methodologies and more like learning in the real world.

For the future, simulations should be developed and applied to develop a teacher's skill in helping students reach important learning goals in specific ways, for example by helping teachers understand and deal with managing learning activities that best match the diverse needs of their students. From management literature, we know that experienced employees learn differently than young employees. Similarly, the distinction between previous education and school education of young woman and men is just as important (Kluge, 2007). The literature on intelligence also supports the idea that there are many ways of knowing and expressing knowledge (Gardner, 2006; Sternberg & Grigorenko, 2000). Learners achieve best when their preferences as well as strengths and interests are taken into account. Educational leaders should set a goal to develop teachers who understand these issues in themselves as well as in their students.

To support this goal, simulation can be used for experiential learning by providing a rich learning environment for both teacher and learner. As implied by Richard et al. (1995), issues of simulation-based learning are as follows: first, relevance, in that the representation of real design and operational issues are the same as those faced by professionals. Second, motivation, which is made possible through the realism of a simulation providing an incentive for students to become and stay more involved in the material they are studying. Third, consolidation and integration, made possible because each simulation requires the application of multiple concepts and techniques in an integrated fashion to address a single set of issues. Finally, transfer, facilitated by the fact that simulations give students experience that can be applied to subsequent cases, other course work, and on-the-job situations.

Next-generation simulation systems for effective learning will include advanced simulation engines, which allow educators to create and modify them with confidence in student's success in simulating real situation (Hoffman, 2003). Squire (2003) said that simulations provide environments where the players can immerse themselves in a stirring education environment. The vision of the future of simulations worth working for is to improve education by stirring people to new heights of learning and performance.

Contributions of Digital Simulation

Simulations, which are used now in a broad range of teaching and training situations that vary tremendously in detail and complexity, can support future learning in several ways.

First, simulation can provide rich learning environments to support constructivist learning for teachers and learners through *learning by doing*. The requirements of the learning situation should derive from the learning context. But a digital learning environment is able to capture and manage information that comes from much more complex learning toolsets such as simulations. In particular, teaching simulations have the potential to engage future teachers in making decisions about student behavior, classroom organization and learning decisions and the impact of these decisions upon individual and collective student learning outcomes. Furthermore, researchers as well as teacher educators and K-12 classroom teachers are able to get close to the learner's experience within simulation learning environments and this allows them to understand how learners feel their way, cognitively and emotionally through learning tasks (Brookfield, 1995; Ferry et al., 2004).

Simulations interwoven with strategies needed to acquire a particular body of knowledge contain elements of traditional apprenticeship processes that encourage student observation and comment, make explicit much of the know-how acquired, and permit the participation of relatively unskilled players. For example, playing a simulation or video game might seem to be focused on the individual, but may also make use of a learning group to support decisions and provide reflection. Such a strategy emphasizes inquiry, skill development, collaboration and reflection (Tan, Turgeon, & Jonassen, 2001). A simulation aim at improving teaching can create a dynamic environment in which future teachers can make decisions concerning human development within a school or classroom setting, and the future teachers will be motivated to learn because the simulation presents an environment that is active and holds their attention (Sottile & Brozik, 2004).

Second, simulation can apply various teaching and learning models to improve higher-order thinking and classroom skills. Problem solving ability concerning student learning is considered an important competence for teachers. Just as an important mission of school education is to help students develop the knowledge, skills, and attitudes needed to deal independently with a problem, a future teacher needs those same skills. For example, practice-based approaches in curriculum design, such as problem-based learning, are being used beyond clinical disciplines and are also being shaped by new technologies, which alter how these approaches are used within curricula. In many respects simulation-based approaches are a step further towards real practice situations. Many institutions are implementing virtual learning environment (VLE) platforms such as WebCT, Blackboard, Moodle, ATutor, ILIAS, to name some of the common methods for delivering e-learning (Maharg & Owen, 2007). Owen (2000) suggests that a professional practice learning system has demands that go beyond traditional VLEs:

- The whole of an activity needs to be considered in the implementation, even if the implementation is not all based in information technology.

- There are needs for conversations to take place at many levels - in role play, in seeking support, in reflection, in assessment, and in feedback.
- There are needs to sustain creative and generative activities as well as responsive ones
- There are needs to work around and share boundary objects (see Brown & Duguid, 2000), and to base the system around human activities and action as well as the need to recall and record.

Third, simulation can help improve teaching skills of pre-service teachers. Much of pre-service teacher education at the undergraduate level is based on indirect links between those preparing to teach and those who spend their days in schools (Sather, 2007). Simulations provide directly linked opportunities for pre-service teacher to explore constructed environments that practice specific skills, and engage in problem solving in which teaching and learning based on virtual environment (Brent, 1997; Brown, 1999; Ferry et al., 2004; Zibit & Gibson, 2005). In this context, Kervin et al. (2005) said that a simulated classroom environment provided teachers with more opportunity to make decisions, to try out different approaches and closely monitor the impact of decisions upon students than would otherwise have been afforded in a regular classroom practicum experience.

Previously, the prevailing teaching methodology was that of classroom lecture and discussion, but today one can find individual analysis of group roles, group dynamics, individual and group decision making, role-playing, and other human development skill exercises used in teacher education college classrooms (Sottile & Brozik, 2004). Teacher education programs need to consider how pre-service teachers can be supported in both their understanding of how children best learn literacy practices and what the teaching may look like in actual classroom practice (Kervin et al., 2005). A well-designed simulation can improve decision-making and critical thinking skills as well as teaching discipline-specific concepts.

Finally, simulation can be used as a tool to support lifelong learning. The European Commission (2003) has defined the meaning of lifelong learning as all learning activity undertaken throughout life, with the aim of improving knowledge, skills and competence, within a personal, civic, social and/or employment-related perspective. Human beings learn throughout their lives and in almost all situations at home, leisure activities and work. We start learning even before birth, and we continue until senility. Some of this learning is incidental and largely unconscious, but a large amount of learning is planned and purposive (Tough, 1971). People need and desire to constantly enhance their knowledge and skills for the sake of their professional or personal development or for problem solving in both areas. Learning is a process of mental and social change of an entire lifetime.

In the future, learners will not be bound to particular locations (Abfalter et al., 2004). New technologies such as simulations can offer the opportunity to learn and study at anytime and anywhere in different ways, according to the user's demands. Simulation can provide a collaborative environment rather than competitive learning environment and can promote cooperative learning as well as the exchange and critical discussion of ideas with others for lifelong learning.

REFERENCES

Abfalter, D., Mirski, P. J., & Hitz, M. (2004). *Mobile learning-knowledge enhancement and vocational training on the move*. Paper presented at the 5th European Conference on Organizational Knowledge, Learning and Capabilities, Innsbruck, Austria

Aldrich, C. (2004). *Clark Aldrich's six criteria of an educational simulation.* Retrieved Feb 1, 2008, from http://www.learningcircuits.org/NR/rdonlyres/F2ED000A-7A59-4108-A6CB- 1BE4F4CC1CA5/ 4719/ clark_e2.pdf

Aldrich, C. (2005). *Learning by doing: A comprehensive guide to simulations, computer games, and pedagogy in e-learning and other educational experiences.* San Francisco: Pfeiffer.

Alessi, S. M., & Trollip, S.R. (2001). *Multimedia for learning: Methods and development* (3rd ed.). Needham Height Massachusett: Allyn and Bacon.

Baek, Y. (2008). Implications of educational digital game structure for use in formal education settings. *Proceedings of the 19th Society for Information Technology & Teacher Education International Conference* (pp. 1613-1619), Las Vegas, USA.

Barles, J., Dras, M., Kavakli, M., Richards, D. & Tychsen, A. (2006). An overview of training simulation research and systems. *Proceedings of the AAMAS 2005 conference workshop on Autonomous Agents & Multi Agent Systems.* Utrecht University, Netherlands.

Barton, K., & Maharg, P. (2006) E-simulations in the wild: interdisciplinary research, design and implementation of simulation environments in legal education, in D. Gibson, C. Aldrich C, M. Prensky (eds), *Games and Simulations in Online Learning: Research and Development Frameworks* (pp. 115-148), Hershey, PA: Idea Group Inc.

Barnett, R. (2000). *Realising the university in an age of Supercomplexity.* Buckingham: Society for Education Research and Open University Press.

Boe, R. (2004). *Starry night enthusiast review.* Retrieved March 31, 2008, from http://www.cloudynights.com/documents/starry.pdf

Brandenburg, D. C. (2006). *Simulation and learning in disaster preparedness: A Research and Theory Review.* Paper presented at the Academy of Human Resource Development International Conference. Columbus, OH.

Brent, R. (1997). Simulated opening of school. *Proceedings of the 8th International Conference of the Society for Information Technology and Teacher Education* (pp.981-983). *Orlando, Fla.*

Brookfield, S. D. (1995). *Becoming a critically reflective teacher.* San Francisco: Jossey-Bass.

Brown, A. (1999). Simulated teaching and artificial students: The potential effects of new technologies on teacher education. *Journal of Research on Computing in Education, 32*(2), 307-318.

Brown, A. H. (1999). Simulated classrooms and artificial students: The potential effects of new technologies on teacher education. *Research on Computing in Education, 32*(2), 307-318.

Brown, J. S., & Duguid, P. (2000). *The social life of information.* Boston, MASS: Harvard Business School Press.

Carstens, A., & Beck, J. (2005). Get ready for the gamer generation. *TechTrends, 49*(3), 22-25.

Choi, H. J., & Johnson, S. D. (2005). The effect of context-based video instruction on learning and motivation in online courses. *The American Journal of Distance Education, 19*(4), 215–227.

Clegg, B. (n.d.). *Software review-starry night: Pro plus.* Retrieved March 31, 2008, from http://www.popularscience.co.uk/software/rev18.htm

D'Augustine, A., & Charks, H. (1973). *Multiple methods of teaching mathematics in elementary schools.* New York: Harper and Row Publishers.

Dede, C., Salzman, M. C., Loftin, R. B., & Sprague, D. (1999). Multisensory immersion as a modeling environment for learning complex scientific concepts. In W. Feurzeig, & N. Roberts (Eds.). *Computer Modeling and Simulation in Science Education* (pp.282-319). NY: Springer-Verlag.

de Jong, T. (1991). Learning and instruction with computer simulations. *Education & Computing, 6,* 217-229.

de Jong, T., & van Joolingen, W. R. (1998). Scientific discovery learning with computer simulations of conceptual domains. *Review of Educational Research, 68*(2), 179-201.

Dwerryhouse, R. (2001). Real work in the 16-19 curriculum: AVCE Business and Young Enterprise. *Education and Training, 43*(3), 153-161.

Eraut, M (1994). *Developing professional knowledge and competence.* London: Falmer Press.

EUROPA (2003). *What Is lifelong learning?.* Retrieved Mar 31, 2008, from http://europa.eu.int/comm/education/policies/lll/life/what_islll_en.html

Felder, R. M., & Silverman, L. K. (1988). Learning and teaching styles in engineering. *Engineer Education, 78*(7), 674–681.

Ferry, B., Kervin, L, Cambourne, B., Turbill, J., Puglisi, S., Jonassen, D., & Hedberg, J. (2004). Online classroom simulation: The next wave for pre-service teacher education? In R. Atkinson, C. McBeath,. D. Jonas-Dwyer, & R. Phillips (Eds.), Beyond the comfort zone: *Proceedings of the 21st ASCILITE Conference* (pp. 294-302). Perth, Western Australia.

Franklin, T. (2008). Panel presentation enriching STEM education through games and simulations. *Proceedings of the 19th Society for Information Technology & Teacher Education International Conference* (pp. 1631-1637), Las Vegas, USA.

Galarneau, L. (2005). *Authentic learning experiences through play: games, simulations and the construction of knowledge.* Retrieved April 1, 2008, from http://www.digra.org/ dl/order_by_author?publication=Changing%20Views:%20Worlds%20in%20Play.

Gardner, H. (2006) *Changing Minds. The art and science of changing our own and other people's minds.* Boston, MA: Harvard Business School Press.

Gredler, M. E. (1996). Educational games and simulations: A technology in search of a paradigm. In D. H. Jonassen (Ed.), *Handbook of research for educational communications and technology* (pp. 521-540). New York: Simon and Schuster Macmillan.

Gredler, M. E. (2004). Games and simulations and their relationships to learning. In Jonassen. D. H. (Ed.), *Handbook of research for educational communications and technology* (2nd ed., pp. 571-581). Mahwah, NJ: Lawrence Erlbaum Associates.

Greenblat, C. S. (1975). Basic concepts and linkages basic concepts and linkages. In C. S. Greenblat & R. D. Duke (Eds.), *Gaming-simulation: Rationale, design and applications: A text with parallel readings for social scientists, educators, and Community Workers* (pp. 180-195). Toronto: Wiley & Sons.

Hartmann, S. (1996). The world as a process: Simulations in the natural and social sciences, in R. Hegselmann et al (Eds.), *Modelling and simulation in the social sciences from the philosophy of science point of view. Theory and decision library* (pp. 77-100), Dordrecht: Kluwer.

Hoffman, T. (2003). Simulations revitalize e-Learning. *Computerworld, 37*, 31.

Hopewell, T. M. (2008). Elaboration, practice, and attention in simulation design. *Proceedings of the 19th Society for Information Technology & Teacher Education International Conference* (pp. 1675-1679), Las Vegas, USA.

Hovenga, E. J., & Bricknell, L. (2004). Current and future trends in teaching and learning. In E. J. S. Hovenga & J. Matas (Eds.). *Global health informatics education* (pp. 131-142), Rockhampton, Australia: IOS Press.

Hubal, R.C., Helms, R.F., & Triplett, S.E. (1997). *Advanced learning environments*. Paper presented at the 19th Interservice/Industry Training, Simulation and Education Conference. Orlando, FL.

Humphreys, P. (1991). Computer simulations. In Fine, M. Forbes & L. Wessels (Eds.), *PSA, 2*, 497-506. East Lansing: Philosophy of Science Association.

Jones, K. (1988). Why gamesters die in space. In C. D., K. J.H.G.., C. A. & A. Delle Piane (Eds.), Simulation-gaming in education and training: *Proceedings of the International Simulation and Gaming Association's 18th International Conference* (pp. 33-37). Toronto: Pergamon Press.

Kervin, L., Turbill, J., Cambourne, B., & Ferry, B. (2005). Simulation technology in pre-service teacher education: Pleasurable learning to inspire passionate teaching. *Proceedings of the Pleasure, Passion Provocation AATE/ALEA National Conference*, Broadbeach, Australia.

Key Skills Trainer. (2005). *Tour guide: Understanding application of number and Communication* (pp.13-48). Retrieved March 24, 2008, from http://www.keyskills4u.com/ files/trainer/ Tutor_guide9_13.pdf

Kim, B. K. (2004). Activity theory as a theoretical framework for understanding activities in educational games. *Korean Journal of Educational Research, 42*(2), 459-483.

Kim, B., & Kim, J. (2008a). Effects of using types of teaching simulation: Teacher-guided use and self-guided use. *Proceedings of the 19th Society for Information Technology & Teacher Education International Conference* (pp. 1721-1725), Las Vegas, USA.

Kim, B., & Kim, J. (2008b). Effects of using types of teaching simulation: Teacher-guided use and self-guided use. *Proceedings of the 19th Society for information Technology and Teacher Education International Conference* (pp. 1721-1725), Las Vegas, USA.

Klein, J. D., & Doran, M. S. (1999). Implementing individual and small group learning structures with a computer simulation. *Educational Technology Research and Development, 47*(1), 97-109.

Kluge, A. (2007). Experiential learning methods, simulation complexity and their effects on different target groups. *Journal of educational computing research, 36*(3), 27.

Kolb, D. A. (1983). *Experiential learning: Experience as the source of learning and development*, Upper Saddle River, NJ: Prentice-Hall PTR.

Law, A.M., & Kelton, W. D. (1991). *Simulation modeling and analysis.* New York: McGraw-Hill.

MacDonald, C. J. (2008). The more the world changes the more the education system remains the same: Addressing the need to use emerging technologies to enhance online learning communities. *Proceedings of the Society for Information Technology and Teacher Education International Conference* (pp. 3047-3054). Las Vegas, NV, USA.

Magee, M. (2006). *State of the field review: Simulation in education final report.* Calgary: Alberta Online Learning Consortium. Retrieved March 15, 2008, from http://www.ccl-cca.ca/NR/rdonlyres/C8CB4C08-F7D3-4915-BDAA-C41250A43516/0/SFRSimulationinEducationJul06REV.pdf

Maharg, P. (2006). Authenticity in learning: Transactional learning in virtual communities. *Innovating E-Learning 2006: Transforming Learning Experiences*, Retrieved March 20, 2008, from http://www.jisc.ac.uk/elp_conference06.html

Maharg, P. (2006). On the edge: ICT and the transformation of professional legal learning. *Web Journal of Current Legal Issues, 3, Special Issue on Legal Education.* Retrieved Mar 31, 2008, from http://webjcli.ncl.ac.uk/2006/issue3/maharg3.html

Maharg, P., & Owen, M. (2007). Simulations, learning and the metaverse: Changing cultures in legal education. *Journal of Information law and technology* (pp 1-15). Retrieved Mar 28, 2008, from http://www2.warwick.ac.uk/fac/soc/law/elj/jilt/2007_1/maharg_owen

Mayes, R. L. (1992). The effects of using software tools on mathematical problem solving in secondary schools. *School Science and Mathematics, 92*(5), 243-248.

Menn, D. (1993). Multimedia in education: Arming our kids for the future. *PC World, 11*(10), 52-28.

Mosley, J. (2000). *Starry night companion.* NY: SPACE.com.

Nurmi, S. (n.d.). *Simulations and learning.* Retrieved March 11, 2008, from http://www.eun. org/insight-pdf/ernist/Q4_2%20_Long_%20answer_Are_%20simulations_%20 useful_%20for_%20learning.pdf

Ormrod, J. E. (2004). *Human learning (4th ed.).* Upper Saddle River, New Jersey: Pearson Education.

Owen, M. (2000). Paradigms for curriculum design: The design of reflective, situated, collaborative professional development supported by virtual learning environments. *European Journal of Open and Distance Learning.* Retrieved Nomber 24, 2006, from http://www.eurodl.org/materials/ contrib/2000/owen/owen.html

Park, H., Jung, J., & Chris, C. (2008). Effect of the activities in the second life® as tool for mutual understanding of gender role. *Proceedings of the 19th Society for information Technology and Teacher Education International Conference* (pp. 1778-1783), Las Vegas, USA.

Prensky, M. (2001). *Digital game-based learning.* New York: McGraw-Hill.

Psotka, J. (1995). Immersive training systems: Virtual reality and education and training. *Instructional Science, 23*(3), 405-431.

Quinn, C. N. (1993). Cognitive skills and computers: Framing the link. *Proceedings of the Fifth International Conference on Thinking*, Townsville, Australia.

Rade, L. (1994). *Simulation and the student engineer.* Paper presented at the Proceedings of the First Scientific Meeting (of the IASE), Universiti di Perugia, Italy.

Ralstone, A., & Reilly, E. D. (Eds.). (1983). *Encyclopedia of computer science and engineering.* New York: Van Nostrand Reinhold.

Reigeluth, C. M., & Schwartz, E. (1989). An instructional theory for the design of computer-based simulations. *Journal of Computer-Based Instruction, 16*(1), 1-10.

Richards, L. G., Gorman, M., Scherer, W. T., & Landel, R. D. (1995). Promoting active learning with cases and instructional modules. *Journal of Engineering Education, 84*(5), 375-381.

Roschelle, J. M., Pea, R. D., Hoadley, C. M., Gordin, D. N., & Means, B. N. (2000). Changing how and what children learn in school with computer-based technologies. *The Future of Children and Computer Technology, 10*(2), 76-101.

Ruben, B. D. (1999). Simulations, games, and experience-based learning: The quest for a new paradigm for teaching and learning. *Simulation & Gaming, 30*(4), 498-505.

Sanders, R. T., & McKeown, L. (2007). *Promoting reflection through action learning in a 3D virtual world.* Paper presented at the Association of Library and Information Science Educators Annual Conference, Seattle, WA.

Sather, A. C. (2007). Direct links: Using e-mail to connect pre service teachers, experienced teachers, and high school students within an undergraduate teacher preparation program. *Journal of Technology and Teacher Education, 15*(1), 11-37.

Schmucker, K. (1999). A taxonomy of simulation software. *Learning Technology Review*, 40-75.

Shank, R., & Cleary, C. (1995). *Engines for education.* Hillsdale, NJ: Erlbaum Associates.

Shaffer, D. W. (2004). Pedagogical praxis: The professions as models for postindustrial education. *Teachers College Record, 106*(7), 1401-1421.

Shaffer, D. W. (2006). *How computer games help children learn.* New York: Palgrave Macmillan.

Shirts, R. G. (1975). Notes on defining simulation. In C. S. Greenblat & R. D. Duke (eds.), *Gaming-simulation: Rationale, design and applications. A text with parallel readings for social scientists, educators, and community workers* (pp. 75-81). Toronto: Wiley & Sons.

Sottile, J. M., & Brozik, D. (2004). *The use of simulations in a teacher education program: The impact on student development.* Paper presented at the 2004 Hawaii International Conference on Education.. Honolulu, HI.

Speelman, P. & Gore, D. (2008). IT Proposal-simulation project as a higher order thinking technique for instruction. In K. McFerrin et al (Eds.), *Proceedings of the 19th Society for Information Technology and Teacher Education International Conference 2008* (pp. 1813-1817). Chesapeake, VA: AACE.

Squire, K. (2003). Video games in education. *International Journal of Simulations and Gaming, 2*(1), 49-62.

Starry Night (n. d.). *Prepare your students for NCLB science testing.* Retrieved March 31, 2008, from http://www.starrynight.com/images/store/edu06brochure.pdf

Sternberg, R. J., & Grigorenko, E. L. (2000). *Teaching for successful intelligence.* Arlington Heights, IL: Skylight Training and Publishing Inc.

Swaak, J., Van Joolingen, W. R., & de Jong, T. (1998). Supporting simulation-based learning: The effects of model progression and assignments on definitional and intuitive knowledge. *Learning and Instruction, 8*(3), 235-252.

Tan , S. C., Turgeon, A. J., & Jonassen, D. H. (2001). Computer-supported collaborative argumentation: An innovative approach to group problem solving. *Journal of Natural Resources and Life Sciences Education, 30*, 54-61.

Tough, A. (1971). *The adult's learning projects.* Toronto, Ontario: Institute for Studies in Education.

Towne, D. (1995). *Learning and instruction in simulation environments.* New York: Springer- Verlag.

van Joolingen, W.R., & de Jong, T. (1991). Characteristics of simulations for instructional settings. *Education & Computing, 6*, 241-262.

Vanlehn, K., Ohlsson, S., & Nason, R. (1994). Applications of simulated students: An exploration. *Journal of Artificial Intelligence in Education, 5*(2), 135-175.

Vogel, J. J., Vogel, D. S., Bowers, J. C., & Bowers, C. A. (2006). Computer gaming and interactive simulations for learning: A meta-analysis. *Journal of Educational Computing Research, 34*(3), 339-243.

Walker, W. (1995). *The use of scenarios and gaming in crisis management planning and training.* Santa Monica: Rand.

Waller, V. (2007). Academics in the arcade. *Journal of British Library Direct,* 35-39.

Wenger, E. (1998). *Communities of practice: Learning, meaning and identity.* Cambridge: Cambridge University Press.

Windschitl, M., & Andre, T. (1998). Using computer simulations to enhance conceptual change: The roles of constructivist instruction and student epistemological beliefs. *Journal of Research in Science Teaching, 35*(2), 145-60.

Winsberg, E. (1999). Sanctioning models: The epistemology of simulation. *Journal of Science in Context, 12*(2), 275-292.

Winsberg, E. (2003). Simulated experiments: Methodology for a virtual world. *Philosophy of Science, 70*, 105-125.

Wolfram, S. (1994). *Cellular automata and complexity.* Boston: Addison-Wesley.

Woolf, B., & Hall, W. (1995). Multimedia pedagogues: Interactive systems for teaching and learning. *IEEE Multimedia*, 74-80.

Zibit, M., & Gibson, D. (2005). simSchool: The game of teaching. *Journal of online education, 1*(6). Retrieved April 1, 2008, from http://www.innovateonline.info

WEBSITES

http://www.aqua-pacific.com/education.htm

http://www.betterstrategy.com/pdfs/Accelerated%20learning%20with%20Strategy%20CoPilot.pdf

http://celsim.com/information/strategy-copilot.html?Itemid=47

http://education.guardian.co.uk/evaluate/story/0,,1255262,00.html

http://www.gamediscovery.com/category/sim/ideas/virtual-school.html

http://www.immersiveeducation.com/krucible/

http://www.imparta.com/__data/assets/pdf_file/2516/Strategy_CoPilot.pdf

http://www.imparta.com/resource_centre/course_areas

http://www.keyskills4u.com/

http://www.keyskills4u.com/trainer/tutorguide/using.htm

http://www.keyskillssupport.net/

http://www.literacytrust.org.uk/Database/keyskills.html

http://mmdesign.co.uk/reach/back_issues/2002/409whtml.htm

http://www.ncsl.org.uk/

http://www.rm.com/_RMVirtual/Media/Downloads/RMSecondaryCatalogue_June.pdf

http://www.sepup.com/TG/SALI/AppendixF.pdf

http://www.simClass.co.kr

http://simSchool.org/

http://skillswise.org.uk/

http://www.wordnet-online.com/simulation.shtml

KEY TERMS

Computer Simulation: A computer simulation is a computer program that attempts to simulate an abstract model of a particular system.

Digital Game: Digital game is a form of game written in a computer language.

Experiential Learning: Experiential learning is a process through which a student develops knowledge, skills, and values from direct experiences. It is a model that views learning as a cyclical process in four stages: concrete experience, reflective observation, abstract conceptualization, and active experimentation. Experiential learning relates to participants' activities and reactions to a training event, in contrast to passive learning.

Fidelity: Fidelity is a notion that at its most abstract level implies a truthful connection to a source or sources.

Game: A game is a structured or semi-structured activity, usually undertaken for enjoyment. Key components of games are goals, rules, challenge, and interactivity.

Game Based Learning: Game based learning (GBL) is a type of learning which uses a game as a tool for students to engage in learning while they are playing.

Guided Discovery Learning: Guided discovery learning is a learner-centered approach that combines didactic instruction with more student-centered and task-based approaches. Guided discovery learning is superior to the less-structured approach of pure discovery learning in promoting learning and knowledge transfer.

Self-Directed Learning: Self-directed learning is a form of self-education in which learners have the primary responsibility for establishing objectives, planning, carrying out, and evaluating their own learning experiences.

Simulation: A simulation is an imitation of some real thing, state of affairs, or process. The act of simulating something generally entails representing certain key characteristics or behaviors of a selected physical or abstract system.

This work was previously published in Digital Simulations for Improving Education: Learning Through Artificial Teaching Environments, edited by D. Gibson and Y. Baek, pp. 25-51, copyright 2009 by Information Science Reference (an imprint of IGI Global).

Appendix A

Barbara A. Soloman
North Carolina State University, USA

Richard M. Felder
North Carolina State University, USA

INDEX OF LEARNING STYLES QUESTIONNAIRE

Directions
Please provide us with your full name. Your name will be printed on the information that is returned to you.

Full Name

For each of the 44 questions below select either "a" or "b" to indicate your answer. Please choose only one answer for each question. If both "a" and "b" seem to apply to you, choose the one that applies more frequently. When you are finished selecting answers to each question please select the submit button at the end of the form.

I understand something better after I

○ **(a)** try it out.
○ **(b)** think it through.

I would rather be considered
- ○ **(a)** realistic.
- ○ **(b)** innovative.

When I think about what I did yesterday, I am most likely to get
- ○ **(a)** a picture.
- ○ **(b)** words.

I tend to
- ○ **(a)** understand details of a subject but may be fuzzy about its overall structure.
- ○ **(b)** understand the overall structure but may be fuzzy about details.

When I am learning something new, it helps me to
- ○ **(a)** talk about it.
- ○ **(b)** think about it.

If I were a teacher, I would rather teach a course
- ○ **(a)** that deals with facts and real life situations.
- ○ **(b)** that deals with ideas and theories.

I prefer to get new information in
- ○ **(a)** pictures, diagrams, graphs, or maps.
- ○ **(b)** written directions or verbal information.

Once I understand
- ○ **(a)** all the parts, I understand the whole thing.
- ○ **(b)** the whole thing, I see how the parts fit.

In a study group working on difficult material, I am more likely to
- ○ **(a)** jump in and contribute ideas.
- ○ **(b)** sit back and listen.

I find it easier
- ○ **(a)** to learn facts.
- ○ **(b)** to learn concepts.

In a book with lots of pictures and charts, I am likely to
- ○ **(a)** look over the pictures and charts carefully.
- ○ **(b)** focus on the written text.

When I solve math problems
- ○ **(a)** I usually work my way to the solutions one step at a time.
- ○ **(b)** I often just see the solutions but then have to struggle to figure out the steps to get to them.

In classes I have taken
- ○ **(a)** I have usually gotten to know many of the students.
- ○ **(b)** I have rarely gotten to know many of the students.

In reading nonfiction, I prefer
- ○ **(a)** something that teaches me new facts or tells me how to do something.
- ○ **(b)** something that gives me new ideas to think about.

I like teachers
- ○ **(a)** who put a lot of diagrams on the board.
- ○ **(b)** who spend a lot of time explaining.

When I'm analyzing a story or a novel
- ○ **(a)** I think of the incidents and try to put them together to figure out the themes.
- ○ **(b)** I just know what the themes are when I finish reading and then I have to go back and find the incidents that demonstrate them.

When I start a homework problem, I am more likely to
- ○ **(a)** start working on the solution immediately.
- ○ **(b)** try to fully understand the problem first.

I prefer the idea of
- ○ **(a)** certainty.
- ○ **(b)** theory.

I remember best
- ○ **(a)** what I see.
- ○ **(b)** what I hear.

It is more important to me that an instructor
- ○ **(a)** lay out the material in clear sequential steps.
- ○ **(b)** give me an overall picture and relate the material to other subjects.

I prefer to study
- ○ **(a)** in a study group.
- ○ **(b)** alone.

I am more likely to be considered
- ○ **(a)** careful about the details of my work.
- ○ **(b)** creative about how to do my work.

When I get directions to a new place, I prefer
- ○ **(a)** a map.
- ○ **(b)** written instructions.

I learn

- **(a)** at a fairly regular pace. If I study hard, I'll "get it."
- **(b)** in fits and starts. I'll be totally confused and then suddenly it all "clicks."

I would rather first

- **(a)** try things out.
- **(b)** think about how I'm going to do it.

When I am reading for enjoyment, I like writers to

- **(a)** clearly say what they mean.
- **(b)** say things in creative, interesting ways.

When I see a diagram or sketch in class, I am most likely to remember

- **(a)** the picture.
- **(b)** what the instructor said about it.

When considering a body of information, I am more likely to

- **(a)** focus on details and miss the big picture.
- **(b)** try to understand the big picture before getting into the details.

I more easily remember

- **(a)** something I have done.
- **(b)** something I have thought a lot about.

When I have to perform a task, I prefer to

- **(a)** master one way of doing it.
- **(b)** come up with new ways of doing it.

When someone is showing me data, I prefer

- **(a)** charts or graphs.
- **(b)** text summarizing the results.

When writing a paper, I am more likely to

- **(a)** work on (think about or write) the beginning of the paper and progress forward.
- **(b)** work on (think about or write) different parts of the paper and then order them.

When I have to work on a group project, I first want to

- **(a)** have "group brainstorming" where everyone contributes ideas.
- **(b)** brainstorm individually and then come together as a group to compare ideas.

I consider it higher praise to call someone

- **(a)** sensible.
- **(b)** imaginative.

When I meet people at a party, I am more likely to remember
- ○ **(a)** what they looked like.
- ○ **(b)** what they said about themselves.

When I am learning a new subject, I prefer to
- ○ **(a)** stay focused on that subject, learning as much about it as I can.
- ○ **(b)** try to make connections between that subject and related subjects.

I am more likely to be considered
- ○ **(a)** outgoing.
- ○ **(b)** reserved.

I prefer courses that emphasize
- ○ **(a)** concrete material (facts, data).
- ○ **(b)** abstract material (concepts, theories).

For entertainment, I would rather
- ○ **(a)** watch television.
- ○ **(b)** read a book.

Some teachers start their lectures with an outline of what they will cover. Such outlines are
- ○ **(a)** somewhat helpful to me.
- ○ **(b)** very helpful to me.

The idea of doing homework in groups, with one grade for the entire group,
- ○ **(a)** appeals to me.
- ○ **(b)** does not appeal to me.

When I am doing long calculations,
- ○ **(a)** I tend to repeat all my steps and check my work carefully.
- ○ **(b)** I find checking my work tiresome and have to force myself to do it.

I tend to picture places I have been
- ○ **(a)** easily and fairly accurately.
- ○ **(b)** with difficulty and without much detail.

When solving problems in a group, I would be more likely to
- ○ **(a)** think of the steps in the solution process.
- ○ **(b)** think of possible consequences or applications of the solution in a wide range of areas.

When you have completed filling out the above form please click on the Submit button below. Your results will be returned to you. If you are not satisfied with your answers above please click on Reset to clear the form.

Submit Reset

Dr. Richard Felder, felder@ncsu.edu

Appendix B

Visual Basic Sample Source Codes – Ver. 1

Login Screen

```
Dim WithEvents Con As ADODB.Connection
Dim WithEvents rst As ADODB.Recordset
Dim Cmd As ADODB.Command
Dim strResponse As String

Private Sub cmdlog_Click()
   Call IdProcess
End Sub

Private Sub IdProcess()
If txtId.Text = "" And txtName.Text = "" Then

   strResponse = MsgBox("Please insert your name and ID first.", vbOKOnly + vbExclamation, _
              "Note")
   Else

   txtId.Text = Trim(txtId.Text)
   txtName.Text = Trim(txtName.Text)
   txtId.Text = UCase(txtId.Text)

   If txtId.Text Like "[A-Z][A-Z][0-9][0-9][0-9][0-9][0-9]" Then

   stra = txtId.Text
   strb = txtName.Text
   ValTime = Time
   ValResult = 0

   Load Q1Objective
   Q1Objective.Visible = True
   Unload Main
   Unload Me
   vStep = 0

   Else

   strResponse = MsgBox("Please insert your ID correctly.(eg AB12345)", vbOKOnly + vbExclamation, _
          "Note")
   End If
End If
End Sub
```

Question 1 – Objective (opening screen)

```
Private Sub cmd1_Click()
   Load Main
   Main.Visible = True
   Unload Me
   vStep = 0
End Sub

Private Sub cmd2_Click()
   Load Q1intro
   Q1intro.Visible = True
   Unload Me
   vStep = 0
End Sub

Private Sub cmdex_Click()
   Dim fx As Form
For Each fx In Forms
  Unload fx
  Set fx = Nothing
Next
End Sub

Private Sub cmdh_Click()
   Me.PopupMenu Q1Menu.helpm, 2
End Sub

Private Sub cmds_Click()
   Dim voice As Boolean
   voice = vOnOff
   Me.PopupMenu Q1Menu.Sound, 2
   If voice <> vOnOff Then
   End If
End Sub

Private Sub cmdt_Click()
   Me.PopupMenu Q1Menu.Tool, 2
End Sub

Private Sub Form_Load()
   CreateRoundRectFromWindow Me
   StopPlaying 2
   SetPos 2, 0
   MediaPlayer1.Open App.Path & "\mov\babbit.avi"
End Sub

Private Sub Form_MouseDown(Button As Integer, Shift As Integer, X As Single, Y As Single)

   Select Case Button
      Case vbLeftButton
         'Allow the form to be moved by dragging
         Call ReleaseCapture
         Call SendMessage(Me.hwnd, WM_NCLBUTTONDOWN, HTCAPTION, 0&)
   End Select

End Sub
```

Question 1 & Solution 1

```
Option Explicit
Dim n, p As Integer
```

Appendix B

```
Dim q As Integer

Private Sub cmd1_Click()
lblc.ForeColor = RGB(0, 0, 0)
lblc.Caption = "The correct answer is a = dv/dt ."
   p = 3
   q = 7
   CkProcess p
   cmd1.Visible = False
   n = 0
   txt1.Text = ""
End Sub

Private Sub cmd2_Click()
   Load Q1Question
   Q1Question.Visible = True
   Unload Me
   vStep = 0
End Sub

Private Sub cmd3_Click()
   Load Q1Sulotion2
   Q1Sulotion2.Visible = True
   Unload Me
End Sub

Private Sub cmda_Click()
Trim (txt1.Text)
If txt1.Text = "" Then
   lblc.ForeColor = RGB(0, 0, 0)
   lblc.Caption = "Please insert the value in first."
ElseIf txt1.Text = "dv/dt" Then
   lblc.ForeColor = RGB(0, 0, 255)
   lblc.Caption = "That's right. a = dv/dt ."
   ValResult = ValResult + 1
   q = 2
   p = 2
   cmd1.Visible = False
   CkProcess p
Else
   n = n + 1
   If (n = 1) Then
   lblc.ForeColor = RGB(255, 0, 0)
   lblc.Caption = "No,this is wrong answer,Do you want some hint."
   q = 3
   ckvoice
   cmdb.Visible = True
   ElseIf (n = 2) Then
   lblc.ForeColor = RGB(255, 0, 0)
   lblc.Caption = "No,the answer is wrong."
   q = 4
   ckvoice
   ElseIf (n >= 3) Then
   lblc.ForeColor = RGB(255, 0, 0)
   lblc.Caption = "No,the answer is wrong, Click'on solve for correct answer"
   q = 5
   ckvoice
   cmd1.Visible = True
   End If
End If
End Sub

Private Sub cmdb_Click()
lblc.ForeColor = RGB(0, 0, 0)
```

```
lblc.Caption = "Plot velocity - time graph ."
q = 6
ckvoice
End Sub

Private Sub cmdex_Click()
   Dim fx As Form
For Each fx In Forms
  Unload fx
  Set fx = Nothing
Next
End Sub

Private Sub cmdh_Click()
 Me.PopupMenu Q1Menu.helpm, 2
End Sub

Private Sub cmdmap_Click()
   Load CharacterMap
   CharacterMap.Visible = True
   CharacterMap.SetFocus
End Sub

Private Sub cmds_Click()
   Dim voice As Boolean
   voice = vOnOff
   Me.PopupMenu Q1Menu.Sound, 2
   If voice <> vOnOff Then
   ckvoice
   End If
End Sub

Private Sub cmdt_Click()
   Me.PopupMenu Q1Menu.Tool, 2
End Sub



Private Sub Form_MouseDown(Button As Integer, Shift As Integer, X As Single, Y As Single)

   Select Case Button
      Case vbLeftButton
         'Allow the form to be moved by dragging
         Call ReleaseCapture
         Call SendMessage(Me.hwnd, WM_NCLBUTTONDOWN, HTCAPTION, 0&)
   End Select

End Sub
Private Sub Form_Load()
 StopPlaying 2
 SetPos 2, 0
 CreateRoundRectFromWindow Me

   ShockwaveFlash1.Loop = False
   ShockwaveFlash1.Zoom (100)
   ShockwaveFlash1.Movie = App.Path & "\swf\new1.swf"



 lbl3.Caption = " STEP1 : To plot v-t graph" & vbCrLf & vbCrLf & _
            " Given"

 lbl4.Caption = " Recall back kinematics"
```

```
 CkProcess p
 lblc.Caption = ""

End Sub

Private Sub CkProcess(int1 As Integer)
   Static Solution As Integer
   Solution = int1
   Select Case Solution
'==================================
      Case Is = 0
      q = 0
      lbl4.Visible = False '0
      img02.Visible = False '0
      cmd3.Visible = False '0
      If vOnOff = False Then
      GoTo Process1
      End If
      ckvoice
      p = 1
      q = 1
'==================================
      Case Is = 1
Process1:
      q = 1
      lbl4.Visible = False '0
      img02.Visible = False '0
      cmd3.Visible = False '0
      fra1.Visible = True '1
      If vOnOff = True Then
       ckvoice
      End If
'==================================
      Case Is = 2
      If vOnOff = True Then
       ckvoice
      Else
      GoTo Process2
      End If
      p = 3
      q = 7
'==================================
      Case Is = 3
Process2:
      q = 7
      If vOnOff = True Then
      ckvoice
      Else
      GoTo Process3
      End If
      lbl4.Visible = True '3
      img02.Visible = True '3
      cmd3.Visible = True '3
      fra1.Visible = False '3
      p = 4
      q = 8
'==================================
      Case Is = 4
Process3:
      q = 8
      If vOnOff = True Then
      ckvoice
      End If
      lbl4.Visible = True '3
```

```
      img02.Visible = True '3
      cmd3.Visible = True '3
      fra1.Visible = False '3
   End Select
End Sub

Private Sub ckvoice()
 Form1.List(2).ListItems.Clear
 If vOnOff = False Then
 StopPlaying 2
 SetPos 2, 0
 Exit Sub
 Else
 Timer1.Enabled = False
 Timer2.Enabled = False
 vTimer1 = 0
 vTimer2 = 0
 Select Case q
 '=================================================
 Case Is = 0
  vTime = 54
  Timer1.Enabled = True
  Form1.AddtoList 2, App.Path & "\wav\lstep1.mp3"
  Form1.PlaySelected 2
 '=================================================
 Case Is = 1
  vTime = 10
  Timer2.Enabled = True
  Form1.AddtoList 2, App.Path & "\wav\lstep1.1.mp3"
  Form1.PlaySelected 2
 '=================================================
 Case Is = 2
  vTime = 7
  Timer1.Enabled = True
  Form1.AddtoList 2, App.Path & "\wav\lright.mp3"
  Form1.PlaySelected 2
 '=================================================
 Case Is = 3
  vTime = 8
  Timer2.Enabled = True
  Form1.AddtoList 2, App.Path & "\wav\l1sttrial.mp3"
  Form1.PlaySelected 2
 '=================================================
 Case Is = 4
  vTime = 7
  Timer2.Enabled = True
  Form1.AddtoList 2, App.Path & "\wav\l2ndtrial.mp3"
  Form1.PlaySelected 2
 '=================================================
 Case Is = 5
  vTime = 11
  Timer2.Enabled = True
  Form1.AddtoList 2, App.Path & "\wav\3rdtrial.mp3"
  Form1.PlaySelected 2
 '=================================================
 Case Is = 6
  'missing
 '=================================================
 Case Is = 7
  vTime = 6
  Timer1.Enabled = True
  Form1.AddtoList 2, App.Path & "\wav\lstep1.3.mp3"
  Form1.PlaySelected 2
 '=================================================
```

```
 Case Is = 8
 vTime = 20
 Timer2.Enabled = True
 Form1.AddtoList 2, App.Path & "\wav\Ibutton3.mp3"
 Form1.PlaySelected 2
 End Select
 End If
End Sub


Private Sub Image1_Click()

End Sub

Private Sub Timer1_Timer()
   Static tmr As Integer
   tmr = vTimer1
   vTimer1 = vTimer1 + 1
   If (tmr = vTime) Or (vOnOff = False) Then
   tmr = 0
   Timer1.Enabled = False
   StopPlaying 2
   SetPos 2, 0
   CkProcess p
   End If
End Sub

Private Sub Timer2_Timer()
   Static tmr As Integer
   tmr = vTimer2
   vTimer2 = vTimer2 + 1
   If (tmr = vTime) Or (vOnOff = False) Then
   tmr = 0
   Timer1.Enabled = False
   StopPlaying 2
   SetPos 2, 0
   End If
End Sub
```

Question 1 – Answer

```
Dim WithEvents Con As ADODB.Connection
Dim WithEvents rst As ADODB.Recordset
Dim Cmd As ADODB.Command
Dim strcol As String
Dim p As Integer
Dim q As Integer

Private Sub cmd2_Click()
   Load Q1Sulotion6
   Q1Sulotion6.Visible = True
   Unload Me
End Sub

Private Sub cmd3_Click()
   Unload Me
End Sub

Private Sub cmdclose_Click()
   Dim fx As Form
```

```
For Each fx In Forms
  Unload fx
  Set fx = Nothing
Next
End Sub

Private Sub cmdd_Click()
   Load Flash3
   Flash3.Visible = True
   Unload Me
End Sub

Private Sub cmdex_Click()
   Dim fx As Form
For Each fx In Forms
  Unload fx
  Set fx = Nothing
Next
End Sub

Private Sub cmdgo_Click()
Select Case cbo1.Text
Case "To main menu"
 Load Main
 Main.Visible = True
 Unload Me

Case "To solve v-t"
 Load Q1Sulotion1
 Q1Sulotion1.Visible = True
 Unload Me

Case "Plot v-t graph"
 Load Flash1
 Flash1.Visible = True
 Unload Me

Case "To solve s-t"
 Load Q1Sulotion5
 Q1Sulotion5.Visible = True
 Unload Me

Case "Plot s-t graph"
 Load Flash2
 Flash2.Visible = True
 Unload Me
End Select

End Sub

Private Sub cmdgraph_Click()
Load Graph
Graph.Visible = True
Unload Me
End Sub

Private Sub cmdh_Click()
 Me.PopupMenu Q1Menu.helpm, 2
End Sub

Private Sub cmds_Click()
   Dim voice As Boolean
   voice = vOnOff
   Me.PopupMenu Q1Menu.Sound, 2
```

```
    If voice <> vOnOff Then
    ckvoice
    End If
End Sub

Private Sub cmdt_Click()
    Me.PopupMenu Q1Menu.Tool, 2
End Sub

Private Sub Form_MouseDown(Button As Integer, Shift As Integer, X As Single, Y As Single)

    Select Case Button
        Case vbLeftButton
            'Allow the form to be moved by dragging
            Call ReleaseCapture
            Call SendMessage(Me.hwnd, WM_NCLBUTTONDOWN, HTCAPTION, 0&)
    End Select

End Sub
Private Sub Form_Load()
     StopPlaying 2
     SetPos 2, 0
    CreateRoundRectFromWindow Me
    Form1.List(2).ListItems.Clear

    strcol = "^|    | 0 <= t <= 10s | 10s < t < 60s    | Units"
    With MSFlexGrid1
        .Cols = 5
        .Rows = 4
        .FormatString = strcol
        .TextMatrix(1, 1) = " a"
        .TextMatrix(2, 1) = " v"
        .TextMatrix(3, 1) = " s"
        .TextMatrix(1, 2) = "    10 "
        .TextMatrix(2, 2) = "    10t "
        .TextMatrix(3, 2) = "     5t "
        .TextMatrix(1, 3) = " 2 "
        .TextMatrix(2, 3) = " -2t + 120 "
        .TextMatrix(3, 3) = " -t" & Chr(178) & " + 120t -600 "
        .TextMatrix(1, 4) = « m/s» & Chr(178)
        .TextMatrix(2, 4) = « m/s»
        .TextMatrix(3, 4) = « m»
    End With
cbo1.Text = «To solve v-t»
p = 0
CkProcess p
CkData
End Sub

Private Sub CkProcess(int1 As Integer)
    Static Solution As Integer
    Solution = int1
    Select Case Solution
        Case Is = 0
         q = 0
    If vOnOff = False Then
         GoTo Process1
    End If
         ckvoice
         p = 1
         q = 1
        '====================
        Case Is = 1
Process1:
```

```
        q = 1
        If vOnOff = True Then
        ckvoice
        End If
    End Select
End Sub

Private Sub ckvoice()
 Form1.List(2).ListItems.Clear
 If vOnOff = False Then
 StopPlaying 2
 SetPos 2, 0
 Else
 Timer1.Enabled = False
 Timer2.Enabled = False
Select Case q
 Case Is = 0
  vTime = 7
  Form1.AddtoList 2, App.Path & «\wav\table.mp3»
  Form1.PlaySelected 2
  vTimer1 = 0
  Timer1.Enabled = True
 Case Is = 1
  vTime = 11
  Form1.AddtoList 2, App.Path & «\wav\graphs.mp3»
  Form1.PlaySelected 2
  vTimer2 = 0
  Timer2.Enabled = True
End Select
  End If
End Sub

Private Sub Timer1_Timer()
    Static tmr As Integer
    tmr = vTimer1
    vTimer1 = vTimer1 + 1
    If (tmr = vTime) Or (vOnOff = False) Then
    tmr = 0
    Timer1.Enabled = False
    StopPlaying 2
    SetPos 2, 0
    CkProcess p
    End If
End Sub

Private Sub Timer2_Timer()
Static tmr As Integer
    tmr = vTimer2
    vTimer2 = vTimer2 + 1
    If (tmr = vTime) Or (vOnOff = False) Then
    tmr = 0
    Timer2.Enabled = False
    StopPlaying 2
    SetPos 2, 0
    End If
End Sub

Private Sub CkData()
Dim ret As Boolean
Dim strtemp As String

Set Con = New ADODB.Connection

Con.CursorLocation = adUseClient
```

```
Con.Open «Provider=Microsoft.Jet.OLEDB.3.51;Persist Security Info=False;Data Source=» & App.Path & «\database.mdb»

DataEnvironment1.Connections(2).Open Con

Set rst = New ADODB.Recordset

   rst.ActiveConnection = Con
   rst.CursorLocation = adUseClient
   rst.CursorType = adOpenDynamic
   rst.LockType = adLockOptimistic
   rst.Open «select * from student «

   On Error GoTo ErrMsg

   ret = FindFirst(rst, «[ID]='» & stra & «'»)

   '==================================================
   If Not ret Then
   'Add New Record
   With rst
   .AddNew
   .Fields(0) = StrConv(stra, vbProperCase)
   .Fields(1) = StrConv(strb, vbProperCase)
   .Fields(2) = StrConv(ValResult, vbProperCase)
   .Fields(3) = ValTime
   .Update
   End With
   GoTo Process1
   End If
   '==================================================

   ret = FindFirst(rst, «[Time]='» & ValTime & «'»)

   '==================================================
   If Not ret Then
   Set rst = Nothing
   'Reset
   Set rst = New ADODB.Recordset
   With rst
   .ActiveConnection = Con
   .CursorLocation = adUseClient
   .CursorType = adOpenDynamic
   .LockType = adLockPessimistic
   .Open «select * from student where ID='» & stra & «'»
   End With
   'Update Record
   strtemp = str(ValResult)
   strtemp = rst.Fields(2).value & «,» & Trim(strtemp)
   With rst
   .Fields(1) = StrConv(strb, vbProperCase)
   .Fields(2) = StrConv(strtemp, vbProperCase)
   .Fields(3) = ValTime
   .Update
   End With
   End If
   '==================================================

Process1:
   rst.Close
   Set rst = Nothing
   Con.Close
   Set Con = Nothing
   DataEnvironment1.Connections(2).Close
   Exit Sub
```

```
ErrMsg:
   MsgBox Err.Description
End Sub
```

Question 1 & Solution 2

```
Dim str1, str2, str3, str4 As String
Dim n, M, p As Integer
Dim q As Single

Private Sub cmd1_Click()
   lblc.ForeColor = RGB(0, 0, 0)
   lblc.Caption = ""
   p = 7
   q = 6
   CkProcess p
   cmd1.Visible = False
End Sub

Private Sub cmd2_Click()
   Load Q1Sulotion1
   Q1Sulotion1.Visible = True
   Unload Me
End Sub

Private Sub cmd3_Click()
   Load Q1Sulotion3
   Q1Sulotion3.Visible = True
   Unload Me
End Sub

Private Sub cmd5_Click()
   lblc.ForeColor = RGB(0, 0, 0)
   lblc.Caption = "The correct answer is v = 100 ."
   p = 6
   CkProcess p
   cmd5.Visible = False
End Sub

Private Sub cmda_Click()
Trim (txtx(0).Text)
Trim (txty(0).Text)
Trim (txtm(0).Text)
Trim (txtn(1).Text)
Trim (txtx(1).Text)
Trim (txty(1).Text)
Trim (txtm(1).Text)
If txtm(0).Text = "" And txtm(1).Text = "" Then
   lblc.ForeColor = RGB(0, 0, 0)
   lblc.Caption = "Please insert the value in first."
ElseIf txtx(0).Text = "v" And (txty(0).Text = "v=0" Or txty(0).Text = "0") _
      And txtm(0).Text = "dv" And txtn(1).Text = "10" And txtx(1).Text = "t" _
      And (txty(1).Text = "t=0" Or txty(1).Text = "0") And txtm(1).Text = "dt" Then
   lblc.ForeColor = RGB(0, 0, 255)
   lblc.Caption = "That's right."
   ValResult = ValResult + 1
   cmd1.Visible = False
   p = 2
   q = 2
   CkProcess p
Else
```

```
    n = n + 1
    If (n = 1) Then
    lblc.ForeColor = RGB(255, 0, 0)
    lblc.Caption = "Try again."
    q = 3
    ckvoice
    ElseIf (n >= 2) Then
    If txtx(0).Text <> "v" Or txty(0).Text <> "v=0" Or txtx(1).Text <> "t" Or txty(1).Text <> "t=0" Then
    lblc.ForeColor = RGB(255, 0, 0)
    lblc.Caption = "The limit bounds are wrong, do you wish to try again or click'on solve for correct answer"
    q = 4
    ckvoice
    Else
    lblc.ForeColor = RGB(255, 0, 0)
    lblc.Caption = "No,the answer is wrong, do you wish to try again or click'on solve for correct answer"
    q = 5
    ckvoice
    End If
    cmd1.Visible = True
    End If
End If
End Sub

Private Sub cmdb_Click()
Trim (txt3.Text)
If txt3.Text = "" Then
    lblc.ForeColor = RGB(0, 0, 0)
    lblc.Caption = "Please insert the value in first."
ElseIf txt3.Text = "100" Then
    lblc.ForeColor = RGB(0, 0, 255)
    lblc.Caption = "That's right. v=100 ."
    ValResult = ValResult + 1
    p = 5
    q = 9
    CkProcess p
Else
    M = M + 1
    If (M = 1) Then
    lblc.ForeColor = RGB(255, 0, 0)
    lblc.Caption = "No,this is wrong answer."
    q = 10
    ckvoice
    ElseIf (M = 2) Then
    lblc.ForeColor = RGB(255, 0, 0)
    lblc.Caption = "No,the answer is wrong."
    q = 11
    ckvoice
    ElseIf (M >= 3) Then
    lblc.ForeColor = RGB(255, 0, 0)
    lblc.Caption = "No,the answer is wrong, do you wish to try again or click'on solve for correct answer"
    p = 12
    ckvoice
    cmd5.Visible = True
    End If
End If
End Sub

Private Sub cmdequ1_Click()
Dim hMenu As Long
Dim hSubMenu As Long
Dim lngID As Long
Dim i As Integer

    hMenu = GetMenu(Q1Menu.hwnd)
```

```
    hSubMenu = GetSubMenu(hMenu, 0)
    hSubMenu = GetSubMenu(hSubMenu, 0)
For i = 0 To 2
    lngID = GetMenuItemID(hSubMenu, i)
    Q1Menu.Pic(i).Picture = Q1Menu.Pic(i).Image
    Call ModifyMenu(hMenu, lngID, 4, lngID, CLng(Q1Menu.Pic(i).Picture))
Next i

Q1Sulotion2.PopupMenu Q1Menu.s21, 2
End Sub

Private Sub cmdex_Click()
    Dim fx As Form
For Each fx In Forms
   Unload fx
   Set fx = Nothing
Next
End Sub

Private Sub cmdequ2_Click()
Dim hMenu As Long
Dim hSubMenu As Long
Dim lngID As Long
Dim i As Integer

    hMenu = GetMenu(Q1Menu.hwnd)
    hSubMenu = GetSubMenu(hMenu, 0)
    hSubMenu = GetSubMenu(hSubMenu, 1)
For i = 0 To 2
    lngID = GetMenuItemID(hSubMenu, i)
    Q1Menu.Pic(i).Picture = Q1Menu.Pic(i).Image
    Call ModifyMenu(hMenu, lngID, 4, lngID, CLng(Q1Menu.Pic(i).Picture))
Next i


Q1Sulotion2.PopupMenu Q1Menu.s22, 2
End Sub

Private Sub cmdh_Click()
 Me.PopupMenu Q1Menu.helpm, 2
End Sub

Private Sub cmdmap_Click()
    Load CharacterMap
    CharacterMap.Visible = True
    CharacterMap.SetFocus
End Sub

Private Sub cmds_Click()
    Dim voice As Boolean
    voice = vOnOff
    Me.PopupMenu Q1Menu.Sound, 2
    If voice <> vOnOff Then
    ckvoice
    End If
End Sub

Private Sub cmdt_Click()
    Me.PopupMenu Q1Menu.Tool, 2
End Sub

Private Sub cmdundo_Click()
    cmdequ1.Visible = True
    cmdequ2.Visible = True
```

```
    imgInt1(0).Visible = False
    imgInt1(1).Visible = False
    txtx(0).Visible = False
    txty(0).Visible = False
    txtm(0).Visible = False
    txtn(0).Visible = False
    txtx(1).Visible = False
    txty(1).Visible = False
    txtm(1).Visible = False
    txtn(1).Visible = False
    txtx(0).Text = «»
    txty(0).Text = «»
    txtm(0).Text = «»
    txtn(0).Text = «»
    txtx(1).Text = «»
    txty(1).Text = «»
    txtm(1).Text = «»
    txtn(1).Text = «»
End Sub

Private Sub Form_MouseDown(Button As Integer, Shift As Integer, X As Single, Y As Single)

    Select Case Button
        Case vbLeftButton
            'Allow the form to be moved by dragging
            Call ReleaseCapture
            Call SendMessage(Me.hwnd, WM_NCLBUTTONDOWN, HTCAPTION, 0&)
    End Select

End Sub
Private Sub Form_Load()
 StopPlaying 2
 SetPos 2, 0
CreateRoundRectFromWindow Me

str1 = «  STEP2 :Consider time interval « & vbCrLf & _
        vbCrLf & vbCrLf & vbCrLf & vbCrLf
str4 = «  Statement : « & vbCrLf & _
        «   In Order to obtain the velocity equation of the car for time « & vbCrLf & _
        «   t=0 to t=10s ; we have to integrate Equation1»
' bold str 1
str2 = vbCrLf & vbCrLf & vbCrLf & vbCrLf & vbCrLf & vbCrLf & _
        «   - This is a linear equation.»

str3 = vbCrLf & vbCrLf & _
        "   When t =  0 ; v =  0" & vbCrLf & _
        "   When t = 10 ; v = 100"
CkProcess p
lbl5.Caption = "   When t =  0 ; v = 0 " & vbCrLf & "   When t = 10 ; v = "
lblc.Caption = """

End Sub

Private Sub CkProcess(int1 As Integer)
    Static Solution As Integer
    Solution = int1
    Select Case Solution
'===================================
        Case Is = 0
         q = 0
         lbl3.Caption = str1 '0
         fra1.Visible = False '0
         cmd3.Visible = False '0
         ShockwaveFlash1.Visible = False '0
```

```
        If vOnOff = False Then
        GoTo Process1
        End If
        ckvoice
        p = 1
        q = 1
`==================================
        Case Is = 1
Process1:
        q = 1
        lbl3.Caption = str1 & str4 `1
        fra1.Visible = True `1
        cmd3.Visible = False `0
        ShockwaveFlash1.Visible = False `0
        If vOnOff = True Then
        ckvoice
        End If
`==================================
        Case Is = 2
        If vOnOff = False Then
        GoTo Process2
        End If
        ckvoice
        p = 7
        q = 6
`==================================
        Case Is = 7
Process2:
        q = 6
        lbl3.Caption = str1 & str4 & str2 `7
        fra1.Visible = False `7
        cmd3.Visible = False `0
        ShockwaveFlash1.Visible = True `7
        ShockwaveFlash1.Loop = False `7
        ShockwaveFlash1.Zoom (100) `7
        ShockwaveFlash1.Movie = App.Path & «\swf\s21.swf» `7
        ShockwaveFlash1.Rewind `7
        ShockwaveFlash1.Play `7
        If vOnOff = False Then
        GoTo Process3
        End If
        ckvoice
        p = 3
        q = 7
`==================================
        Case Is = 3
Process3:
        q = 7
        lbl3.Caption = str1 & str4 & str2 `7
        fra1.Visible = False `7
        cmd3.Visible = False `0
        ShockwaveFlash1.Visible = True `7
        ShockwaveFlash1.Loop = False `7
        ShockwaveFlash1.Zoom (100) `7
        ShockwaveFlash1.Movie = App.Path & «\swf\s21.swf» `7
        ShockwaveFlash1.Rewind `7
        ShockwaveFlash1.Play `7
        If vOnOff = False Then
        GoTo Process4
        End If
        ckvoice
        p = 4
        q = 8
`==================================
```

```
        Case Is = 4
Process4:
        q = 8
        lbl3.Caption = str1 & str4 & str2
        lbl3.Caption = str1 & str4 & str2 '7
        fra1.Visible = False '7
        fra2.Visible = True '4
        cmd3.Visible = False '0
        ShockwaveFlash1.Visible = True '7
        ShockwaveFlash1.Loop = False '7
        ShockwaveFlash1.Zoom (100) '7
        ShockwaveFlash1.Movie = App.Path & «\swf\s21.swf» '7
        ShockwaveFlash1.Rewind '7
        ShockwaveFlash1.Play '7
        If vOnOff = True Then
        ckvoice
        End If
'==================================
        Case Is = 5
        If vOnOff = False Then
        GoTo Process5
        End If
        ckvoice
        p = 6
'==================================
        Case Is = 6
Process5:
        lbl3.Caption = str1 & str4 & str2 & str3 '6
        fra1.Visible = False '7
        fra2.Visible = False '6
        cmd3.Visible = True
        ShockwaveFlash1.Visible = True '7
        ShockwaveFlash1.Loop = False '7
        ShockwaveFlash1.Zoom (100) '7
        ShockwaveFlash1.Movie = App.Path & «\swf\s21.swf» '7
        ShockwaveFlash1.Rewind '7
        ShockwaveFlash1.Play '7
    End Select

End Sub

Private Sub ckvoice()
 Form1.List(2).ListItems.Clear
 If vOnOff = False Then
 StopPlaying 2
 SetPos 2, 0
 Else
 Timer1.Enabled = False
 Timer2.Enabled = False
 vTimer1 = 0
 vTimer2 = 0
Select Case q
 '==================================================
 Case Is = 0
  vTime = 23
  Form1.AddtoList 2, App.Path & «\wav\Istep2.mp3»
  Form1.PlaySelected 2
  Timer1.Enabled = True
 '==================================================
 Case Is = 1
  vTime = 28
  Timer2.Enabled = True
  Form1.AddtoList 2, App.Path & «\wav\Istatement.mp3»
  Form1.PlaySelected 2
```

```
`=================================================
Case Is = 2
 vTime = 7
 Form1.AddtoList 2, App.Path & «\wav\lright.mp3»
 Form1.PlaySelected 2
 Timer1.Enabled = True
`=================================================
Case Is = 3
 vTime = 5
 Timer2.Enabled = True
 Form1.AddtoList 2, App.Path & «\wav\wro.mp3»
 Form1.PlaySelected 2
`=================================================
Case Is = 4
 vTime = 15
 Timer2.Enabled = True
 Form1.AddtoList 2, App.Path & «\wav\ltry3.mp3»
 Form1.PlaySelected 2
`=================================================
Case Is = 5
 vTime = 11
 Timer2.Enabled = True
 Form1.AddtoList 2, App.Path & «\wav\3rdtrial.mp3»
 Form1.PlaySelected 2
`=================================================
Case Is = 6
 vTime = 22
 Form1.AddtoList 2, App.Path & «\wav\lstep2.3.mp3»
 Form1.PlaySelected 2
 Timer1.Enabled = True
`=================================================
Case Is = 7
 vTime = 8
 Form1.AddtoList 2, App.Path & «\wav\lstep2.4.mp3»
 Form1.PlaySelected 2
 Timer1.Enabled = True
`=================================================
Case Is = 8
 vTime = 15
 Timer2.Enabled = True
 Form1.AddtoList 2, App.Path & «\wav\lstep2.6.mp3»
 Form1.PlaySelected 2
`=================================================
Case Is = 9
 vTime = 7
 Form1.AddtoList 2, App.Path & «\wav\lright.mp3»
 Form1.PlaySelected 2
 Timer1.Enabled = True
`=================================================
Case Is = 10
 vTime = 4
 Timer2.Enabled = True
 Form1.AddtoList 2, App.Path & «\wav\l1stchoice.mp3»
 Form1.PlaySelected 2
`=================================================
Case Is = 11
 vTime = 7
 Timer2.Enabled = True
 Form1.AddtoList 2, App.Path & «\wav\l2ndtrial.mp3»
 Form1.PlaySelected 2
`=================================================
Case Is = 12
 vTime = 11
 Timer2.Enabled = True
```

```
  Form1.AddtoList 2, App.Path & «\wav\3rdtrial.mp3»
  Form1.PlaySelected 2
End Select
  End If
End Sub

Private Sub Timer1_Timer()
    Static tmr As Integer
    tmr = vTimer1
    vTimer1 = vTimer1 + 1
    If (tmr = vTime) Or (vOnOff = False) Then
    tmr = 0
    Timer1.Enabled = False
    StopPlaying 2
    SetPos 2, 0
    CkProcess p
    End If
End Sub

Private Sub Timer2_Timer()
    Static tmr As Integer
    tmr = vTimer2
    vTimer2 = vTimer2 + 1
    If (tmr = vTime) Or (vOnOff = False) Then
    tmr = 0
    Timer2.Enabled = False
    StopPlaying 2
    SetPos 2, 0
    End If
End Sub

Private Sub Timer3_Timer()
    Static tmr As Integer
    tmr = vTimer3
    vTimer3 = vTimer3 + 1
    If (tmr = vTime) Or (vOnOff = False) Then
    tmr = 0
    Timer3.Enabled = False
    StopPlaying 2
    SetPos 2, 0
    CkProcess p
    End If
End Sub
```

Question 1 & Solution 3

```
Dim str1, str2, str3 As String
Dim n, M, p As Integer
Dim q As Long

Private Sub cmd1_Click()
    p = 5
    CkProcess p
    lblc.ForeColor = RGB(0, 0, 0)
    lblc.Caption = ""
    cmd1.Visible = False
End Sub

Private Sub cmd2_Click()
    Load Q1Sulotion2
    Q1Sulotion2.Visible = True
    Unload Me
```

```
End Sub

Private Sub cmd3_Click()
   Load Q1Sulotion4
   Q1Sulotion4.Visible = True
   Unload Me
End Sub

Private Sub cmda_Click()
If Opt1(0).value = False And Opt1(1).value = False And Opt1(2).value = False Then
 lblc.ForeColor = RGB(0, 0, 0)
 lblc.Caption = "Please select a value."
Else
 If Opt1(0).value = True Then
  lblc.ForeColor = RGB(0, 0, 255)
  lblc.Caption = "That's Right."
  ValResult = ValResult + 1
  q = 2
 Else
  lblc.ForeColor = RGB(255, 0, 0)
  lblc.Caption = "No,this is wrong answer,we suppose to intergrate."
  q = 3
 End If
   p = 2
   CkProcess p
End If

End Sub

Private Sub cmdb_Click()
Trim (txt1.Text)
Trim (txt2.Text)
Trim (txt3.Text)
Trim (txt4.Text)
If txt1.Text = "" Or txt2.Text = "" Or txt3.Text = "" Or txt4.Text = "" Then
   lblc.ForeColor = RGB(0, 0, 0)
   lblc.Caption = "Please insert the value in first."
ElseIf txt1.Text = "v" And (txt2.Text = "v=100" Or txt2.Text = "100") And _
              txt3.Text = "t" And (txt4.Text = "t=10" Or txt4.Text = "10") Then
   lblc.ForeColor = RGB(0, 0, 255)
   lblc.Caption = "That's right. "
   ValResult = ValResult + 1
   p = 4
   q = 4
   CkProcess p
   cmd1.Visible = False
Else
   M = M + 1
   If (M = 1) Then
   lblc.ForeColor = RGB(255, 0, 0)
   lblc.Caption = "The upper and lower limit for dv and dt is wrong"
   q = 5
   ckvoice
   ElseIf (M >= 2) Then
   lblc.ForeColor = RGB(255, 0, 0)
   lblc.Caption = "Do you with to try again or click on solve for answer"
   cmd1.Visible = True
   q = 6
   ckvoice
   End If
End If
End Sub
```

Appendix B

```
Private Sub cmdex_Click()
  Dim fx As Form
For Each fx In Forms
  Unload fx
  Set fx = Nothing
Next
End Sub

Private Sub cmdh_Click()
 Me.PopupMenu Q1Menu.helpm, 2
End Sub

Private Sub cmdmap_Click()
  Load CharacterMap
  CharacterMap.Visible = True
  CharacterMap.SetFocus
End Sub

Private Sub cmds_Click()
  Dim voice As Boolean
  voice = vOnOff
  Me.PopupMenu Q1Menu.Sound, 2
  If voice <> vOnOff Then
  ckvoice
  End If
End Sub

Private Sub cmdt_Click()
  Me.PopupMenu Q1Menu.Tool, 2
End Sub

Private Sub Form_MouseDown(Button As Integer, Shift As Integer, X As Single, Y As Single)

  Select Case Button
    Case vbLeftButton
      'Allow the form to be moved by dragging
      Call ReleaseCapture
      Call SendMessage(Me.hwnd, WM_NCLBUTTONDOWN, HTCAPTION, 0&)
  End Select

End Sub
Private Sub Form_Load()
 StopPlaying 2
 SetPos 2, 0
CreateRoundRectFromWindow Me



str1 = "  STEP3 - 1 :" & vbCrLf & _
    "  Consider " & vbCrLf & vbCrLf & _
    "     Given              ; as found when " & _
    vbCrLf & vbCrLf
str3 = "   Kinematic :"
'bold str1
str2 = vbCrLf & vbCrLf & vbCrLf & vbCrLf & _
    "  What will be the next step ..?" & vbCrLf & _
    "    -Integration."

CkProcess p
lblc.Caption = ""

End Sub
```

```
Private Sub CkProcess(int1 As Integer)
  Static Solution As Integer
  Solution = int1
  Select Case Solution
'==================================
    Case Is = 0
    q = 0
    lbl3.Caption = str1 '0
    img3.Visible = False '0
    img5.Visible = False '0
    fra1.Visible = False '0
    cmd3.Visible = False '0
    If vOnOff = False Then
     GoTo Process1
     End If
     ckvoice
     p = 1
     q = 1
'==================================
    Case Is = 1
Process1:
    q = 1
    lbl3.Caption = str1 & str3 '1
    img3.Visible = True '1
    img5.Visible = False '0
    fra1.Visible = True '1
    cmd3.Visible = False '0
    If vOnOff = True Then
    ckvoice
    End If
'==================================
    Case Is = 2
     If vOnOff = False Then
     GoTo Process2
     End If
     ckvoice
     p = 3
'==================================
    Case Is = 3
Process2:
    lbl3.Caption = str1 & str3 & str2 '3
    img3.Visible = True '1
    img5.Visible = False '0
    fra1.Visible = False '3
    fra2.Visible = True '3
    cmd3.Visible = False '0
'==================================
    Case Is = 4
     If vOnOff = False Then
     GoTo Process3
     End If
     ckvoice
     q = 7
     p = 5
'==================================
    Case Is = 5
Process3:
    q = 7
    lbl3.Caption = str1 & str3 & str2 '3
    img3.Visible = True '1
    img5.Visible = True '5
    fra1.Visible = False '3
    fra2.Visible = False '5
    cmd3.Visible = True '5
```

```
      If vOnOff = True Then
       ckvoice
      End If
   End Select

End Sub

Private Sub ckvoice()
 Form1.List(2).ListItems.Clear
 If vOnOff = False Then
 StopPlaying 2
 SetPos 2, 0
 Else
 Timer1.Enabled = False
 Timer2.Enabled = False
 vTimer1 = 0
 vTimer2 = 0
Select Case q
 Case Is = 0
  vTime = 23
  Form1.AddtoList 2, App.Path & "\wav\lstep3.mp3"
  Form1.PlaySelected 2
  Timer1.Enabled = True
 '==================================
 Case Is = 1
  vTime = 21
  Timer2.Enabled = True
  Form1.AddtoList 2, App.Path & "\wav\lstep3.1.mp3"
  Form1.PlaySelected 2
  '==================================
 Case Is = 2
  vTime = 5
  Form1.AddtoList 2, App.Path & "\wav\correct.mp3"
  Form1.PlaySelected 2
  Timer1.Enabled = True
  '==================================
 Case Is = 3
  vTime = 5
  Form1.AddtoList 2, App.Path & "\wav\wro.mp3"
  Form1.PlaySelected 2
  Timer1.Enabled = True
  '==================================
  Case Is = 4
  vTime = 7
  Form1.AddtoList 2, App.Path & "\wav\lright.mp3"
  Form1.PlaySelected 2
  Timer1.Enabled = True
  '==================================
 Case Is = 5
  vTime = 5
  Timer2.Enabled = True
  Form1.AddtoList 2, App.Path & "\wav\lwrong3.mp3"
  Form1.PlaySelected 2
  '==================================
 Case Is = 6
  vTime = 9
  Timer2.Enabled = True
  Form1.AddtoList 2, App.Path & "\wav\lcorrect3.mp3"
  Form1.PlaySelected 2
  '==================================
 Case Is = 7
  vTime = 26
  Timer2.Enabled = True
  Form1.AddtoList 2, App.Path & "\wav\lright1.mp3"
```

```
 Form1.PlaySelected 2
 '=================================
End Select
 End If
End Sub

Private Sub Timer1_Timer()
   Static tmr As Integer
   tmr = vTimer1
   vTimer1 = vTimer1 + 1
   If (tmr = vTime) Or (vOnOff = False) Then
   tmr = 0
   Timer1.Enabled = False
   StopPlaying 2
   SetPos 2, 0
   CkProcess p
   End If
End Sub

Private Sub Timer2_Timer()
   Static tmr As Integer
   tmr = vTimer2
   vTimer2 = vTimer2 + 1
   If (tmr = vTime) Or (vOnOff = False) Then
   tmr = 0
   Timer2.Enabled = False
   StopPlaying 2
   SetPos 2, 0
   End If
End Sub

Private Sub Timer3_Timer()
   Static tmr As Integer
   tmr = vTimer3
   vTimer3 = vTimer3 + 1
   If (tmr = vTime) Or (vOnOff = False) Then
   CkProcess p
   tmr = 0
   Timer3.Enabled = False
   StopPlaying 2
   SetPos 2, 0
   End If
End Sub
```

Question 1 & Solution 4

```
Dim n, M, O, p As Integer
Dim str1, str2, str3, str4  As String
Dim q As Integer

Private Sub cmd2_Click()
   Load Q1Sulotion3
   Q1Sulotion3.Visible = True
   Unload Me
End Sub

Private Sub cmd3_Click()
   Load Q1Sulotion5
   Q1Sulotion5.Visible = True
   Unload Me
End Sub
```

Appendix B

```
Private Sub cmda_Click()
Trim (txt1.Text)
Trim (txt2.Text)
If txt1.Text = "" Or txt2.Text = "" Then
   lblc.ForeColor = RGB(0, 0, 0)
   lblc.Caption = "Please insert the value in first."
ElseIf txt1.Text = "v-100" And txt2.Text = "-2(t-10)" Then
   lblc.ForeColor = RGB(0, 0, 255)
   lblc.Caption = "That's right. v-100=-2(t-100)."
   ValResult = ValResult + 1
   p = 1
   q = 1
   CkProcess p
Else
   n = n + 1
   If (n = 1) Then
   lblc.ForeColor = RGB(255, 0, 0)
   lblc.Caption = "Please do revision on how to intergrate"
   q = 2
   ckvoice
   ElseIf (n = 2) Then
   lblc.ForeColor = RGB(255, 0, 0)
   lblc.Caption = "No,the answer is wrong."
   q = 3
   ckvoice
   ElseIf (n >= 3) Then
   lblc.ForeColor = RGB(255, 0, 0)
   lblc.Caption = "The correct answer is v-100=-2(t-10)."
   q = 4
   p = 2
   CkProcess p
   End If
End If
End Sub


Private Sub cmdc_Click()
Trim (txt4.Text)
If txt4.Text = "" Then
   lblc.ForeColor = RGB(0, 0, 0)
   lblc.Caption = "Please insert the value in first."
ElseIf txt4.Text = "0" Then
   lblc.ForeColor = RGB(0, 0, 255)
   lblc.Caption = "That's right. v = 0 ."
   ValResult = ValResult + 1
   p = 4
   q = 1
   CkProcess p
Else
   O = O + 1
   If (O = 1) Then
   lblc.ForeColor = RGB(255, 0, 0)
   lblc.Caption = "Wrong answer."
   q = 2
   ckvoice
   ElseIf (O = 2) Then
   lblc.ForeColor = RGB(255, 0, 0)
   lblc.Caption = "Wrong answer , click on Hint"
   q = 6
   ckvoice
   cmdd.Visible = True
   ElseIf (O >= 3) Then
   lblc.ForeColor = RGB(255, 0, 0)
```

```
   lblc.Caption = "The correct answer is v = 0 ."
   q = 7
   p = 5
   CkProcess p
   End If
End If
End Sub

Private Sub cmdd_Click()
   lblc.ForeColor = RGB(0, 0, 0)
   lblc.Caption = "Car stops, velocity=?"
   q = 8
   ckvoice
End Sub

Private Sub cmde_Click()
   Load Flash1
   Flash1.Visible = True
   q = 9
   ckvoice
   Unload Me
End Sub

Private Sub cmdex_Click()
   Dim fx As Form
For Each fx In Forms
  Unload fx
  Set fx = Nothing
Next
End Sub

Private Sub cmdh_Click()
 Me.PopupMenu Q1Menu.helpm, 2
End Sub

Private Sub cmdmap_Click()
   Load CharacterMap
   CharacterMap.Visible = True
   CharacterMap.SetFocus
End Sub

Private Sub cmds_Click()
   Dim voice As Boolean
   voice = vOnOff
   Me.PopupMenu Q1Menu.Sound, 2
   If voice <> vOnOff Then
   ckvoice
   End If
End Sub

Private Sub cmdt_Click()
   Me.PopupMenu Q1Menu.Tool, 2
End Sub

Private Sub Form_MouseDown(Button As Integer, Shift As Integer, X As Single, Y As Single)

   Select Case Button
      Case vbLeftButton
         `Allow the form to be moved by dragging
         Call ReleaseCapture
         Call SendMessage(Me.hwnd, WM_NCLBUTTONDOWN, HTCAPTION, 0&)
   End Select

End Sub
```

```
Private Sub Form_Load()
 StopPlaying 2
 SetPos 2, 0
CreateRoundRectFromWindow Me

str1 = "  STEP3-2 :    From  "

str2 = vbCrLf & vbCrLf & vbCrLf & vbCrLf & vbCrLf & vbCrLf & _
     "   After rearrange the equation , we found that" & vbCrLf & vbCrLf & _
     "                                   ( This is a linear equation )"

str3 = "When t = 10 ; v = 100m/s" & Chr(178) & vbCrLf & vbCrLf & _
     "When t = t' ; v = "

str4 = vbCrLf & vbCrLf & _
     "   When " & vbCrLf & vbCrLf & _
     "   When "

CkProcess p
lbl7.Caption = str3
lblc.Caption = ""
End Sub

Private Sub CkProcess(int1 As Integer)
   Static Solution As Integer
   Solution = int1
   Select Case Solution
'====================================
      Case Is = 0
      q = 0
      lbl3.Caption = str1 '0
      ShockwaveFlash1.Visible = False '0
      ShockwaveFlash2.Visible = False '0
      img4.Visible = False '0
      cmd3.Visible = False '0
      cmde.Visible = False '0
      fra1.Visible = True '0
      If vOnOff = True Then
      ckvoice
      End If
'====================================
      Case Is = 1
      If vOnOff = True Then
       ckvoice
       Else
       GoTo Process1
       End If
       q = 4
       p = 2
'====================================
      Case Is = 2
Process1:
      q = 4
      lbl3.Caption = str1 & str2 '2
      ShockwaveFlash1.Visible = True '2
      ShockwaveFlash2.Visible = True '2
      img4.Visible = False '0
      cmd3.Visible = False '0
      cmde.Visible = False '0
      fra1.Visible = True '0
      ShockwaveFlash1.Loop = False '2
      ShockwaveFlash1.Zoom (100) '2
      ShockwaveFlash1.Movie = App.Path & "\swf\s41.swf" '2
      ShockwaveFlash1.Rewind '2
```

```
    ShockwaveFlash1.Play '2
    ShockwaveFlash2.Loop = False '2
    ShockwaveFlash2.Zoom (100) '2
    ShockwaveFlash2.Movie = App.Path & "\swf\s42.swf" '2
    ShockwaveFlash2.Rewind '2
    ShockwaveFlash2.Play '2
    fra1.Visible = False
    If vOnOff = True Then
     ckvoice
    Else
    GoTo Process2
    End If
    q = 0
    p = 3
'================================
    Case Is = 3
Process2:
    q = 0
    lbl3.Caption = str1 & str2 '2
    ShockwaveFlash1.Visible = True '2
    ShockwaveFlash2.Visible = True '2
    img4.Visible = False '0
    cmd3.Visible = False '0
    cmde.Visible = False '0
    fra1.Visible = False '3
    fra3.Visible = True '3
    ShockwaveFlash1.Loop = False '2
    ShockwaveFlash1.Zoom (100) '2
    ShockwaveFlash1.Movie = App.Path & «\swf\s41.swf» '2
    ShockwaveFlash1.Rewind '2
    ShockwaveFlash1.Play '2
    ShockwaveFlash2.Loop = False '2
    ShockwaveFlash2.Zoom (100) '2
    ShockwaveFlash2.Movie = App.Path & «\swf\s42.swf» '2
    ShockwaveFlash2.Rewind '2
    ShockwaveFlash2.Play '2
    If vOnOff = True Then
     ckvoice
    End If
'================================
    Case Is = 4
    If vOnOff = True Then
     ckvoice
    Else
    GoTo Process3
    End If
    p = 5
    q = 7
'================================
    Case Is = 5
Process3:
    q = 7
    lbl3.Caption = str1 & str2 & str4 '5
    ShockwaveFlash1.Visible = True '2
    ShockwaveFlash2.Visible = True '2
    img4.Visible = True '5
    cmd3.Visible = True '5
    cmde.Visible = True '5
    fra1.Visible = False '3
    fra3.Visible = False '5
    ShockwaveFlash1.Loop = False '2
    ShockwaveFlash1.Zoom (100) '2
    ShockwaveFlash1.Movie = App.Path & "\swf\s41.swf" '2
    ShockwaveFlash1.Rewind '2
```

```
        ShockwaveFlash1.Play '2
        ShockwaveFlash2.Loop = False '2
        ShockwaveFlash2.Zoom (100) '2
        ShockwaveFlash2.Movie = App.Path & "\swf\s42.swf" '2
        ShockwaveFlash2.Rewind '2
        ShockwaveFlash2.Play '2
        If vOnOff = True Then
         ckvoice
        End If

    End Select

End Sub

Private Sub ckvoice()
 Form1.List(2).ListItems.Clear
 If vOnOff = False Then
 StopPlaying 2
 SetPos 2, 0
 Else
 Timer1.Enabled = False
 Timer2.Enabled = False
 vTimer1 = 0
 vTimer2 = 0
Select Case q
 Case Is = 0
  vTime = 11
  Timer2.Enabled = True
  Form1.AddtoList 2, App.Path & "\wav\lstate.mp3"
  Form1.PlaySelected 2
  '=====================================
 Case Is = 1
  vTime = 7
  Form1.AddtoList 2, App.Path & "\wav\lright.mp3"
  Form1.PlaySelected 2
  Timer1.Enabled = True
  '=====================================
 Case Is = 2
  vTime = 10
  Timer2.Enabled = True
  Form1.AddtoList 2, App.Path & "\wav\lagain.mp3"
  Form1.PlaySelected 2
  '=====================================
 Case Is = 3
  vTime = 6
  Timer2.Enabled = True
  Form1.AddtoList 2, App.Path & "\wav\wrong2.mp3"
  Form1.PlaySelected 2
  '=====================================
  Case Is = 4
  vTime = 39
  Form1.AddtoList 2, App.Path & "\wav\ans2.mp3"
  Form1.PlaySelected 2
  Timer1.Enabled = True
  '=====================================
 Case Is = 5
  vTime = 7
  Timer2.Enabled = True
  Form1.AddtoList 2, App.Path & "\wav\lright.mp3"
  Form1.PlaySelected 2
  '=====================================
 Case Is = 6
  vTime = 9
  Timer2.Enabled = True
```

```
 Form1.AddtoList 2, App.Path & "\wav\hint.mp3"
 Form1.PlaySelected 2
 '==================================
Case Is = 7
 vTime = 17
 Form1.AddtoList 2, App.Path & "\wav\finans.mp3"
 Form1.PlaySelected 2
 Timer1.Enabled = True
 '==================================
Case Is = 8
 vTime = 5
 Timer2.Enabled = True
 Form1.AddtoList 2, App.Path & "\wav\hint1.mp3"
 Form1.PlaySelected 2
 '==================================
 Case Is = 9
 vTime = 8
 Timer2.Enabled = True
 Form1.AddtoList 2, App.Path & "\wav\graph.mp3"
 Form1.PlaySelected 2
End Select
 End If
End Sub

Private Sub Timer1_Timer()
   Static tmr As Integer
   tmr = vTimer1
   vTimer1 = vTimer1 + 1
   If (tmr = vTime) Or (vOnOff = False) Then
   tmr = 0
   Timer1.Enabled = False
   StopPlaying 2
   SetPos 2, 0
   CkProcess p
   End If
End Sub

Private Sub Timer2_Timer()
   Static tmr As Integer
   tmr = vTimer2
   vTimer2 = vTimer2 + 1
   If (tmr = vTime) Or (vOnOff = False) Then
   tmr = 0
   Timer2.Enabled = False
   StopPlaying 2
   SetPos 2, 0
   End If
End Sub

Private Sub Timer3_Timer()
   Static tmr As Integer
   tmr = vTimer3
   vTimer3 = vTimer3 + 1
   If (tmr = vTime) Or (vOnOff = False) Then
   CkProcess p
   tmr = 0
   Timer3.Enabled = False
   StopPlaying 2
   SetPos 2, 0
   End If
End Sub
```

Appendix B

Question 1 & Solution 5

```
Dim n, M, O, p As Integer
Dim str1, str2 As String
Dim q As Integer

Private Sub cmd1_Click()
    p = 3
    q = 6
   lblc.ForeColor = RGB(0, 0, 0)
   lblc.Caption = ""
   CkProcess p
   cmd1.Visible = False
End Sub

Private Sub cmd2_Click()
   Load Q1Sulotion4
   Q1Sulotion4.Visible = True
   Unload Me
End Sub

Private Sub cmd3_Click()
   Load Q1Sulotion6
   Q1Sulotion6.Visible = True
   Unload Me
End Sub

Private Sub cmda_Click()
Trim (txtx(0).Text)
Trim (txty(0).Text)
Trim (txtm(0).Text)
Trim (txtn(1).Text)
Trim (txtx(1).Text)
Trim (txty(1).Text)
Trim (txtm(1).Text)
If txtm(0).Text = "" And txtm(1).Text = "" Then
   lblc.ForeColor = RGB(0, 0, 0)
   lblc.Caption = "Please insert the value in first."
ElseIf txtx(0).Text = "s" And (txty(0).Text = "s=0" Or txty(0).Text = "0") _
      And txtm(0).Text = "ds" And txtn(1).Text = "10" And txtx(1).Text = "t" _
      And (txty(1).Text = "t=0" Or txty(1).Text = "0") And txtm(1).Text = "tdt" Then
   lblc.ForeColor = RGB(0, 0, 255)
   lblc.Caption = "That's right."
   ValResult = ValResult + 1
   p = 2
   q = 2
   cmd1.Visible = False
   CkProcess p
Else
   n = n + 1
   If (n = 1) Then
   lblc.ForeColor = RGB(255, 0, 0)
   lblc.Caption = "Try again."
   q = 3
   ckvoice
   ElseIf (n >= 2) Then
   If txtx(0).Text <> "s" Or txty(0).Text <> "s=0" Or txtx(0).Text <> "t" Or txty(1).Text <> "t=0" Then
   lblc.ForeColor = RGB(255, 0, 0)
   lblc.Caption = "The limit bounds are wrong, do you wish to try again or click'on solve for correct answer"
   q = 4
   ckvoice
   Else
```

```
   lblc.ForeColor = RGB(255, 0, 0)
   lblc.Caption = "No,the answer is wrong, do you wish to try again or click'on solve for correct answer"
   q = 5
   ckvoice
   End If
   cmd1.Visible = True
   End If
End If
End Sub

Private Sub cmdavi_Click()
Load Q1avi1
Q1avi1.Visible = True
End Sub

Private Sub cmdb_Click()
Trim (txt2.Text)
If txt2.Text = "" Then
   lblc.ForeColor = RGB(0, 0, 0)
   lblc.Caption = "Please insert the value in first."
ElseIf txt2.Text = "s=(10t" & Chr(178) & ")/2" Then
   lblc.ForeColor = RGB(0, 0, 255)
   lblc.Caption = "That's right. s=(10t" & Chr(178) & ")/2"
   ValResult = ValResult + 1
   q = 2
   p = 5
   CkProcess p
Else
   M = M + 1
   If (M = 1) Then
   lblc.ForeColor = RGB(255, 0, 0)
   lblc.Caption = "No,this is wrong answer."
   q = 3
   ckvoice
   ElseIf (M = 2) Then
   lblc.ForeColor = RGB(255, 0, 0)
   lblc.Caption = "No,the answer is wrong."
   q = 9
   ckvoice
   ElseIf (M >= 3) Then
   lblc.ForeColor = RGB(255, 0, 0)
   lblc.Caption = "The correct answer is s=(10t" & Chr(178) & ")/2"
   q = 7
   p = 6
   CkProcess p
   End If
End If
End Sub

Private Sub cmdc_Click()
Trim (txt3.Text)
If txt3.Text = "" Then
   lblc.ForeColor = RGB(0, 0, 0)
   lblc.Caption = "Please insert the value in first."
ElseIf txt3.Text = "s=5t" & Chr(178) Then
   lblc.ForeColor = RGB(0, 0, 255)
   lblc.Caption = "That's right. s=5t" & Chr(178)
   ValResult = ValResult + 1
   p = 8
   q = 2
   CkProcess p
Else
   O = O + 1
   If (O = 1) Then
```

```
    lblc.ForeColor = RGB(255, 0, 0)
    lblc.Caption = "No,this is wrong answer."
    q = 3
    ckvoice
    ElseIf (O = 2) Then
    lblc.ForeColor = RGB(255, 0, 0)
    lblc.Caption = "No,the answer is wrong."
    q = 9
    ckvoice
    ElseIf (O >= 3) Then
    lblc.ForeColor = RGB(255, 0, 0)
    lblc.Caption = "The correct answer is s=5t" & Chr(178)
    q = 8
    p = 9
    CkProcess p
    End If
End If
End Sub

Private Sub cmdequ1_Click()
Dim hMenu As Long
Dim hSubMenu As Long
Dim lngID As Long
Dim i As Integer

    hMenu = GetMenu(Q1Menu.hwnd)
    hSubMenu = GetSubMenu(hMenu, 0)
    hSubMenu = GetSubMenu(hSubMenu, 2)
For i = 0 To 2
    lngID = GetMenuItemID(hSubMenu, i)
    Q1Menu.Pic(i).Picture = Q1Menu.Pic(i).Image
    Call ModifyMenu(hMenu, lngID, 4, lngID, CLng(Q1Menu.Pic(i).Picture))
Next i

Q1Sulotion5.PopupMenu Q1Menu.s51, 2
End Sub

Private Sub cmdequ2_Click()
    Dim hMenu As Long
    Dim hSubMenu As Long
    Dim lngID As Long
    Dim i As Integer

    hMenu = GetMenu(Q1Menu.hwnd)
    hSubMenu = GetSubMenu(hMenu, 0)
    hSubMenu = GetSubMenu(hSubMenu, 3)
For i = 0 To 2
    lngID = GetMenuItemID(hSubMenu, i)
    Q1Menu.Pic(i).Picture = Q1Menu.Pic(i).Image
    Call ModifyMenu(hMenu, lngID, 4, lngID, CLng(Q1Menu.Pic(i).Picture))
Next i

Q1Sulotion5.PopupMenu Q1Menu.s52, 2
End Sub

Private Sub cmdex_Click()
    Dim fx As Form
For Each fx In Forms
   Unload fx
   Set fx = Nothing
Next
End Sub

Private Sub cmdh_Click()
```

```
 Me.PopupMenu Q1Menu.helpm, 2
End Sub

Private Sub cmdmap_Click()
   Load CharacterMap
   CharacterMap.Visible = True
   CharacterMap.SetFocus
End Sub

Private Sub cmds_Click()
   Dim voice As Boolean
   voice = vOnOff
   Me.PopupMenu Q1Menu.Sound, 2
   If voice <> vOnOff Then
   ckvoice
   End If
End Sub

Private Sub cmdt_Click()
   Me.PopupMenu Q1Menu.Tool, 2
End Sub

Private Sub cmdundo_Click()
   cmdequ1.Visible = True
   cmdequ2.Visible = True
   imgInt1(0).Visible = False
   imgInt1(1).Visible = False
   txtx(0).Visible = False
   txty(0).Visible = False
   txtm(0).Visible = False
   txtn(0).Visible = False
   txtx(1).Visible = False
   txty(1).Visible = False
   txtm(1).Visible = False
   txtn(1).Visible = False
   txtx(0).Text = «»
   txty(0).Text = «»
   txtm(0).Text = «»
   txtn(0).Text = «»
   txtx(1).Text = «»
   txty(1).Text = «»
   txtm(1).Text = «»
   txtn(1).Text = «»
End Sub

Private Sub Form_MouseDown(Button As Integer, Shift As Integer, X As Single, Y As Single)

   Select Case Button
      Case vbLeftButton
         `Allow the form to be moved by dragging
         Call ReleaseCapture
         Call SendMessage(Me.hwnd, WM_NCLBUTTONDOWN, HTCAPTION, 0&)
   End Select

End Sub
Private Sub Form_Load()
 StopPlaying 2
 SetPos 2, 0
CreateRoundRectFromWindow Me
ShockwaveFlash1.Loop = False
   ShockwaveFlash1.Zoom (100)
   ShockwaveFlash1.Movie = App.Path & «\swf\lat2.swf»

str1 = «STEP4 : To plot s - t graph» & _
```

```
    vbCrLf & vbCrLf & vbCrLf & vbCrLf & vbCrLf & vbCrLf & vbCrLf & _
    «   Defination of velocity:»

str2 = «When» & vbCrLf & vbCrLf & _
    «When»

CkProcess p
lblc.Caption = «»
lbl3.Caption = str1
lbl4.Caption = str2
End Sub

Private Sub CkProcess(int1 As Integer)
  Static Solution As Integer
  Solution = int1
  Select Case Solution
'==================================
     Case Is = 0
     q = 0
     lbl4.Visible = False '0
     img3.Visible = False '0
     img4.Visible = False '0
     img5.Visible = False '0
     img6.Visible = False '0
     cmd3.Visible = False '0
     cmdavi.Visible = False '0
     If vOnOff = False Then
      GoTo Process1
     End If
      ckvoice
      p = 1
      q = 1
'==================================
     Case Is = 1
Process1:
     q = 1
     lbl4.Visible = False '0
     img3.Visible = False '0
     img4.Visible = False '0
     img5.Visible = False '0
     img6.Visible = False '0
     cmd3.Visible = False '0
     cmdavi.Visible = False '0
     fra1.Visible = True '1
     If vOnOff = True Then
     ckvoice
     End If
'==================================
     Case Is = 2
     If vOnOff = True Then
      ckvoice
      Else
      GoTo Process2
      End If
      q = 6
      p = 3
'==================================
     Case Is = 3
Process2:
     q = 6
     lbl4.Visible = False '0
     img3.Visible = True '3
     img4.Visible = False '0
     img5.Visible = False '0
```

```
        img6.Visible = False '0
        cmd3.Visible = False '0
        cmdavi.Visible = False '0
        fra1.Visible = False '3
        If vOnOff = True Then
        ckvoice
        Else
        GoTo Process3
        End If
        p = 4
        q = 1
'================================
        Case Is = 4
Process3:
        q = 1
        lbl4.Visible = False '0
        img3.Visible = True '3
        img4.Visible = False '0
        img5.Visible = False '0
        img6.Visible = False '0
        cmd3.Visible = False '0
        cmdavi.Visible = False '0
        fra1.Visible = False '3
        fra2.Visible = True '4
        If vOnOff = True Then
        ckvoice
        End If
'================================
        Case Is = 5
        If vOnOff = True Then
         ckvoice
         Else
         GoTo Process4
         End If
         q = 7
         p = 6
'================================
        Case Is = 6
Process4:
        q = 7
        lbl4.Visible = False '0
        img3.Visible = True '3
        img5.Visible = False '0
        img6.Visible = False '0
        cmd3.Visible = False '0
        cmdavi.Visible = False '0
        fra1.Visible = False '3
        fra2.Visible = False '6
        If vOnOff = True Then
        ckvoice
        Else
        GoTo Process5
        End If
        p = 7
        q = 1
'================================
        Case Is = 7
Process5:
        q = 1
        lbl4.Visible = False '0
        img3.Visible = True '3
        img4.Visible = True '7
        img5.Visible = False '0
        img6.Visible = False '0
```

```
        cmd3.Visible = False '0
        cmdavi.Visible = False '0
        fra1.Visible = False '3
        fra2.Visible = False '6
        fra3.Visible = True '7
        If vOnOff = True Then
        ckvoice
        End If
'==================================
         Case Is = 8
        If vOnOff = True Then
         ckvoice
         Else
         GoTo Process6
         End If
         q = 8
         p = 9
'==================================
        Case Is = 9
Process6:
        q = 8
        lbl4.Visible = False '0
        img3.Visible = True '3
        img4.Visible = False '9
        img5.Visible = True '9
        img6.Visible = False '0
        cmd3.Visible = True '9
        cmdavi.Visible = False '0
        fra1.Visible = False '3
        fra2.Visible = False '6
        fra3.Visible = False '9
        If vOnOff = True Then
        ckvoice
        Else
        GoTo Process7
        End If
        q = 10
        p = 10
'==================================
    Case Is = 10
Process7:
        q = 10
        If vOnOff = True Then
        ckvoice
        End If
        lbl4.Visible = True '10
        img3.Visible = True '3
        img4.Visible = False '9
        img5.Visible = True '9
        img6.Visible = True '10
        cmd3.Visible = True '9
        cmdavi.Visible = True '0
        fra1.Visible = False '3
        fra2.Visible = False '6
        fra3.Visible = False '9
    End Select

End Sub

Private Sub ckvoice()
 Form1.List(2).ListItems.Clear
 If vOnOff = False Then
 StopPlaying 2
 SetPos 2, 0
```

```
Else
Timer1.Enabled = False
Timer2.Enabled = False
vTimer1 = 0
vTimer2 = 0
Select Case q
Case Is = 0
 vTime = 57
 Form1.AddtoList 2, App.Path & «\wav\2ndportion.mp3»
 Form1.PlaySelected 2
 Timer1.Enabled = True
 '==================================
Case Is = 1
 vTime = 12
 Timer2.Enabled = True
 Form1.AddtoList 2, App.Path & «\wav\cont.mp3»
 Form1.PlaySelected 2
 '==================================
Case Is = 2
 vTime = 5
 Form1.AddtoList 2, App.Path & «\wav\correct.mp3»
 Form1.PlaySelected 2
 Timer1.Enabled = True
 '==================================
Case Is = 3
 vTime = 4
 Timer2.Enabled = True
 Form1.AddtoList 2, App.Path & «\wav\incorrect.mp3»
 Form1.PlaySelected 2
 '==================================
Case Is = 4
 vTime = 15
 Timer2.Enabled = True
 Form1.AddtoList 2, App.Path & «\wav\ltry3.mp3»
 Form1.PlaySelected 2
 '==================================
Case Is = 5
 vTime = 11
 Timer2.Enabled = True
 Form1.AddtoList 2, App.Path & «\wav\3rdtrial.mp3»
 Form1.PlaySelected 2
 '==================================
Case Is = 6
 vTime = 25
 Form1.AddtoList 2, App.Path & «\wav\int1.mp3»
 Form1.PlaySelected 2
 Timer1.Enabled = True
 '==================================
Case Is = 7
 vTime = 7
 Form1.AddtoList 2, App.Path & «\wav\int2.mp3»
 Form1.PlaySelected 2
 Timer1.Enabled = True
 '==================================
Case Is = 8
 vTime = 8
 Form1.AddtoList 2, App.Path & «\wav\int3.mp3»
 Form1.PlaySelected 2
 Timer1.Enabled = True
 '==================================
Case Is = 9
 vTime = 7
 Timer2.Enabled = True
 Form1.AddtoList 2, App.Path & «\wav\l2ndtrial.mp3»
```

```
 Form1.PlaySelected 2
 '====================================
Case Is = 10
 vTime = 16
 Timer2.Enabled = True
 Form1.AddtoList 2, App.Path & «\wav\explain.mp3»
 Form1.PlaySelected 2
End Select
 End If
End Sub

Private Sub Timer1_Timer()
   Static tmr As Integer
   tmr = vTimer1
   vTimer1 = vTimer1 + 1
   If (tmr = vTime) Or (vOnOff = False) Then
   tmr = 0
   Timer1.Enabled = False
   StopPlaying 2
   SetPos 2, 0
   CkProcess p
   End If
End Sub

Private Sub Timer2_Timer()
   Static tmr As Integer
   tmr = vTimer2
   vTimer2 = vTimer2 + 1
   If (tmr = vTime) Or (vOnOff = False) Then
   tmr = 0
   Timer2.Enabled = False
   StopPlaying 2
   SetPos 2, 0
   End If
End Sub

Private Sub Timer3_Timer()
   Static tmr As Integer
   tmr = vTimer3
   vTimer3 = vTimer3 + 1
   If (tmr = vTime) Or (vOnOff = False) Then
   CkProcess p
   tmr = 0
   Timer3.Enabled = False
   StopPlaying 2
   SetPos 2, 0
   End If
End Sub
```

Question 1 & Solution 6

```
Dim n, M, O, p As Integer
Dim str1, str2 As String
Dim q As Integer

Private Sub cmd1_Click()
     p = 3
     q = 6
   lblc.ForeColor = RGB(0, 0, 0)
   lblc.Caption = ""
   CkProcess p
```

```
   cmd1.Visible = False
End Sub

Private Sub cmdavi_Click()
Load Q1avi2
Q1avi2.Visible = True
End Sub

Private Sub cmdequ1_Click()
Dim hMenu As Long
Dim hSubMenu As Long
Dim lngID As Long
Dim i As Integer

   hMenu = GetMenu(Q1Menu.hwnd)
   hSubMenu = GetSubMenu(hMenu, 0)
   hSubMenu = GetSubMenu(hSubMenu, 4)
For i = 0 To 2
   lngID = GetMenuItemID(hSubMenu, i)
   Q1Menu.Pic(i).Picture = Q1Menu.Pic(i).Image
   Call ModifyMenu(hMenu, lngID, 4, lngID, CLng(Q1Menu.Pic(i).Picture))
Next i

Q1Sulotion6.PopupMenu Q1Menu.s61, 2
End Sub

Private Sub cmdequ2_Click()
   Dim hMenu As Long
   Dim hSubMenu As Long
   Dim lngID As Long
   Dim i As Integer

   hMenu = GetMenu(Q1Menu.hwnd)
   hSubMenu = GetSubMenu(hMenu, 0)
   hSubMenu = GetSubMenu(hSubMenu, 5)
For i = 0 To 2
   lngID = GetMenuItemID(hSubMenu, i)
   Q1Menu.Pic(i).Picture = Q1Menu.Pic(i).Image
   Call ModifyMenu(hMenu, lngID, 4, lngID, CLng(Q1Menu.Pic(i).Picture))
Next i

Q1Sulotion6.PopupMenu Q1Menu.s62, 2
End Sub
Private Sub cmd2_Click()
   Load Q1Sulotion5
   Q1Sulotion5.Visible = True
   Unload Me
End Sub

Private Sub cmd3_Click()
   Load Q1Answer
   Q1Answer.Visible = True
   Unload Me
End Sub

Private Sub cmda_Click()
Trim (txtx(0).Text)
Trim (txty(0).Text)
Trim (txtm(0).Text)
Trim (txtx(1).Text)
Trim (txty(1).Text)
Trim (txtm(1).Text)
If txtm(0).Text = «» And txtm(1).Text = «» Then
   lblc.ForeColor = RGB(0, 0, 0)
```

```
    lblc.Caption = «Please insert the value in first.»
ElseIf txtx(0).Text = «s» And (txty(0).Text = «500» Or txty(0).Text = «s=500») _
      And txtm(0).Text = «ds» And _
      txtx(1).Text = «t» And (txty(1).Text = «10» Or txty(1).Text = «t=10») _
And txtm(1).Text = «(-2t+120)dt» Then
    lblc.ForeColor = RGB(0, 0, 255)
    lblc.Caption = «That's right.»
    ValResult = ValResult + 1
    p = 2
    q = 2
    cmd1.Visible = False
    CkProcess p
Else
    n = n + 1
    If (n = 1) Then
    lblc.ForeColor = RGB(255, 0, 0)
    lblc.Caption = «Try again.»
    q = 3
    ckvoice
    ElseIf (n >= 2) Then
    If txtx(0).Text <> «s» Or txty(0).Text <> «s=500» Or txtx(0).Text <> «t» Or txty(1).Text <> «t=10» Then
    lblc.ForeColor = RGB(255, 0, 0)
    lblc.Caption = «The limit bounds are wrong, do you wish to try again or click'on solve for correct answer»
    q = 4
    ckvoice
    Else
    lblc.ForeColor = RGB(255, 0, 0)
    lblc.Caption = «No,the answer is wrong, do you wish to try again or click'on solve for correct answer»
    q = 5
    ckvoice
    End If
    cmd1.Visible = True
    End If
End If
End Sub

Private Sub cmdb_Click()
Trim (txt2.Text)
If txt2.Text = «» Then
    lblc.ForeColor = RGB(0, 0, 0)
    lblc.Caption = «Please insert the value in first.»
ElseIf txt2.Text = «s-500=-t» & Chr(178) & «+120t-[-(10)» & Chr(178) & «+120(10)]» Then
    lblc.ForeColor = RGB(0, 0, 255)
    lblc.Caption = «That's right.s-500=-t» & Chr(178) & «+120t-[-(10)» & Chr(178) & «+120(10)]»
    ValResult = ValResult + 1
    q = 2
    p = 5
    CkProcess p
Else
    M = M + 1
    If (M = 1) Then
    lblc.ForeColor = RGB(255, 0, 0)
    lblc.Caption = "No,this is wrong answer."
    q = 3
    ckvoice
    ElseIf (M = 2) Then
    lblc.ForeColor = RGB(255, 0, 0)
    lblc.Caption = "No,the answer is wrong."
     q = 9
    ckvoice
    ElseIf (M >= 3) Then
    lblc.ForeColor = RGB(255, 0, 0)
    lblc.Caption = "The correct answer is s-500=-t" & Chr(178) & "+120t-[-(10)" & Chr(178) & "+120(10)]"
    q = 7
```

```
    p = 6
    CkProcess p
    End If
End If
End Sub

Private Sub cmdc_Click()
Trim (txt3.Text)
If txt3.Text = "" Then
    lblc.ForeColor = RGB(0, 0, 0)
    lblc.Caption = "Please insert the value in first."
ElseIf txt3.Text = "s=-t" & Chr(178) & "+120t-600" Then
    lblc.ForeColor = RGB(0, 0, 255)
    lblc.Caption = "That's right.s=-t" & Chr(178) & "+120t-600"
    ValResult = ValResult + 1
    p = 8
    q = 2
    CkProcess p
Else
    O = O + 1
    If (O = 1) Then
    lblc.ForeColor = RGB(255, 0, 0)
    lblc.Caption = "No,this is wrong answer."
    q = 3
    ckvoice
    ElseIf (O = 2) Then
    lblc.ForeColor = RGB(255, 0, 0)
    lblc.Caption = "No,the answer is wrong."
    q = 9
    ckvoice
    ElseIf (O >= 3) Then
    lblc.ForeColor = RGB(255, 0, 0)
    lblc.Caption = "The correct answer is s=-t" & Chr(178) & "+120t-600"
    q = 8
    p = 9
    CkProcess p
    End If
End If
End Sub

Private Sub cmdd_Click()
    Load Flash2
    Flash2.Visible = True
    q = 11
    ckvoice
    Unload Me
End Sub

Private Sub cmdex_Click()
    Dim fx As Form
For Each fx In Forms
    Unload fx
    Set fx = Nothing
Next
End Sub

Private Sub cmdh_Click()
 Me.PopupMenu Q1Menu.helpm, 2
End Sub

Private Sub cmdmap_Click()
    Load CharacterMap
    CharacterMap.Visible = True
    CharacterMap.SetFocus
```

```
End Sub

Private Sub cmds_Click()
   Dim voice As Boolean
   voice = vOnOff
   Me.PopupMenu Q1Menu.Sound, 2
   If voice <> vOnOff Then
   ckvoice
   End If
End Sub

Private Sub cmdt_Click()
   Me.PopupMenu Q1Menu.Tool, 2
End Sub

Private Sub cmdundo_Click()
   cmdequ1.Visible = True
   cmdequ2.Visible = True
   imgInt1(0).Visible = False
   imgInt1(1).Visible = False
   txtx(0).Visible = False
   txty(0).Visible = False
   txtm(0).Visible = False
   txtn(0).Visible = False
   txtx(1).Visible = False
   txty(1).Visible = False
   txtm(1).Visible = False
   txtx(0).Text = ""
   txty(0).Text = ""
   txtm(0).Text = ""
   txtn(0).Text = ""
   txtx(1).Text = ""
   txty(1).Text = ""
   txtm(1).Text = ""
End Sub

Private Sub Form_MouseDown(Button As Integer, Shift As Integer, X As Single, Y As Single)

   Select Case Button
      Case vbLeftButton
         'Allow the form to be moved by dragging
         Call ReleaseCapture
         Call SendMessage(Me.hwnd, WM_NCLBUTTONDOWN, HTCAPTION, 0&)
   End Select

End Sub
Private Sub Form_Load()
 StopPlaying 2
 SetPos 2, 0
CreateRoundRectFromWindow Me



str1 = "STEP 5 :   Using initial condition" & _
      vbCrLf & vbCrLf & vbCrLf & _
      "   When " & vbCrLf & vbCrLf & _
      "   When " & vbCrLf & vbCrLf & _
      "   Kinematic: "

str2 = "When " & vbCrLf & vbCrLf & _
      "   The position is ,"

CkProcess p
```

```
lbl3.Caption = str1
lbl4.Caption = str2
lblc.Caption = ""
End Sub

Private Sub CkProcess(int1 As Integer)
   Static Solution As Integer
   Solution = int1
   Select Case Solution
'================================
      Case Is = 0
      q = 0
      img3.Visible = False '0
      img4.Visible = False '0
      img5.Visible = False '0
      img6.Visible = False '0
      img7.Visible = False '0
      lbl4.Visible = False '0
      cmd3.Visible = False '0 Next
      cmdd.Visible = False '0 Graph
      cmdavi.Visible = False '0 Avi
      If vOnOff = False Then
       GoTo Process1
      End If
       ckvoice
       p = 1
       q = 1
'================================
      Case Is = 1
Process1:
      q = 1
      img3.Visible = False '0
      img4.Visible = False '0
      img5.Visible = False '0
      img6.Visible = False '0
      img7.Visible = False '0
      lbl4.Visible = False '0
      cmd3.Visible = False '0 Next
      cmdd.Visible = False '0 Graph
      cmdavi.Visible = False '0 Avi
      fra1.Visible = True '1
      If vOnOff = True Then
      ckvoice
      End If
'================================
      Case Is = 2
      If vOnOff = True Then
       ckvoice
       Else
       GoTo Process2
       End If
       q = 6
       p = 3
'================================
      Case Is = 3
Process2:
      img3.Visible = True '3
      img4.Visible = False '0
      img5.Visible = False '0
      img6.Visible = False '0
      img7.Visible = False '0
      lbl4.Visible = False '0
      cmd3.Visible = False '0 Next
      cmdd.Visible = False '0 Graph
```

```
      cmdavi.Visible = False '0 Avi
      fra1.Visible = False '3
      If vOnOff = True Then
      ckvoice
      Else
      GoTo Process3
      End If
      p = 4
      q = 1
'==================================
      Case Is = 4
Process3:
      q = 1
      img3.Visible = True '3
      img4.Visible = False '0
      img5.Visible = False '0
      img6.Visible = False '0
      img7.Visible = False '0
      lbl4.Visible = False '0
      cmd3.Visible = False '0 Next
      cmdd.Visible = False '0 Graph
      cmdavi.Visible = False '0 Avi
      fra1.Visible = False '3
      fra2.Visible = True '4
      If vOnOff = True Then
      ckvoice
      End If
'==================================
       Case Is = 5
      If vOnOff = True Then
       ckvoice
       Else
       GoTo Process4
       End If
       q = 7
       p = 6
'==================================
      Case Is = 6
Process4:
      q = 7
      img3.Visible = True '3
      img4.Visible = True '6
      img5.Visible = False '0
      img6.Visible = False '0
      img7.Visible = False '0
      lbl4.Visible = False '0
      cmd3.Visible = False '0 Next
      cmdd.Visible = False '0 Graph
      cmdavi.Visible = False '0 Avi
      fra1.Visible = False '3
      fra2.Visible = False '6
      If vOnOff = True Then
      ckvoice
      Else
      GoTo Process5
      End If
      p = 7
      q = 1
'==================================
      Case Is = 7
Process5:
      q = 1
      img3.Visible = True '3
      img4.Visible = True '6
```

```
      img5.Visible = False '0
      img6.Visible = False '0
      img7.Visible = False '0
      lbl4.Visible = False '0
      cmd3.Visible = False '0 Next
      cmdd.Visible = False '0 Graph
      cmdavi.Visible = False '0 Avi
      fra1.Visible = False '3
      fra2.Visible = False '6
      fra3.Visible = True '7
      If vOnOff = True Then
      ckvoice
      End If
'==================================
      Case Is = 8
      If vOnOff = True Then
      ckvoice
      Else
      GoTo Process6
      End If
      q = 8
      p = 9
'==================================
      Case Is = 9
Process6:
      q = 8
      img3.Visible = True '3
      img4.Visible = True '6
      img5.Visible = True '8
      img6.Visible = True '8
      img7.Visible = True '8
      lbl4.Visible = True '8
      cmd3.Visible = True '8
      cmdd.Visible = True '8
      cmdavi.Visible = False '0 Avi
      fra1.Visible = False '3
      fra2.Visible = False '6
      fra3.Visible = False '8
      If vOnOff = True Then
      ckvoice
      Else
      GoTo Process7
      End If
      q = 10
      p = 10
'==================================
      Case Is = 10
Process7:
      q = 10
      If vOnOff = True Then
      ckvoice
      End If
      img3.Visible = True '3
      img4.Visible = True '6
      img5.Visible = True '8
      img6.Visible = True '8
      img7.Visible = True '8
      lbl4.Visible = True '8
      cmd3.Visible = True '8
      cmdd.Visible = True '8
      cmdavi.Visible = True '10
      fra1.Visible = False '3
      fra2.Visible = False '6
      fra3.Visible = False '8
```

```
    If vOnOff = True Then
    ckvoice
    End If
  End Select

End Sub


Private Sub ckvoice()
 Form1.List(2).ListItems.Clear
 If vOnOff = False Then
 StopPlaying 2
 SetPos 2, 0
 Else
 Timer1.Enabled = False
 Timer2.Enabled = False
 vTimer1 = 0
 vTimer2 = 0
 Select Case q
 Case Is = 0
  vTime = 47
  Form1.AddtoList 2, App.Path & «\wav\explain2.mp3»
  Form1.PlaySelected 2
  Timer1.Enabled = True
'==================================================
 Case Is = 1
  vTime = 12
  Timer2.Enabled = True
  Form1.AddtoList 2, App.Path & «\wav\cont.mp3»
  Form1.PlaySelected 2
'==================================================
 Case Is = 2
  vTime = 5
  Form1.AddtoList 2, App.Path & «\wav\correct.mp3»
  Form1.PlaySelected 2
  Timer1.Enabled = True
'==================================================
 Case Is = 3
  vTime = 4
  Timer2.Enabled = True
  Form1.AddtoList 2, App.Path & «\wav\incorrect.mp3»
  Form1.PlaySelected 2
'==================================================
 Case Is = 4
  vTime = 15
  Timer2.Enabled = True
  Form1.AddtoList 2, App.Path & «\wav\ltry3.mp3»
  Form1.PlaySelected 2
'==================================================
 Case Is = 5
  vTime = 11
  Timer2.Enabled = True
  Form1.AddtoList 2, App.Path & «\wav\3rdtrial.mp3»
  Form1.PlaySelected 2
'==================================================
 Case Is = 6
  vTime = 29
  Form1.AddtoList 2, App.Path & «\wav\int4.mp3»
  Form1.PlaySelected 2
  Timer1.Enabled = True
'==================================================
 Case Is = 7
  vTime = 23
  Form1.AddtoList 2, App.Path & «\wav\int5.mp3»
```

```
 Form1.PlaySelected 2
 Timer1.Enabled = True
'================================================================
Case Is = 8
 vTime = 12
 Form1.AddtoList 2, App.Path & «\wav\int6.mp3»
 Form1.PlaySelected 2
 Timer1.Enabled = True
'================================================================
Case Is = 9
 vTime = 7
 Timer2.Enabled = True
 Form1.AddtoList 2, App.Path & «\wav\l2ndtrial.mp3»
 Form1.PlaySelected 2
Case Is = 10
 vTimer2 = 0
 vTime = 13
 Timer2.Enabled = True
 Form1.AddtoList 2, App.Path & «\wav\explain4.mp3»
 Form1.PlaySelected 2
'================================================================
Case Is = 11
 vTime = 8
 Timer2.Enabled = True
 Form1.AddtoList 2, App.Path & «\wav\graphc.mp3»
 Form1.PlaySelected 2
End Select
 End If
End Sub

Private Sub Timer1_Timer()
   Static tmr As Integer
   tmr = vTimer1
   vTimer1 = vTimer1 + 1
   If (tmr = vTime) Or (vOnOff = False) Then
   tmr = 0
   Timer1.Enabled = False
   StopPlaying 2
   SetPos 2, 0
   CkProcess p
   End If
End Sub

Private Sub Timer2_Timer()
   Static tmr As Integer
   tmr = vTimer2
   vTimer2 = vTimer2 + 1
   If (tmr = vTime) Or (vOnOff = False) Then
   tmr = 0
   Timer2.Enabled = False
   StopPlaying 2
   SetPos 2, 0
   End If
End Sub

Private Sub Timer3_Timer()
   Static tmr As Integer
   tmr = vTimer3
   vTimer3 = vTimer3 + 1
   If (tmr = vTime) Or (vOnOff = False) Then
   CkProcess p
   tmr = 0
   Timer3.Enabled = False
   StopPlaying 2
```

```
   SetPos 2, 0
   End If
End Sub
```

Question 1 & Menu

```
Private Sub a33_Click(index As Integer)

   If index = 1 Then
      Q1Sulotion2.imgInt1(0).Visible = True
      Q1Sulotion2.txtx(0).Visible = True
      Q1Sulotion2.txty(0).Visible = True
      Q1Sulotion2.txtm(0).Visible = True
      Q1Sulotion2.cmdequ1.Visible = False
   Else
      Q1Sulotion2.lblc.ForeColor = RGB(255, 0, 0)
      Q1Sulotion2.lblc.Caption = "You selected wrong symbol ,Please select again."
   End If
End Sub

Private Sub a34_Click(index As Integer)
   If index = 2 Then
      Q1Sulotion2.imgInt1(1).Visible = True
      Q1Sulotion2.txtx(1).Visible = True
      Q1Sulotion2.txty(1).Visible = True
      Q1Sulotion2.txtm(1).Visible = True
      Q1Sulotion2.txtn(1).Visible = True
      Q1Sulotion2.cmdequ2.Visible = False
   Else
      Q1Sulotion2.lblc.ForeColor = RGB(255, 0, 0)
      Q1Sulotion2.lblc.Caption = "You selected wrong symbol ,Please select again."
   End If
End Sub

Private Sub a35_Click(index As Integer)
   If index = 1 Then
      Q1Sulotion5.imgInt1(0).Visible = True
      Q1Sulotion5.txtx(0).Visible = True
      Q1Sulotion5.txty(0).Visible = True
      Q1Sulotion5.txtm(0).Visible = True
      Q1Sulotion5.cmdequ1.Visible = False
   Else
    Q1Sulotion5.lblc.ForeColor = RGB(255, 0, 0)
    Q1Sulotion5.lblc.Caption = "You selected wrong symbol ,Please select again."
   End If
End Sub

Private Sub a36_Click(index As Integer)
   If index = 2 Then
      Q1Sulotion5.imgInt1(1).Visible = True
      Q1Sulotion5.txtx(1).Visible = True
      Q1Sulotion5.txty(1).Visible = True
      Q1Sulotion5.txtm(1).Visible = True
      Q1Sulotion5.txtn(1).Visible = True
      Q1Sulotion5.cmdequ2.Visible = False
   Else
      Q1Sulotion5.lblc.ForeColor = RGB(255, 0, 0)
      Q1Sulotion5.lblc.Caption = "You selected wrong symbol ,Please select again."
   End If
End Sub

Private Sub a37_Click(index As Integer)
```

```
   If index = 1 Then
      Q1Sulotion6.imgInt1(0).Visible = True
      Q1Sulotion6.txtx(0).Visible = True
      Q1Sulotion6.txty(0).Visible = True
      Q1Sulotion6.txtm(0).Visible = True
      Q1Sulotion6.cmdequ1.Visible = False
      Else
       Q1Sulotion6.lblc.ForeColor = RGB(255, 0, 0)
       Q1Sulotion6.lblc.Caption = "You selected wrong symbol ,Please select again."
      End If
End Sub

Private Sub a38_Click(index As Integer)
   If index = 1 Then
      Q1Sulotion6.imgInt1(1).Visible = True
      Q1Sulotion6.txtx(1).Visible = True
      Q1Sulotion6.txty(1).Visible = True
      Q1Sulotion6.txtm(1).Visible = True
      Q1Sulotion6.cmdequ2.Visible = False
      Else
      Q1Sulotion6.lblc.ForeColor = RGB(255, 0, 0)
      Q1Sulotion6.lblc.Caption = "You selected wrong symbol ,Please select again."
      End If
End Sub


Private Sub Calc_Click()
   Calc = Shell(App.Path & "\calc.exe", 1)
End Sub

Private Sub Form_Load()
   'laod mixer
   Load Form1
   'Load file
   Form1.AddtoList 1, App.Path & "\wav\01.mp3"
   'Play
   Form1.PlaySelected 1
   vTimer3 = 0
   vTimeM = 111
   Timer1.Enabled = True
   vOnOff = True
End Sub

Private Sub Glossory_Click()
   Load Glossary
   Glossary.Visible = True
End Sub

Private Sub help_Click()
   Load HelpFile
   HelpFile.Visible = True
   HelpFile.WebBrowser1.Navigate App.Path & "\help\help_file.htm"
End Sub

Private Sub Music_Click()
   If Music.Checked = True Then
   Music.Checked = False
   StopPlaying 1
   SetPos 1, 0
   Timer1.Enabled = False
   Else
   Music.Checked = True
   Form1.PlaySelected 1
   Timer1.Enabled = True
```

```
    End If
End Sub

Private Sub Note_Click()
    Note = Shell(App.Path & "\notepad.exe", 1)
End Sub

Private Sub Timer1_Timer()
    Static tmr As Integer
    tmr = vTimer3
    vTimer3 = vTimer3 + 1
    If (tmr = vTimeM) Then
    Form1.List(1).ListItems.Clear
    'Load file
    Form1.AddtoList 1, App.Path & "\wav\01.mp3"
    'Play
    Form1.PlaySelected 1
    tmr = 0
    End If
End Sub

Private Sub voice_Click()
    If voice.Checked = True Then
    voice.Checked = False
    vOnOff = False
    Else
    voice.Checked = True
    vOnOff = True
    End If
End Sub
```

Question 1 – Audio / Video

```
Private Sub Form_Load()
 StopPlaying 2
 SetPos 2, 0
End Sub

Private Sub Form_MouseDown(Button As Integer, Shift As Integer, X As Single, Y As Single)

    Select Case Button
        Case vbLeftButton
            'Allow the form to be moved by dragging
            Call ReleaseCapture
            Call SendMessage(Me.hwnd, WM_NCLBUTTONDOWN, HTCAPTION, 0&)
    End Select

End Sub

Private Sub Timer1_Timer()
    MediaPlayer1.Open App.Path & "\mov\metal.avi"
    Timer1.Enabled = False
End Sub
```

Introduction Screen

```
Private Sub cmd1_Click()
```

```
   Load Q1Objective
   Q1Objective.Visible = True
   Unload Me
   vStep = 0
End Sub

Private Sub cmd2_Click()
   Load Q1Question
   Q1Question.Visible = True
   Unload Me
   vStep = 0
End Sub

Private Sub cmdex_Click()
   Dim fx As Form
For Each fx In Forms
  Unload fx
  Set fx = Nothing
Next
End Sub

Private Sub cmdh_Click()
   Me.PopupMenu Q1Menu.helpm, 2
End Sub

Private Sub cmds_Click()
   Dim voice As Boolean
   voice = vOnOff
   Me.PopupMenu Q1Menu.Sound, 2
   If voice <> vOnOff Then
   voiceck
   End If
End Sub

Private Sub cmdt_Click()
   Me.PopupMenu Q1Menu.Tool, 2
End Sub

Private Sub Form_MouseDown(Button As Integer, Shift As Integer, X As Single, Y As Single)

   Select Case Button
      Case vbLeftButton
         'Allow the form to be moved by dragging
         Call ReleaseCapture
         Call SendMessage(Me.hwnd, WM_NCLBUTTONDOWN, HTCAPTION, 0&)
   End Select

End Sub
Private Sub Form_Load()
 StopPlaying 2
 SetPos 2, 0
CreateRoundRectFromWindow Me
voiceck

txt1.Text = "Kinematics of a Particle  : " & vbCrLf & _
        "Kinematics treat only the geometric aspects of motion. " & _
        "Kinetics is the analysis of the forces causing the motion. " & _
        "Rectilinear kinematics states that a particle can move along either a straight or a curved path." & _
        "The kinematics of the motion is characterized by specifying,at any given instant," & _
        "the particle's position,velocity and acceleration."
End Sub

Private Sub voiceck()
Form1.List(2).ListItems.Clear
```

```
If vOnOff = False Then
StopPlaying 2
SetPos 2, 0
Else
 If vStep = 0 Then
  vTimer1 = 0
  vTime = 47
  Timer1.Enabled = True
  Form1.AddtoList 2, App.Path & "\wav\lintro2.mp3"
  Form1.PlaySelected 2
 ElseIf vStep = 1 Then
  vTimer2 = 0
  vTime = 22
  Timer2.Enabled = True
  Form1.AddtoList 2, App.Path & "\wav\lbutton.mp3"
  Form1.PlaySelected 2
 End If
End If
End Sub


Private Sub Timer1_Timer()
Static tmr As Integer
    tmr = vTimer1
    vTimer1 = vTimer1 + 1
    If tmr = vTime Then
    vStep = 1
    tmr = 0
    Timer1.Enabled = False
    StopPlaying 2
    SetPos 2, 0
    voiceck
    End If
End Sub

Private Sub Timer2_Timer()
    Static tmr As Integer
    tmr = vTimer2
    vTimer2 = vTimer2 + 1
    If tmr = vTime Then
    StopPlaying 2
    SetPos 2, 0
    tmr = 0
    Timer2.Enabled = False
    End If
End Sub
```

Glossary of Commands

```
Dim WithEvents Con As ADODB.Connection
Dim WithEvents rst As ADODB.Recordset
Dim Cmd As ADODB.Command
Dim strResponse As String


Private Sub cmd1_Click()

chec:
On Error GoTo errh

Set rst = New ADODB.Recordset
```

```
With rst

    .ActiveConnection = Con
    .CursorLocation = adUseClient
    .CursorType = adOpenDynamic
    .LockType = adLockOptimistic
    .Open "glossary" 'opening glossary table

    'adding records from textbox to recordset
    .AddNew
    .Fields(0) = StrConv(txt1, vbProperCase)
    .Fields(1) = StrConv(txt2, vbProperCase)
    .Update

End With

    ` closing the recordset
    rst.Close
    Set rst = Nothing


    lbl1.Caption = txt2.Text
    txt1.Text = ""
    txt2.Text = ""
    Call dload ` calling private procedure to fill the flexgrid

errh:       'in case of error, informing the user

If Err.Description <> vbNullString Then
    MsgBox Err.Description
End If

End Sub

Private Sub cmd2_Click()

    strResponse = MsgBox("Do you want to update this record?", vbYesNo + vbQuestion, _
                "Conform Update")

    If strResponse = 7 Then
    Exit Sub
    End If

    Dim str As String
    str = List1.Text

On Error GoTo errhan

    Set rst = New ADODB.Recordset

With rst

    .ActiveConnection = Con
    .CursorLocation = adUseClient
    .CursorType = adOpenDynamic
    .LockType = adLockPessimistic
    .Open "select * from glossary where Title='" & str & "'"

    .Fields(0) = StrConv(txt1, vbProperCase)
    .Fields(1) = StrConv(txt2, vbProperCase)
    .Update ` updating the recordset

End With
```

```
    Set rst = Nothing

    lbl1.Caption = txt2.Text
    txt1.Text = ""
    txt2.Text = ""
    Call dload

errhan:

If Err.Description <> vbNullString Then
    MsgBox Err.Description
End If
End Sub

Private Sub cmd3_Click()

    strResponse = MsgBox("Do you want to delete this record?", vbYesNo + vbQuestion, _
                "Conform Delete")

    If strResponse = 7 Then
    Exit Sub
    End If

    Dim str As String

    str = List1.Text

    Set Cmd = New ADODB.Command

With Cmd

    .ActiveConnection = Con
    .CommandType = adCmdText
    .CommandText = "delete from glossary where Title = '" & str & "'"
    .Execute

End With

    Set Cmd = Nothing

    lbl1.Caption = ""
    txt1.Text = ""
    txt2.Text = ""

Call dload
End Sub

Private Sub cmdmap_Click()
    Load CharacterMap
    CharacterMap.Visible = True
    CharacterMap.SetFocus
End Sub

Private Sub List1_Click()
Call dload2
End Sub

Private Sub Form_Load()
    Call connect

    Set rst = New ADODB.Recordset

    rst.ActiveConnection = Con
    rst.CursorLocation = adUseClient
```

```
    rst.CursorType = adOpenDynamic
    rst.LockType = adLockOptimistic
    rst.Open "select * from glossary where Title='Displacement'"

    strtmp = rst.Fields(0).value
    lbl1.Caption = rst.Fields(1).value
    txt1.Text = rst.Fields(0).value
    txt2.Text = rst.Fields(1).value

    Set rst = Nothing

End Sub

Private Sub Form_Unload(Cancel As Integer)
Con.Close
Set Con = Nothing
DataEnvironment1.Connections(1).Close
End Sub

Private Sub connect()

Set Con = New ADODB.Connection

Con.CursorLocation = adUseClient

Con.Open "Provider=Microsoft.Jet.OLEDB.3.51;Persist Security Info=False;Data Source=" & App.Path & "\database.mdb"

DataEnvironment1.Connections(1).Open Con

Call dload

End Sub

Private Sub dload()
Dim S As String
List1.Clear

Set rst = New ADODB.Recordset

    rst.ActiveConnection = Con
    rst.CursorLocation = adUseClient
    rst.CursorType = adOpenDynamic
    rst.LockType = adLockOptimistic
    rst.Source = "Glossary"
    rst.Open

    rst.MoveFirst

While Not rst.EOF()
    S = ""
    S = rst.Fields(0).value
    List1.AddItem S
    rst.MoveNext
Wend

Set rst = Nothing

End Sub

Private Sub dload2()
    Dim str As String

    str = List1.Text
```

```
    Set rst = New ADODB.Recordset

    rst.ActiveConnection = Con
    rst.CursorLocation = adUseClient
    rst.CursorType = adOpenDynamic
    rst.LockType = adLockOptimistic
    rst.Open "select * from glossary where Title='" & str & "'"

    lbl1.Caption = rst.Fields(1).value
    txt1.Text = rst.Fields(0).value
    txt2.Text = rst.Fields(1).value

    Set rst = Nothing

End Sub
```

Appendix C

GENERAL CONCEPT OF TRUSS ANALYSIS

A 2-D plane structure, also known as truss, is an engineering structure made of two force members all pin connected to each other. As an example the schematic of a plane truss is shown in Figure C-1.

The truss in Figure 1 consists of 9 members and 6 joints. For example, there is a member from joint *A* to *B*, another from joint *B* to *C*, and a third from joint *C* to *D*. Due to the external loading (i.e. P_1 and R_2) acting on the truss, each member of the truss will be subjected to internal forces either of the tensile or compressive nature. These internal forces can be calculated by applying any of the three equation of equilibrium with respect to the *x-y* reference system as follows:

$\sum \mathbf{F}_x = 0 \qquad \sum \mathbf{F}_y = 0 \qquad \text{and } \sum \mathbf{M} = 0$

These equations of equilibrium states that in order for the structure to remain in the state of equilibrium then (i) the sum of all forces in the *x*-direction must be equal to zero ($\sum \mathbf{F}_x = 0$), (ii) the sum of all forces in the *y*-direction must be equal to zero ($\sum \mathbf{F}_y = 0$) and (iii) the sum of all moments about any point on the structure must also be equal to zero ($\sum \mathbf{M} = 0$).

Figure C-1. A typical loaded plane truss that is supported at A and D.

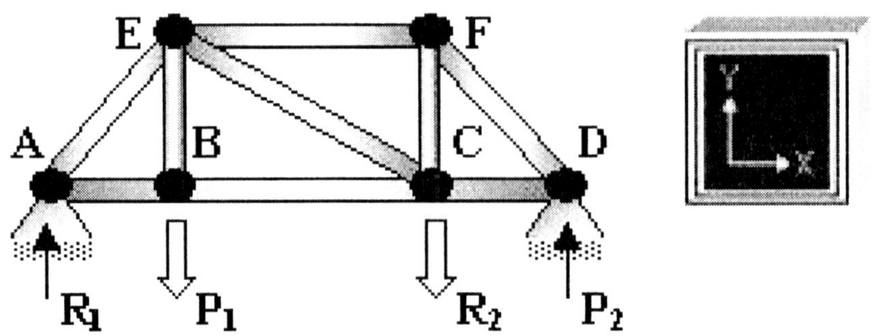

Prior to solving the problem, it is normal practice to adopt a reference system such as the *x-y* system as shown in the top right hand corner in Figure 1. All forces must first be resolved into their components acting in the *x*- and *y*-directions. According to the reference system, force acting in the same direction of the *x*-axis is taken as positive whereas force acting in the opposite direction of the *x*-axis is taken as negative. A similar concept is applied to forces acting in the *y*-direction. As for taking moments about a chosen point, the choice is arbitrary; one can assume that counterclockwise moment is positive and clockwise moment is negative or vice-versa. Thus, the components of any unknown forces can then be determined by applying the equations of equilibrium with respect to this reference system.

As an example for the truss problem shown in Figure C-2 (a), the students are required to determine the force in each member of the truss and indicate whether the member is in tension or compression.

A free body diagram showing all forces acting on the truss is first drawn as shown in Figure C-2 (b). The two supports at *A* and *C*, are represented by the internal reactions A_y, C_x and C_y as shown in Figure C-2 (b). These unknown reaction forces can be determine by applying the equations of equilibrium as follow:

$\sum \mathbf{F}_x = 0$: $600 - C_x = 0$ $\therefore C_x = 600$ N

$\sum \mathbf{M}_C = 0$: $-A_y(6) + 600(4) = 0$ $\therefore A_y = 600$ N

$\sum \mathbf{F}_y = 0$: $A_y - 400 - C_y = 0$

$600 - 400 - C_y = 0$ $\therefore C_y = 200$ N

The analysis for the aforementioned problem can now start at joint *A* or *C*. The choice is arbitrary, since there is one known and two unknown member forces acting on the pin of these joints.

Figure C-2 (a). A typical loaded truss and C-2 (b). Free body diagram of the truss.

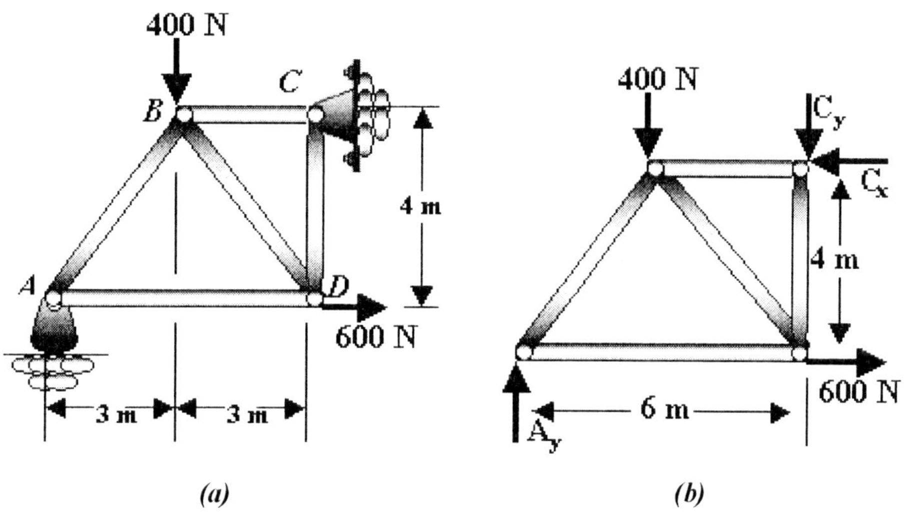

Joint A. As shown in the free-body diagram in Figure C-3, there are three forces that are acting at joint *A*. The inclination of $\mathbf{F}_{AB}$ and $\mathbf{F}_{AD}$ are determined from the geometry of the truss. By inspection, students must apply knowledge gained during lectures on how to assume the nature of the unknown forces. Alternatively, the student can initially assume that all the members are under tension (taken as positive) and if the computed value gives a negative, then the student will understand that the member in question is not under tension but rather under compression. For argument sake, let's assume initially that member *AB* is under compression whereas member *AD* is under tension. These assumptions are represented diagrammatically in the free body diagram by the sense of direction of the arrows representing the forces $\mathbf{F}_{AB}$ and $\mathbf{F}_{AD}$ as shown in Figure C-3. If the assumptions are correct, the computed values for both forces will be positive.

Thus applying the equations of equilibrium, we have:

$$\sum \mathbf{F}_y = 0\text{: } 600 - 4/5\ \mathbf{F}_{AB} = 0 \qquad \therefore\ \mathbf{F}_{AB} = 750 \text{ N (C)}$$

$$\sum \mathbf{F}_x = 0\text{: } \mathbf{F}_{AD} - 3/5\ (750) = 0 \qquad \therefore\ \mathbf{F}_{AD} = 450 \text{ N (T)}$$

The letters (T) and (C) indicate tensile and compressive in nature, respectively. Since, both answers are positive, therefore the initial assumption was correct.

Joint D. In a similar fashion, as shown in the free-body diagram in Figure C-4, joint *D* is chosen since by inspection of Figure C-2a, the force in member *AD* (i.e. $\mathbf{F}_{AD}$) is already known. As such, the two unknown forces in member *DB* ($\mathbf{F}_{DB}$) and member *DC* ($\mathbf{F}_{DC}$) can be determined. Let's assume initially that both members are under compression. As such, the sense of direction of both internal forces (i.e. $\mathbf{F}_{DB}$ and $\mathbf{F}_{DC}$) is as shown schematically in Figure C-4. By summing all the forces in the *x*-direction, we have:

$$\sum \mathbf{F}_x = 0 – 450 + (3/5)\ \mathbf{F}_{DB} + 600 = 0 \qquad \therefore\ \mathbf{F}_{DB} = –250 \text{ N}$$

The negative sign indicates that $\mathbf{F}_{DB}$ acts in the opposite sense to that shown in Figure C-4 and is therefore tensile in nature (i.e. member *DB* is under tension). Hence,

Figure C-3. Free body diagram showing all forces acting at joint A.

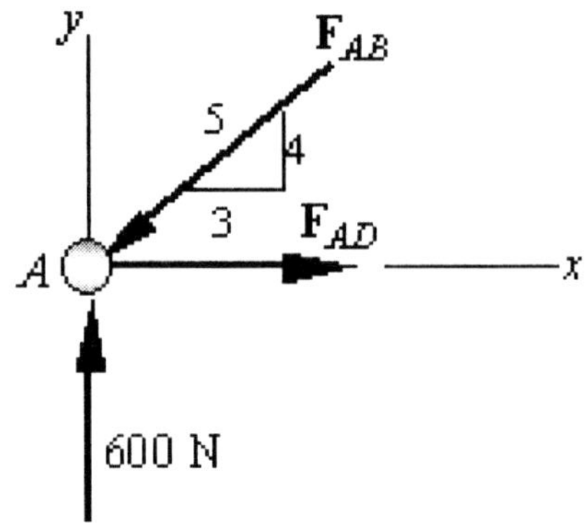

$\therefore F_{DB} = 250$ N (T)

To determine $\mathbf{F}_{DC}$, the student can either correct the sense of $\mathbf{F}_{DB}$ or then apply $\sum \mathbf{F}_y = 0$, or apply this equation of equilibrium and retain the negative value for $\mathbf{F}_{DB}$. Assuming we choose the latter option, we have:

$$\sum \mathbf{F}_y = 0: -\mathbf{F}_{DC} - (4/5)\,\mathbf{F}_{DB} = 0$$

$$-\mathbf{F}_{DC} - (4/5)(-250) = 0 \quad \therefore \mathbf{F}_{DC} = 200 \text{ N (C)}$$

Joint C. The free body diagram of all forces acting on joint C is as shown in Figure C-5. As can be noted from Figure C-5, the only unknown force is the reaction force in member *CB*. If the student assumes that that the member *CB* is under compression, as indicated by the sense of direction of $\mathbf{F}_{CB}$ in Figure C-5, by applying the equation of equilibrium in the *x*-direction, we have:

$$\sum \mathbf{F}_x = 0: \mathbf{F}_{CB} - 600 = 0 \quad \therefore \mathbf{F}_{CB} = 600 \text{ N (C)}$$

COACH BASED EXAMPLE: MOTION OF PROJECTILE IN MECHANICS DYNAMICS

In general, most multimedia-based learning software packages does not support learning by discovery but instead provide the basic capabilities for learning-by-doing. Adding interactivity by integrating multiple media elements such as audio, video, image and motion could provide the dynamic assessment, coaching, and feedback capabilities that are essential for positive learning. Some useful learning by doing approach taken from (Manjit and Ramesh, 2005) for mechanical engineering problem solving module that has a coach-based virtual learning environment incorporated is presented in Table C-1. The main interface of the engineering TAPS package is shown in Figure C-7 and Figure C-8, which highlights the configuration, interface and problem solving steps environment, developed using a variety of multimedia authoring software.

Figure C-4. Free body diagram showing all forces acting at joint D.

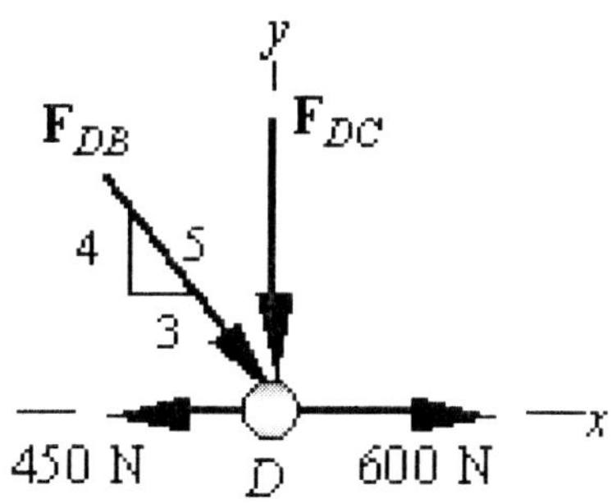

Figure C-5. Free body diagram showing all forces acting at joint C.

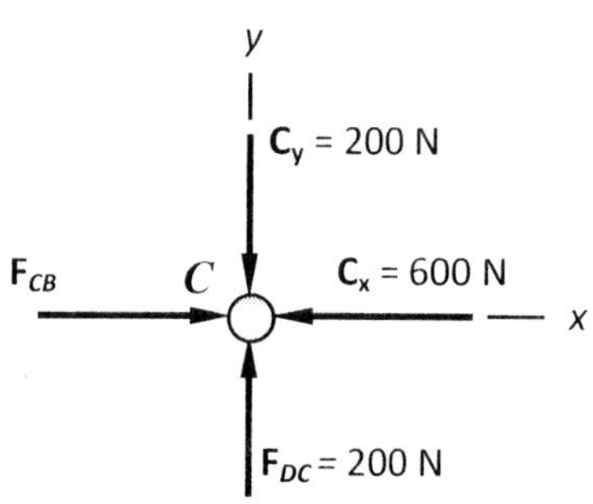

Figure C-6. The interface of the TAPS Package for (Experiment 1)

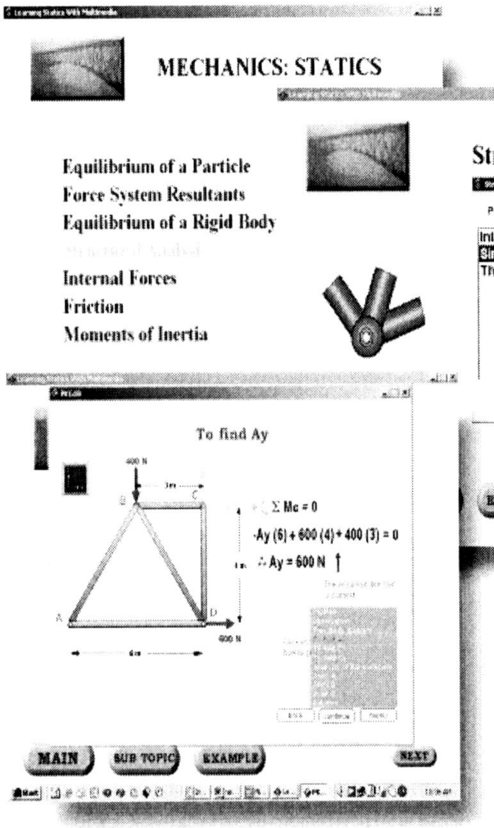

Table C-1. Learning by doing approach with a coach-based virtual learning environment.

Learning by Doing Approach	User Activities	Coach Virtual Learning Environment
Interaction	*The user interacts and observes meaningful tasks, e.g. the motion of a rider jumping of a platform.*	• Animated video files are integrated with audio files and graphics. • During the motion, the question is read out to the user.
Steps & Solutions	*A sequence of steps and solutions of the problem is presented to the user. The user moves forward to the next step or back to the previous step or solution.*	• Animated page showing steps and solutions are created and integrated with the TAPS package. • The TAPS package guides the user to manage the sequence of steps the user should perform to solve the problem and control the 2D animated mechanisms i.e. play, stop, reset and pause.
Simulations	*The user experiences a problem-solving environment in a virtual manner through the accumulation of his actions and the behavior of the animated mechanisms in a 2D environment.*	• The simulations are integrated with 2D graphics that are embedded with audio files. • The TAPS package manages the state of the 2D animated mechanisms and the user's interactions. • The TAPS package further provides graph for users to view data and interpret in a pictorial form.

Figure C-7. Main interface screen snap shot of the mechanics dynamics TAPS package.

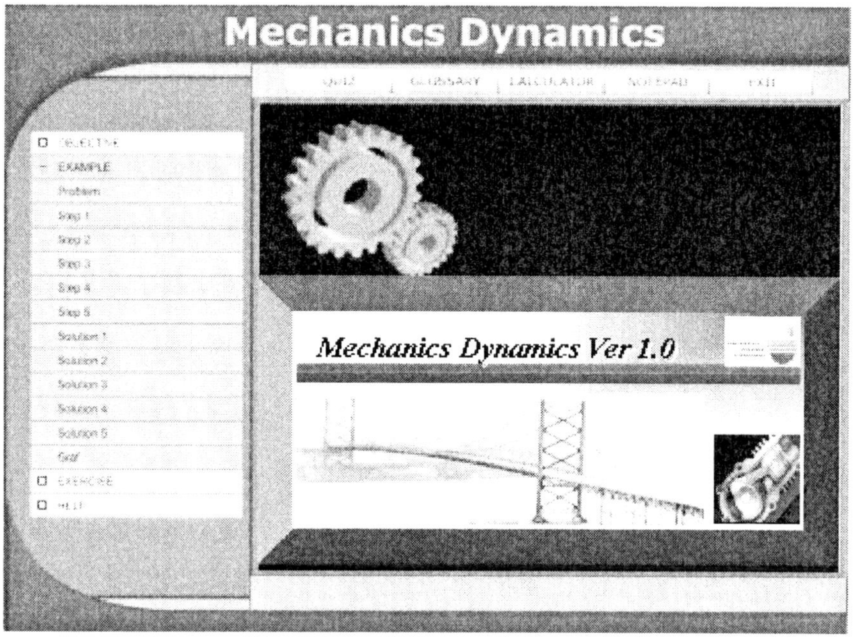

The engineering TAPS package implemented and presented in this example is used to illustrate a 2D virtual environment to allow users to understand as well as to visualize and ultimately to solve problems pertaining to motion of projectile in mechanics dynamics course.

By definition, a projectile has only one force acting upon it, i.e. the weight (W) of the projectile due to the gravitational attraction. All other forces such as wind resistance and the effect of drag are neglected. Thus, the free body diagram of a projectile would show a force W acting downwards.

Our past experiences in UNITEN have shown that many students have difficulty with the concept that the only force acting upon an upwardly moving projectile is W (i.e. $W = m \times g$, where m is the mass of the projectile and g is the acceleration due to gravity, normally taken as 9.81 m/s^2). Furthermore, their (students) conception of motion prompts them to think that if an object is moving upward, then there *must* be an upward force lifting the projectile or if an object is moving upward and rightward, there *must* be both an upward and rightward force acting on the projectile.

To explain further, lets consider the projectile problem given in Figure C-9. The rider leaves the 30° platform with an initial velocity (V_A) and in the absence of gravity (i.e., supposing that the "gravity switch could be turned off") the rider would then travel in a straight-line path in the direction of motion. The rider would continue in motion at a constant speed (V_A) in the same direction of motion provided there is no unbalanced force acting on him. This is the case for an object moving through space in the absence of gravity. However, if now the "gravity switch could be turned on" then upon leaving the platform the rider is treated as a projectile and the rider would be under *free-fall*. Under this circumstance, gravity pull takes effect and the path of motion of the projectile would no longer be a straight-line motion. In fact, the projectile would travel with a *parabolic trajectory* as shown in Figure C-10. As such, the downward force due to gravity effect, i.e. W, will act upon the rider to cause a vertical motion having a downward acceleration of $a_y = -9.81$ m/s^2 (the negative sign indicate that the motion is downward).

Figure C-8. Configuration, interface and problem solving steps environment.

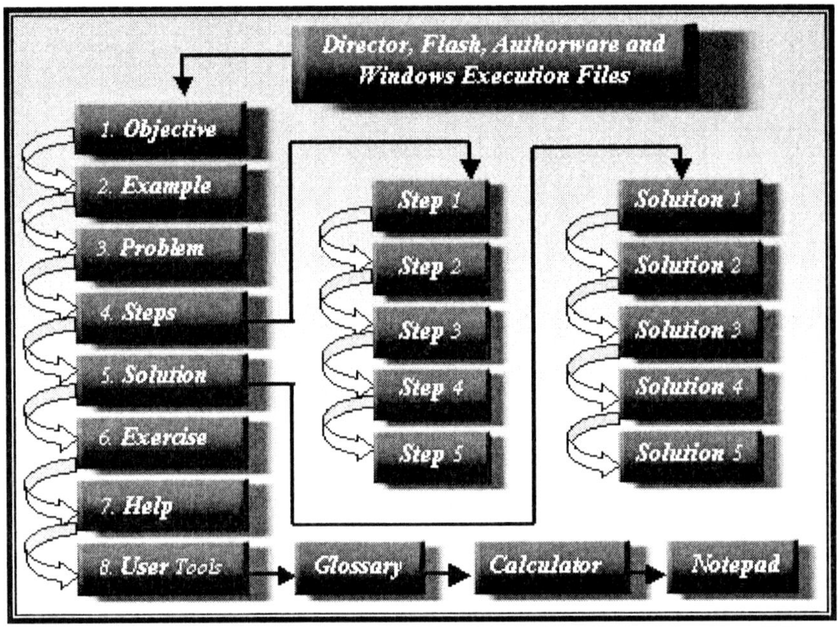

Figure C-9. Typical projectile problem in mechanics dynamics TAPS package.

The track for this racing event was designed so that riders jump off the platform at 30°, from a height of 1 m. During a race it was observed that the rider remained in mid air for 1.5 seconds. Determine the speed (V_A) at which he was traveling off the slope, and the maximum height (h) he attain. Neglect the size of the bike and the rider.

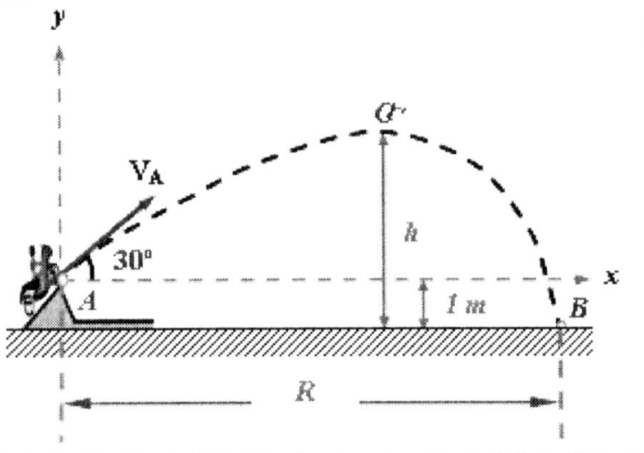

The presence of gravity, however does not affect the horizontal motion of the projectile. The projectile still moves the same horizontal distance in each second of travel as it did when the "gravity switch was turned off." Since, the force due to gravity (W) is a vertical force acting in the y-direction, it does not affect the horizontal motion in the x-direction. According to Newton's second law of motion, since there is no unbalance force acting on the projectile in the x-direction, then the projectile will not experience acceleration in the x-direction and the horizontal component of acceleration i.e. $a_x = 0$. As such, the projectile moves with a constant horizontal velocity in the x-direction (i.e. V_x = constant) throughout the flight.

Figure C-10. Motion of a projectile

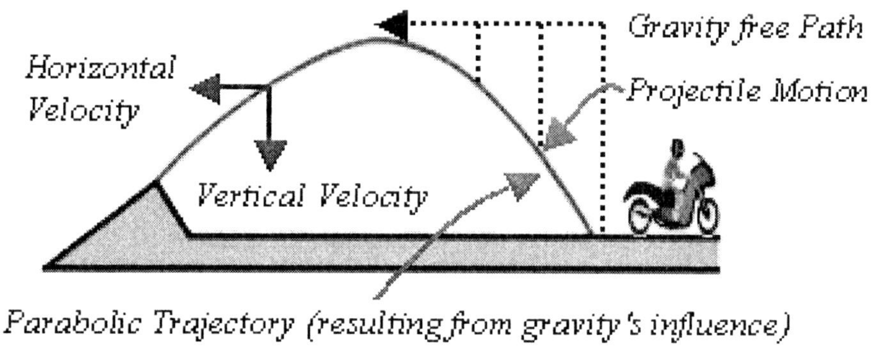

In the problem shown in Figure C-11, as the rider leaves the 30° platform, he undergoes an upward acceleration. However, as the rider strikes the ground, he undergoes a downward acceleration. A downwardly moving rider that is gaining speed is said to have a downward acceleration. In the animation, the downward acceleration is depicted by a change in the vertical component of velocity. This downward acceleration is attributed to the downward force of gravity that acts upon the rider. If the rider motion can be approximated as projectile motion (that is, if the influence of air resistance can be assumed negligible), then there will be no horizontal acceleration (i.e. $a_x = 0$ as shown in Figure C-11). In the absence of horizontal forces, the horizontal component of velocity at any instant will remain constant, i.e. $V_x = (V_A)_x$. This is illustrated graphically in Figure C-11 where the horizontal velocity component remains the same size throughout the entire motion of the rider.

Despite the aforementioned facts, many students would insist there should be horizontal force acting on the rider since he has a horizontal motion. However, this is simply not the case. The horizontal motion of the rider is the result of its own inertia. Inertia is the tendency of an object to resist changes in its state of motion. When jumped from the slope, the rider already possessed a horizontal motion, and thus will maintain this state of horizontal motion unless acted upon by a horizontal force. Therefore, the rider will continue in motion with the same horizontal velocity.

In the conventional method this sort of problems are usually presented to the student as a combination of schematic diagrams and text descriptions. The user/student must immediately apply learned knowledge in order to form an internal model of what the problem means. In addition, in mechanical engineering, the shapes and lines that make up the schematic diagram have very specific engineering meanings. Furthermore, the words accompanying the diagram normally provide additional hints to the problem in question and the user is expected to understand this before applying the appropriate theories in solving the question (Manjit and Ramesh, 2005).

On the contrary, with multimedia technology, such information and theories could be explained clearly through various media, which is not possible via conventional method of classroom teaching.

The key benefits that users could gained by using the multimedia TAPS package to study the motion of projectile in Figure C-9 are summarized as follows:

- The user could see the motion of the projectile at any instant or over a period of time.
- The velocity components (i.e. V_x and V_y) at any instant could be analyzed. This is clearly illustrated in Figure C-11. For example, when time $t = 0$, the rider is at the start point A on the slope of the

Figure C-11. A snap shot of the animation showing the change in the length of V_y with time depicting the change in the vertical velocity component of the rider during the motion. Note also that the length of the arrow representing V_x is constant throughout the motion to indicate that the horizontal velocity component is constant since $a_x = 0$.

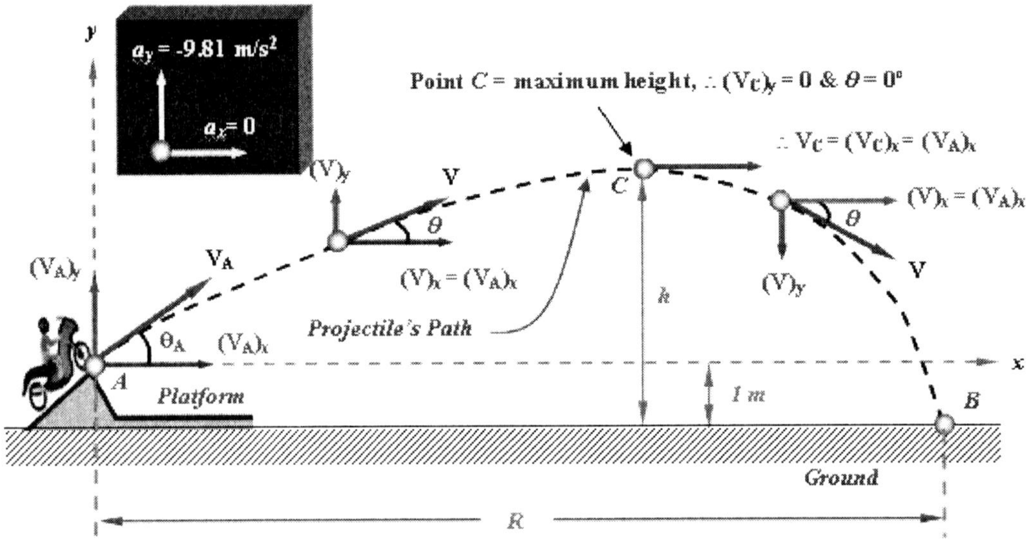

platform. At this instant, the rider is moving with a velocity of V_A measured at an angle of 30° from the *x*-axis. In the analysis, the velocity V_A can be represented by its components, i.e. $(V_A)_y$ measured along the y-axis and $(V_A)_x$ measured along the x-axis. Since the motion of the projectile at *A* is known, point *A* is taken as the reference point and the origin of the *x-y* axes is located at *A*.

- Enhanced visualization and understanding of the problem. The trace of the path taken by the rider can be visualized in a step-by-step manner as shown in Figure C-12. The user can then observe that the rider has taken a parabolic path, and as such is treated as a projectile motion. As the rider moves in a parabolic trajectory after leaving the platform, the velocity (V) of the rider changes with time and this is illustrated by the change in the length of the arrows representing the velocity components, i.e. $(V)_x$ and $(V)_y$. However, since there is no acceleration in the *x*-direction (i.e. $a_x = 0$), the horizontal component of velocity (V_x) will always be the same as the initial $(V_A)_x$. This is shown in the TAPS package by keeping the length V_x equal to $(V_A)_x$. To explain further on the theory, during the motion the user is narrated to reinforce learning and understanding. When the rider reached the maximum height at point *C*, the user can observe that at this instant, $(V_C)_y = 0$ and thus the velocity of the rider at *C* will be $V_C = (V_C)_x = (V_A)_x$.
- Promote learning. The incorporation of multimedia technology in this TAPS package could give a better understanding of the underlying projectile theory i.e. projectile travel with a parabolic trajectory due to the fact that the downward force of gravity accelerates the rider downward from his otherwise straight-line, gravity-free trajectory. This downward force and acceleration results in a downward displacement from the position that the object would be if there were no gravity. The force of gravity does not affect the horizontal component of velocity; a projectile maintains a

Figure C-12. Multimedia animation showing the sequence of motion of the rider with respect to time.

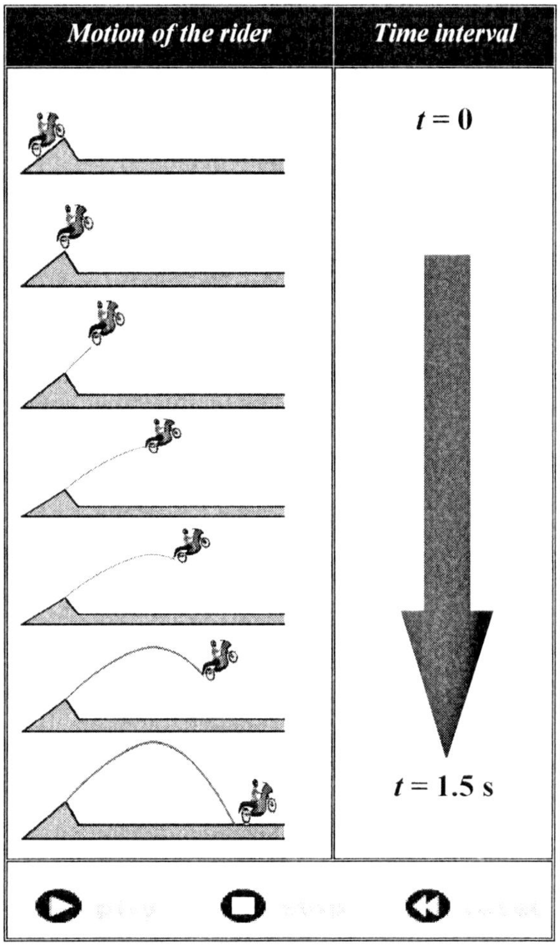

constant horizontal velocity since there are no horizontal forces acting upon it as clearly illustrated in Figure C-11. For slow learner, the motion of the projectile can be replay as many time necessary and in any order until the user understands the underlying principles.

- In addition, animated graphs could be shown to interpret the velocity (V) versus time (t) and distance (R) versus time (t) simultaneously of the projectile motion from $t = 0$ (point A) to $t = 1.5$ s (point B) as shown in Figure C-13.

The interactive multimedia TAPS package in the above example has achieved its objectives, as users were able to describe the position, velocity and acceleration as two-dimensional vectors, recognize two-dimensional projectile motion as simultaneous one-dimensional motion in two directions. Users were able to apply the relevant mechanics theories and kinematics equations to solve projectile motion. Furthermore, users could retained more information through interacting with the multimedia TAPS packages then they do from reading books and attending lectures.

This study has shown that multimedia technology is a powerful learning aid that could help learners/users to understand the underlying mechanics principles, visualize the motion of projectile in a dynamic

Figure C-13. Animated graphs describing the motion of the rider with respect to time.

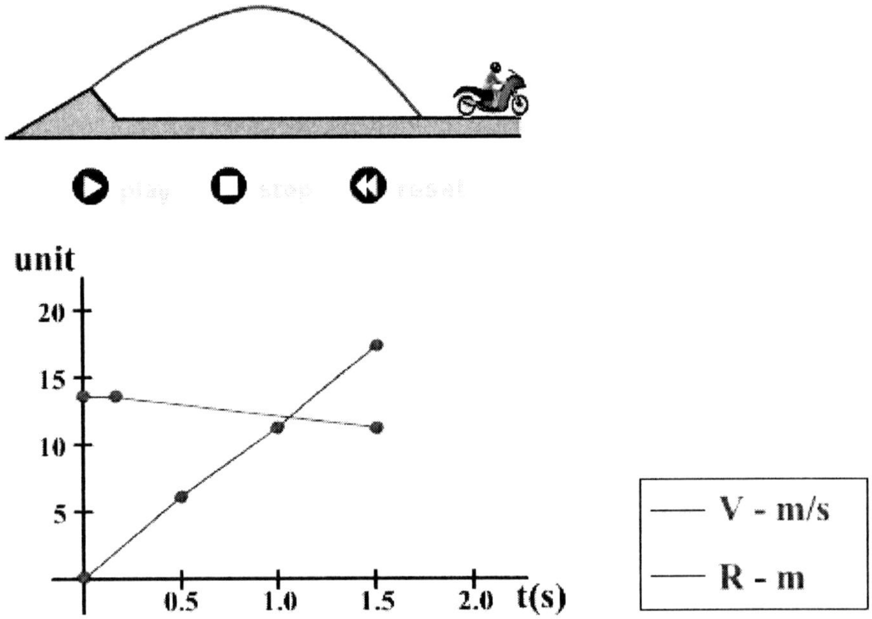

manner and more importantly, to promote deep learning. The interactivity that was incorporated in the multimedia TAPS package allowed users to manage and control the delivery of the material and act as a guide in problem solving. The potential benefits offered by effective virtual learning environment and the approach of learning by doing was identified and discussed.

Appendix D

A typical animation sequence for analyzing forces at joint D of the loaded truss.

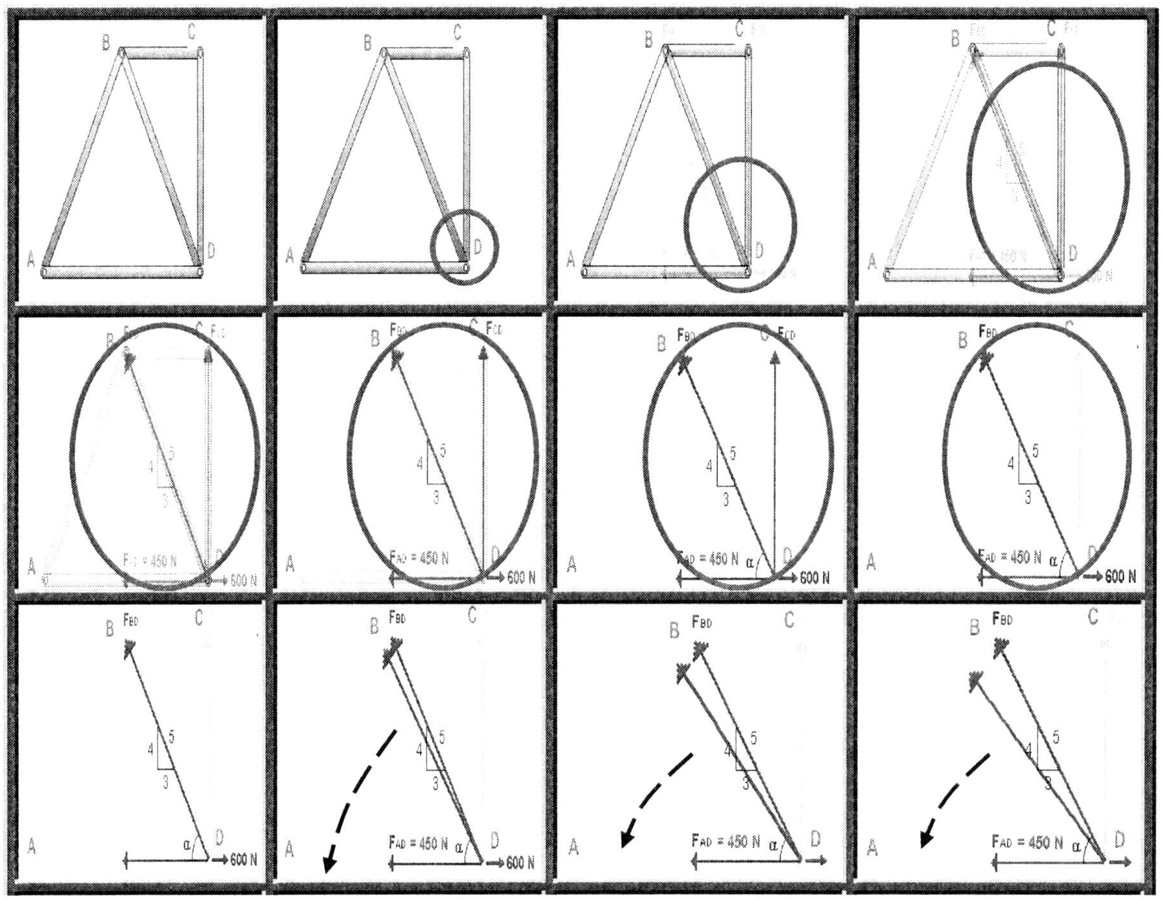

continued on the following page

A typical animation sequence for analyzing forces at joint D of the loaded truss, continued

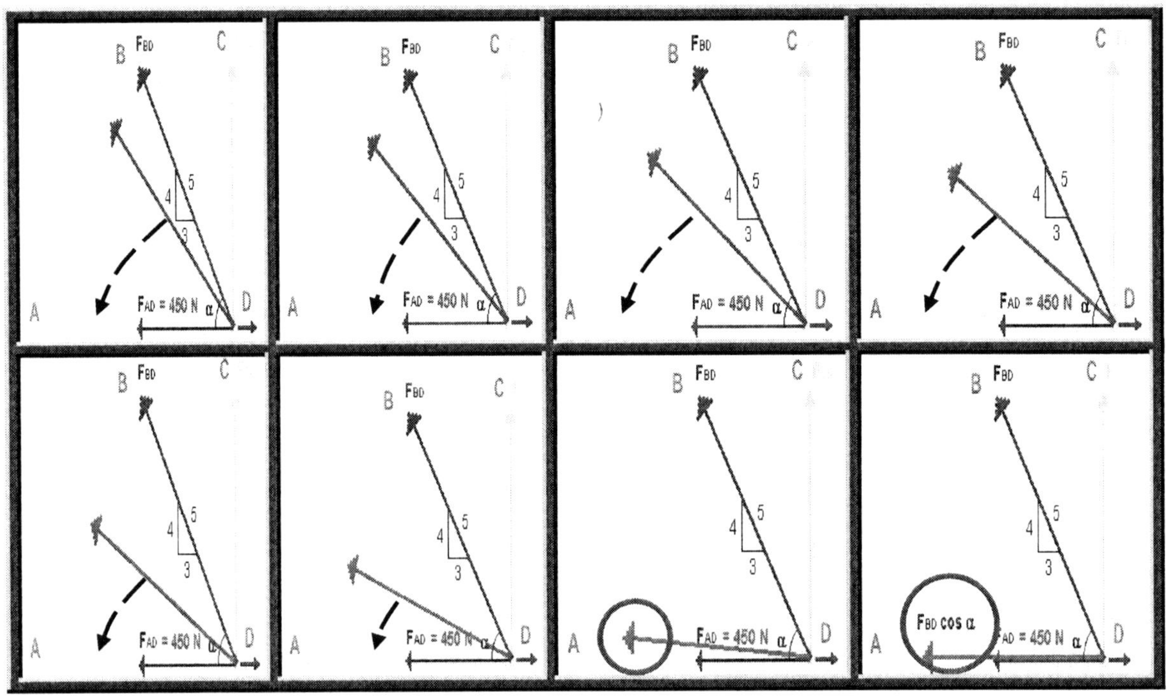

Appendix E

CURVILINEAR MOTION: *CYLINDRICAL COMPONENTS*

The rod *OA* in Figure, 1a is rotating in the horizontal plane such that $\theta = (t^3)$ rad. At the same time the collar *B* is sliding outward along *OA* so that $r = (100t^2)$ mm. If in both cases $t = 1$ s.

Coordinate System. Since time-parametric equations of the path are given, it is not necessary to relate r to θ.

Velocity and Acceleration. Determining the time derivations and evaluating when $t = 1$ s, we have

$r = 100t^2|_{t=1\text{s}} = 100$ mm $\theta = t^3|_{t=1\text{s}} = 1$ rad $= 57.3^{\circ}$

$\dot{r} = 200t|_{t=1\text{s}} = 200$ mm/s $\theta = 3t^2|_{t=1\text{s}} = 3$ rad/s

$\ddot{r} = 200|_{t=1\text{s}} = 200$ mm/s^2 $\theta = 6t|_{t=1\text{s}} = 6$ rad/s^2

Figure 1a.

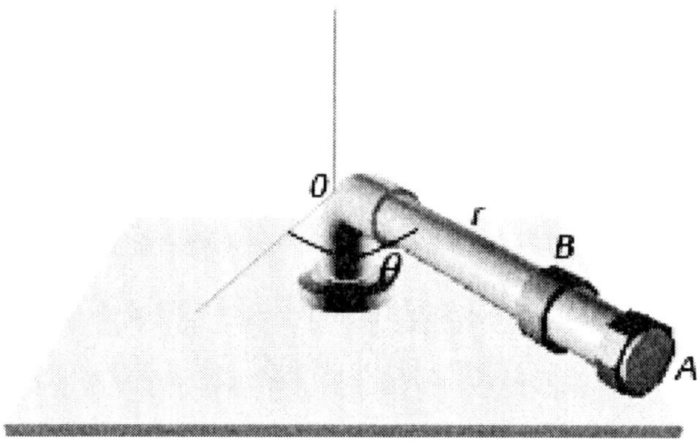

$$\mathbf{v} = \dot{r}\mathbf{u}_r + r\dot{\theta}\mathbf{u}_\theta$$
$$= 200\mathbf{u}_r + 100(3)\mathbf{u}_\theta$$
$$= \{200\mathbf{u}_r + 300\mathbf{u}_\theta\} \text{ mm/s}$$

The magnitude of **v** is:

$$v = \sqrt{(200)^2 + (300)^2} = 361 \text{ mm/s}$$

$$\delta = \tan^{-1}\left(\frac{300}{200}\right) = 56.3° \qquad \delta + 57.3° = 114°$$

$$\mathbf{a} = \left(\ddot{r} - r\dot{\theta}^2\right)\mathbf{u}_r + \left[100(6) + 2(200)3\right]\mathbf{u}_\theta$$
$$= \{-700\mathbf{u}_r + 1800\mathbf{u}_\theta\} \text{ mm/s}^2$$

The magnitude of **a** is:

$$a = \sqrt{(700)^2 + (1800)^2} = 1930 \text{ mm/s}^2$$

$$\varphi = \tan^{-1}\left(\frac{1800}{700}\right) = 68.7° \qquad (\ 80°\text{-}\varphi) + 57.3° = 169°$$

REFERENCE

Hibbler, R.C. (2002). *Engineering Mechanics Dynamics.* Upper Saddle River, NJ: Prentice Hall.

Appendix F

Figure F-1. 2-D animated example 1 illustrating curvilinear motion

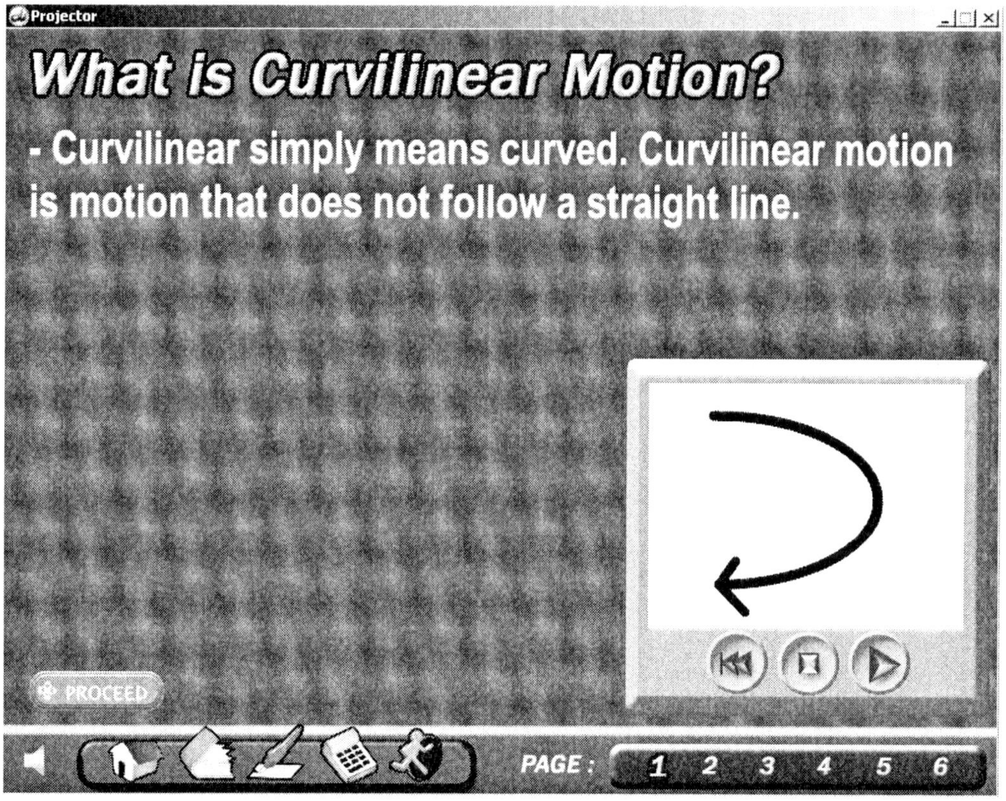

Figure F-2. 2-D animated example 2 illustrating curvilinear motion

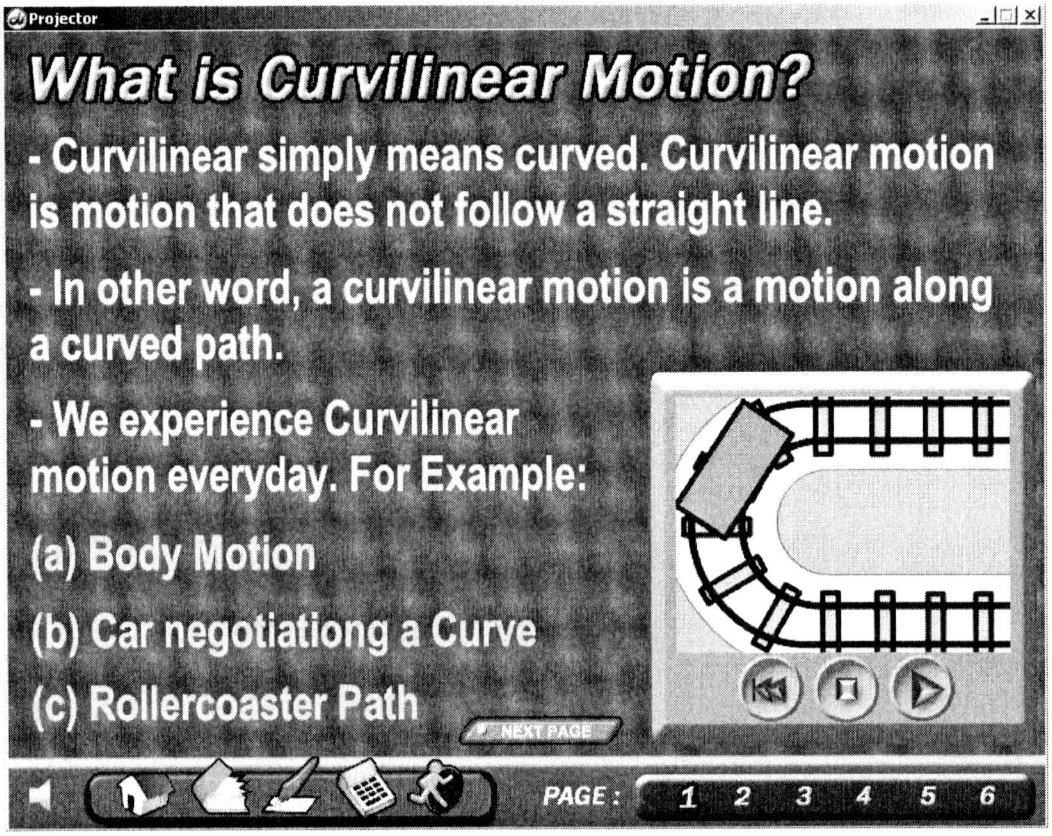

Appendix G

Figure G-1. 3-D animated example showing linear motion

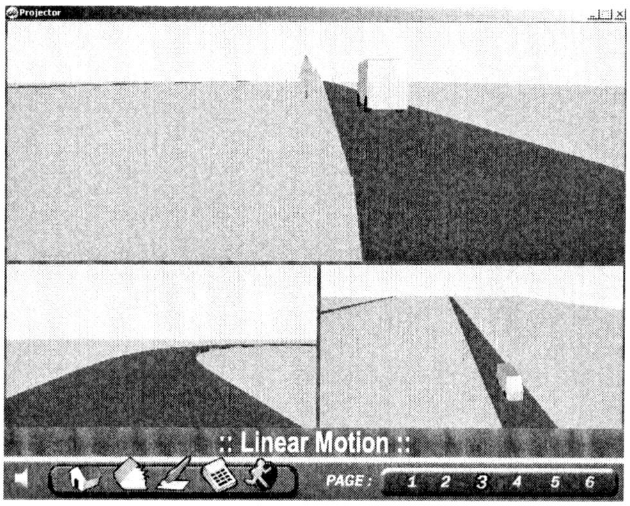

Figure G-2. 3-D animated example showing curvilinear motion

Figure G-3. 3-D animated example showing curvilinear and linear motion

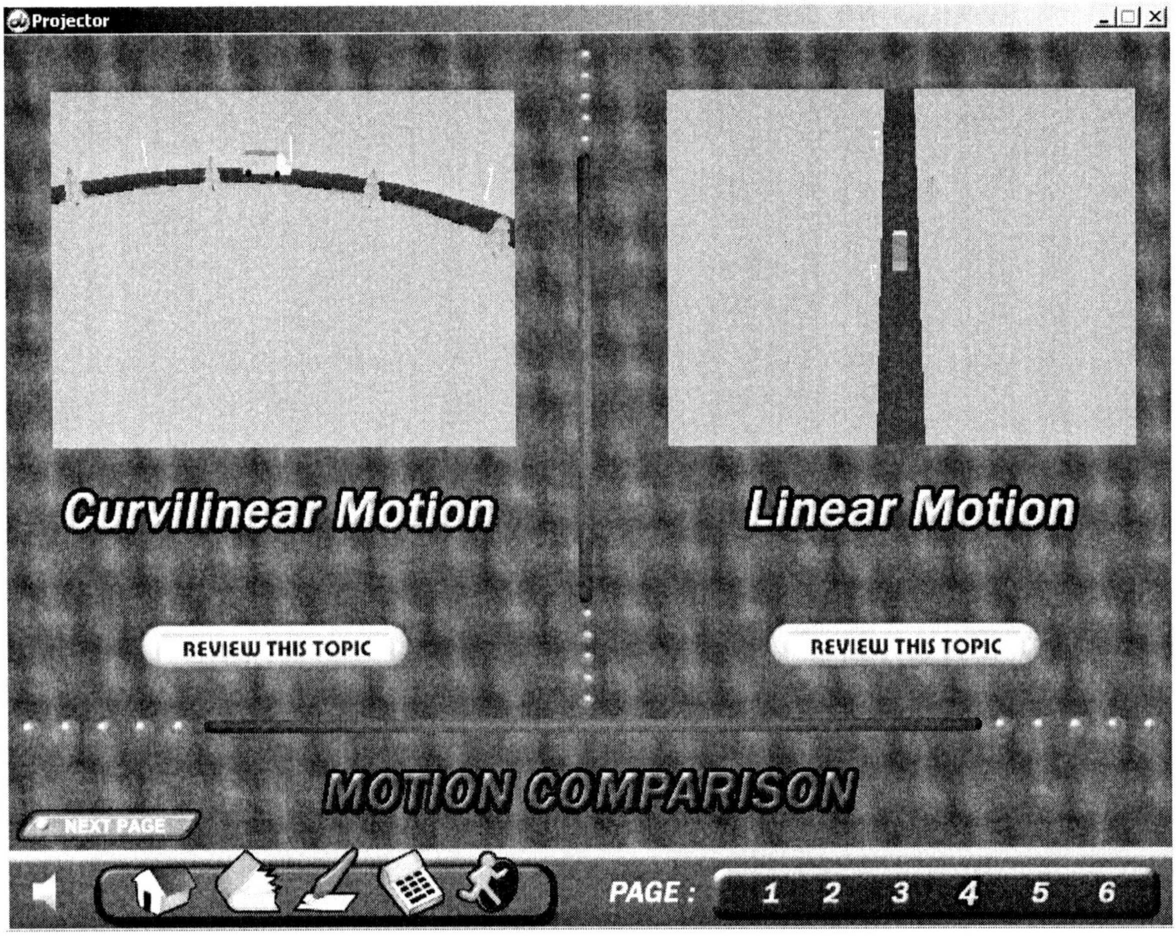

Appendix H

Figure H-1. Components of a particle (Velocity) that experiences curvilinear motion

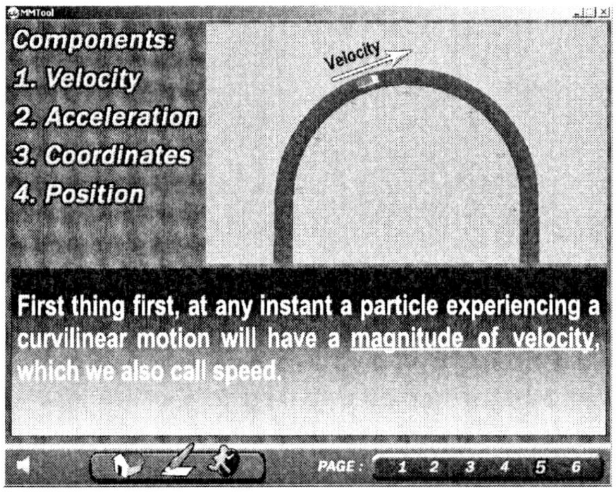

Figure H-2. Components of a particle (Acceleration) that experiences curvilinear motion

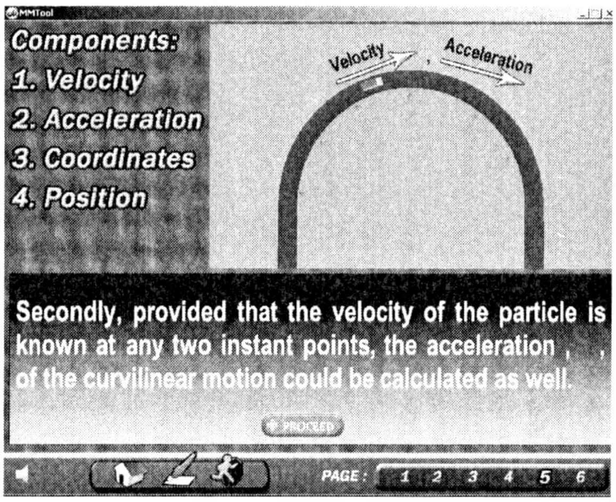

Figure H-3. Components of a particle (Coordinates – Part 1) that experiences curvilinear motion

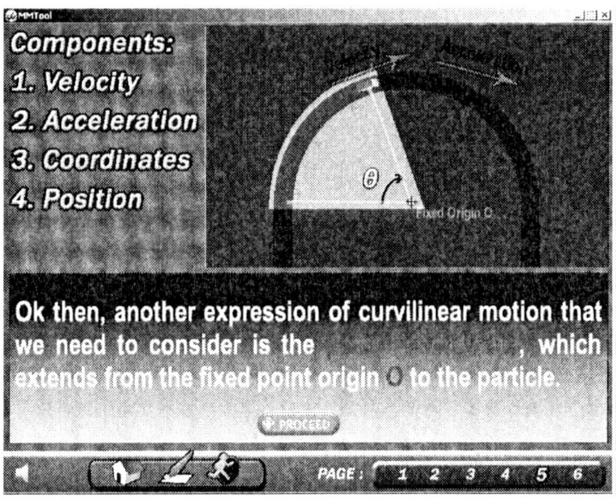

Figure H-4. Components of a particle (Coordinates – Part 2) that experiences curvilinear motion

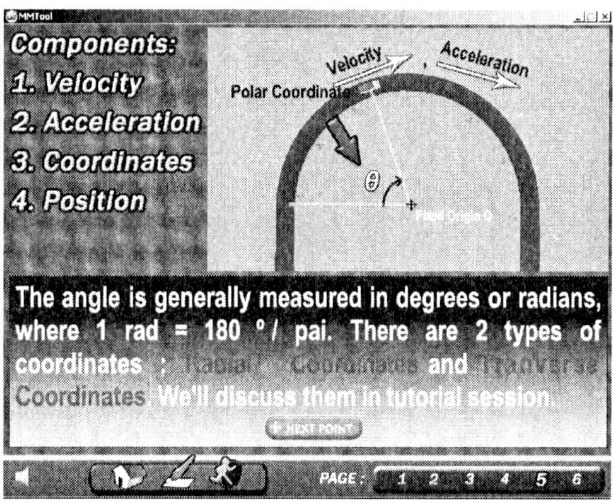

Figure H-5. Components of a particle (Position) that experiences curvilinear motion

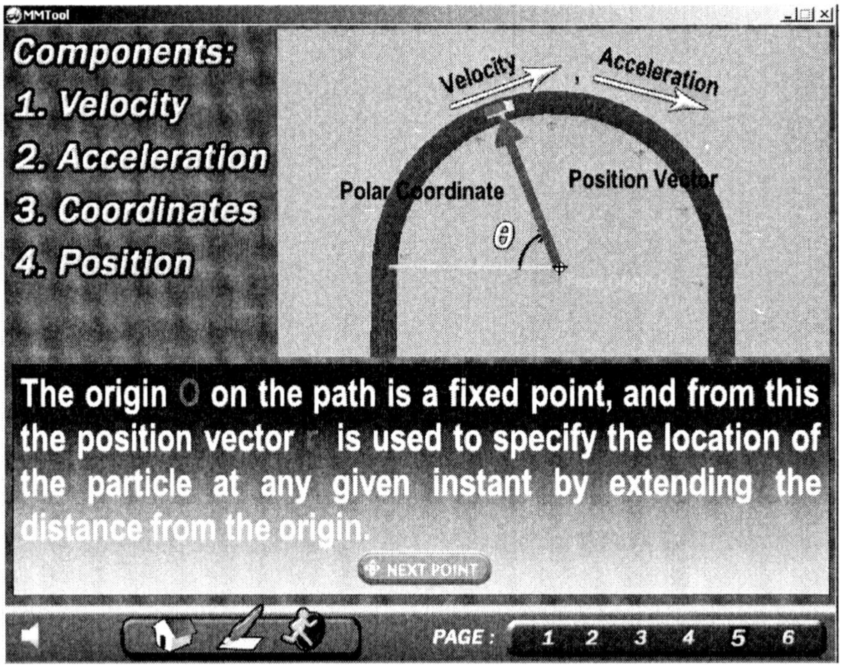

Appendix I

Figure I –1. Problem-solving models of engineering posed in the DVR TAPS package

Appendix J

Figure J-1. A typical problem solving question in the DVR TAPS Package

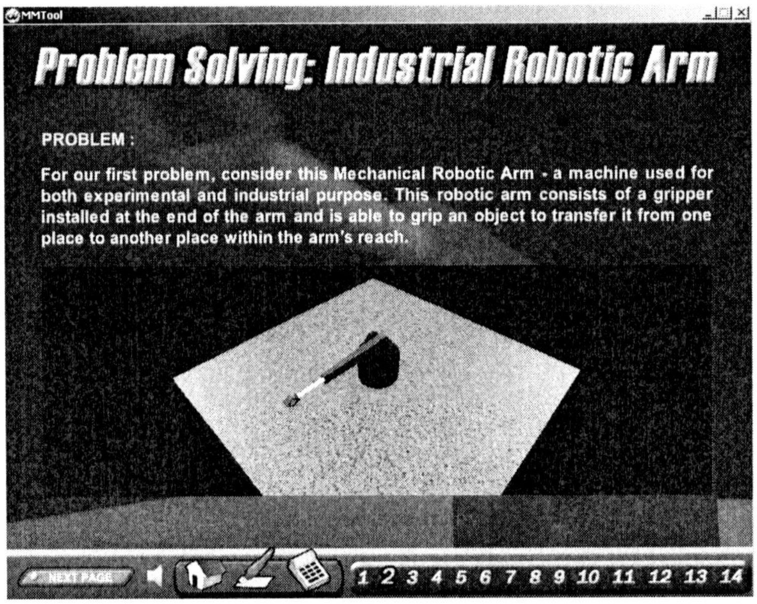

Figure J-2. Robotic arm movements i.e. arm extension, vertical & horizontal and rotational are animated and shown on the screen

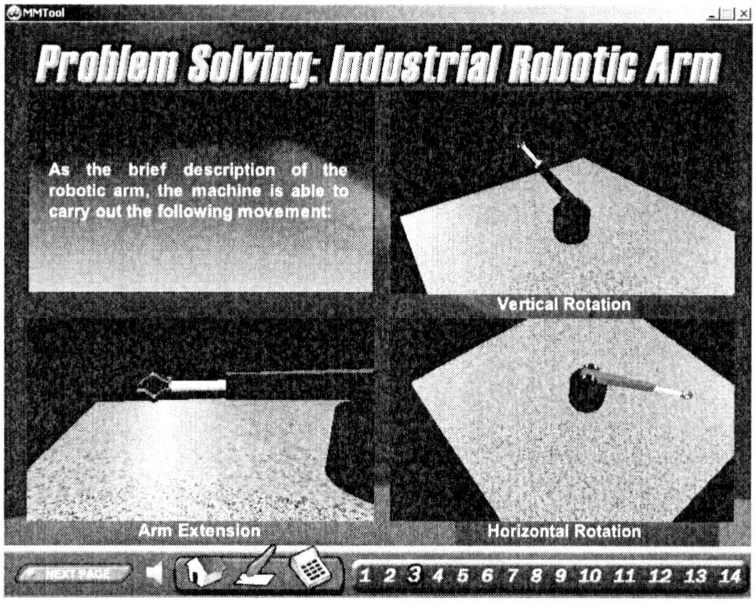

Figure J-3. Robotic arm gripping an object (e.g. ball) while the arm is extending outwards at a constant rate

Figure J-4. Student prompted to determine the magnitudes of the velocity and acceleration of the ball

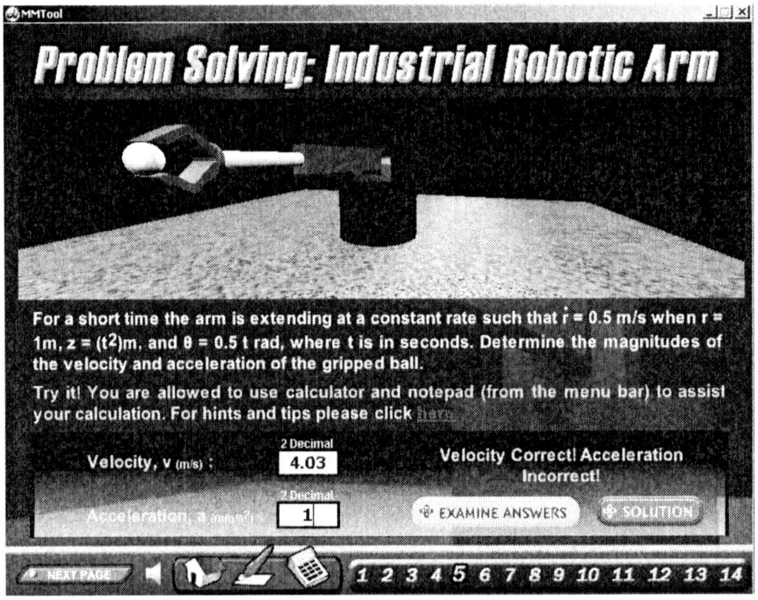

Appendix K

Figure K-1. Complete solution is provided and narrated leading from a series of steps to the answer (screen 1)

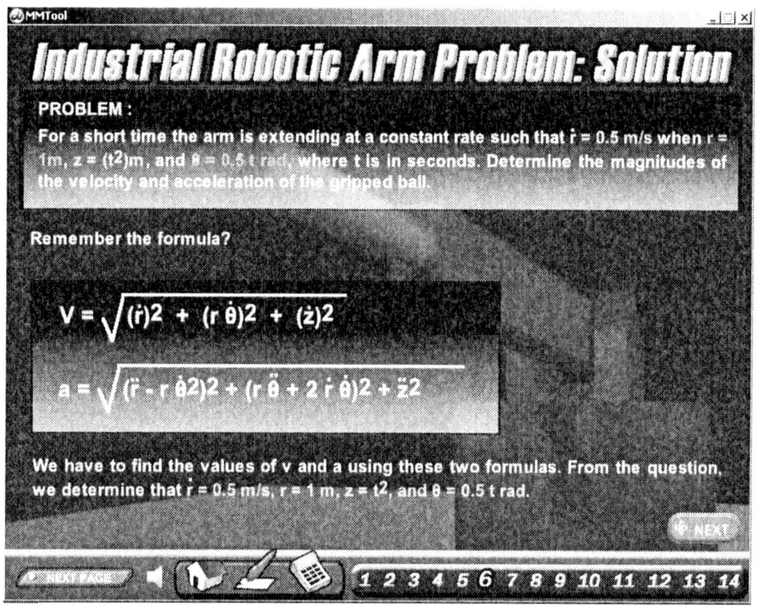

Figure K-2. Complete solution is provided and narrated leading from a series of steps to the answer (screen 2)

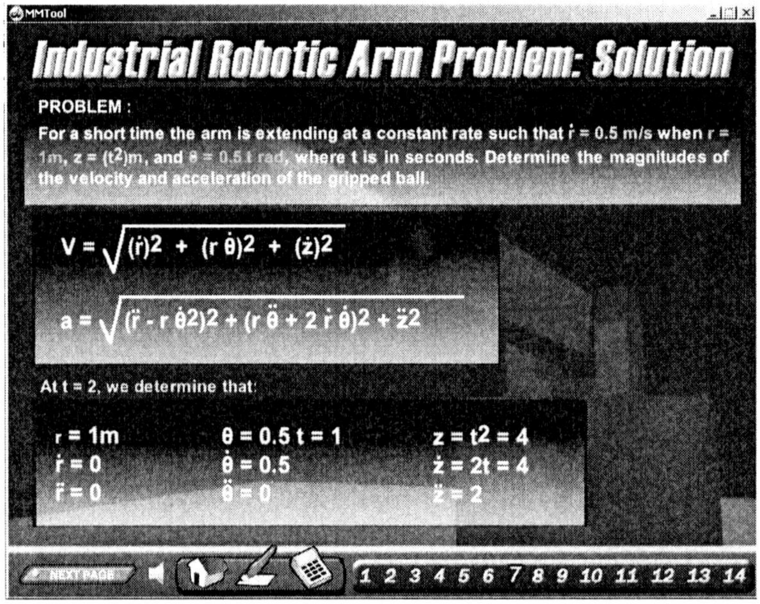

Figure K-3. Complete solution is provided and narrated leading from a series of steps to the answer (screen 3)

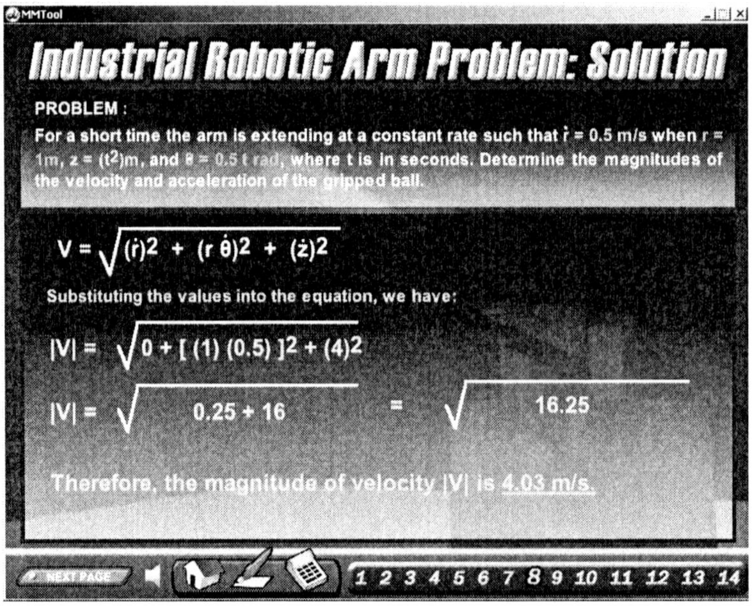

Figure K-4. Complete solution is provided and narrated leading from a series of steps to the answer (screen 4)

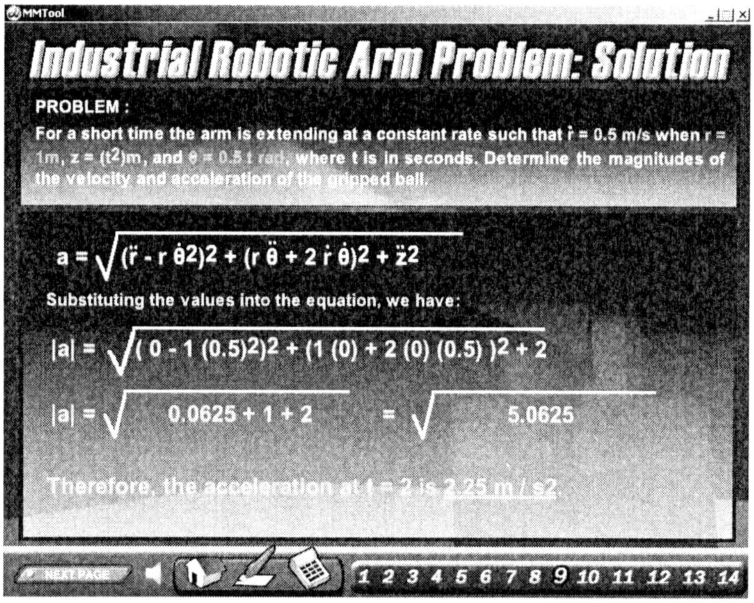

Appendix L

The animated motion of the robotic arm based on the answers shown in Appendix K:

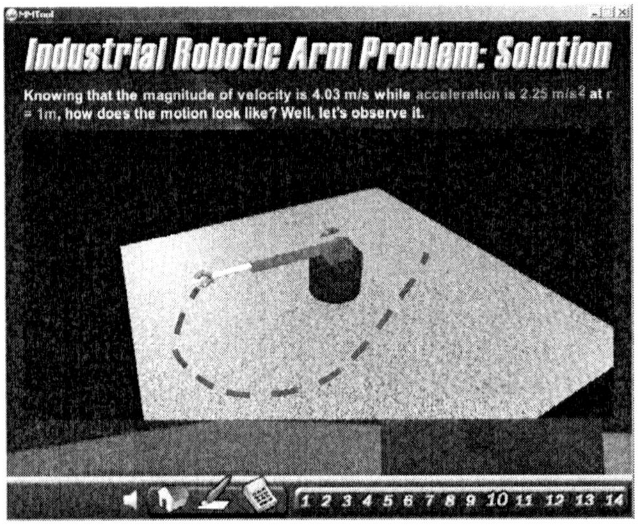

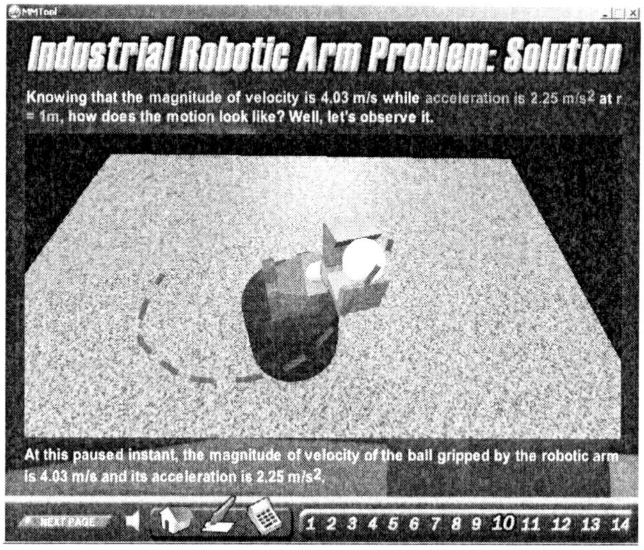
MMTool
Industrial Robotic Arm Problem: Solution
Knowing that the magnitude of velocity is 4.03 m/s while acceleration is 2.25 m/s2 at r = 1m, how does the motion look like? Well, let's observe it.
At this paused instant, the magnitude of velocity of the ball gripped by the robotic arm is 4.03 m/s and its acceleration is 2.25 m/s2.
NEXT PAGE
1 2 3 4 5 6 7 8 9 10 11 12 13 14

Appendix M

Figure M-1. The 3-D industrial robotic arm (normal mode)

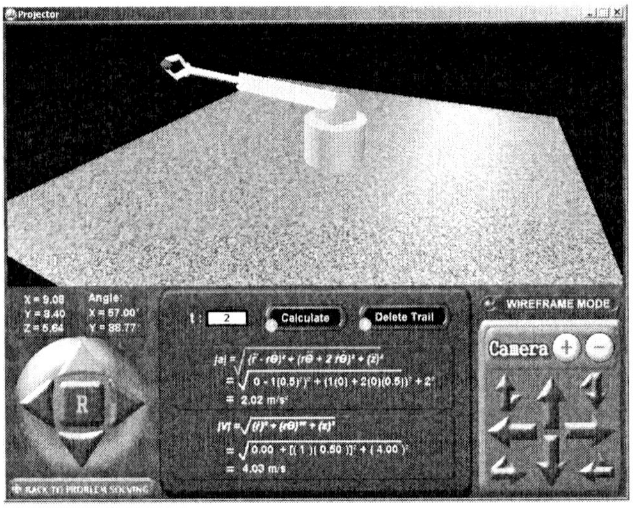

Figure M-2. The 3-D industrial robotic arm (wire-frame mode)

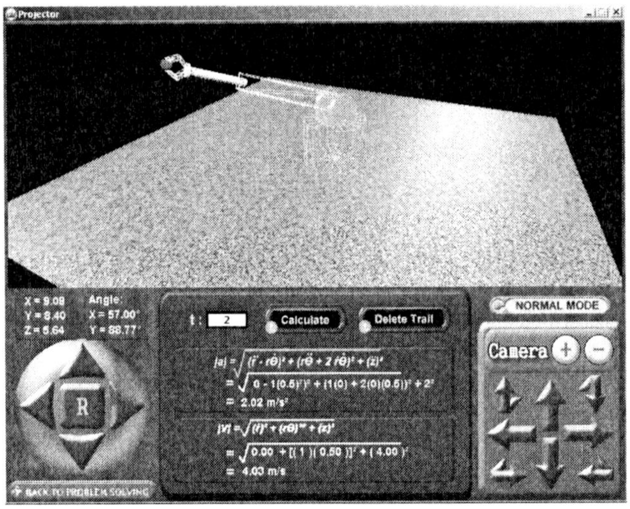

Appendix N

Figure N-1. Normal animation mode

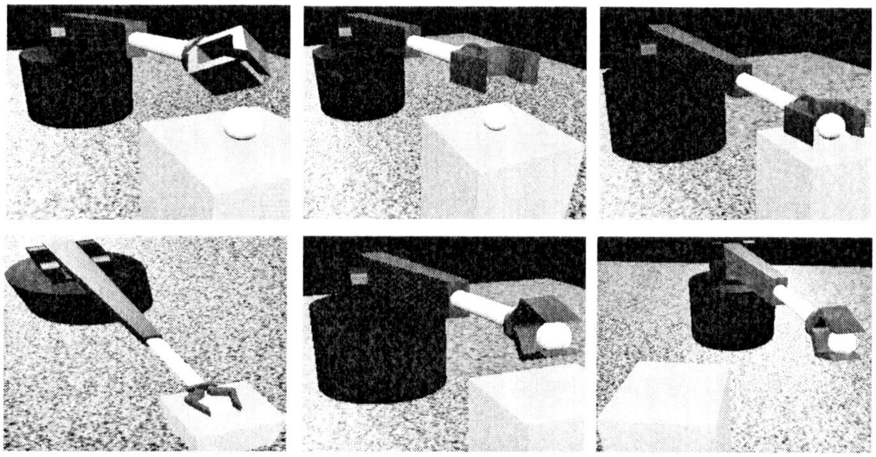

Figure N-2. Desktop virtual reality mode (animated stereoscopic view)

Figure N-3. Typical rotational sequences of the robotic arm and output showing the trail plotted by the DVR TAPS package on time input by user. The trails plotted clearly shows that the robotic arm has a curvilinear motion.

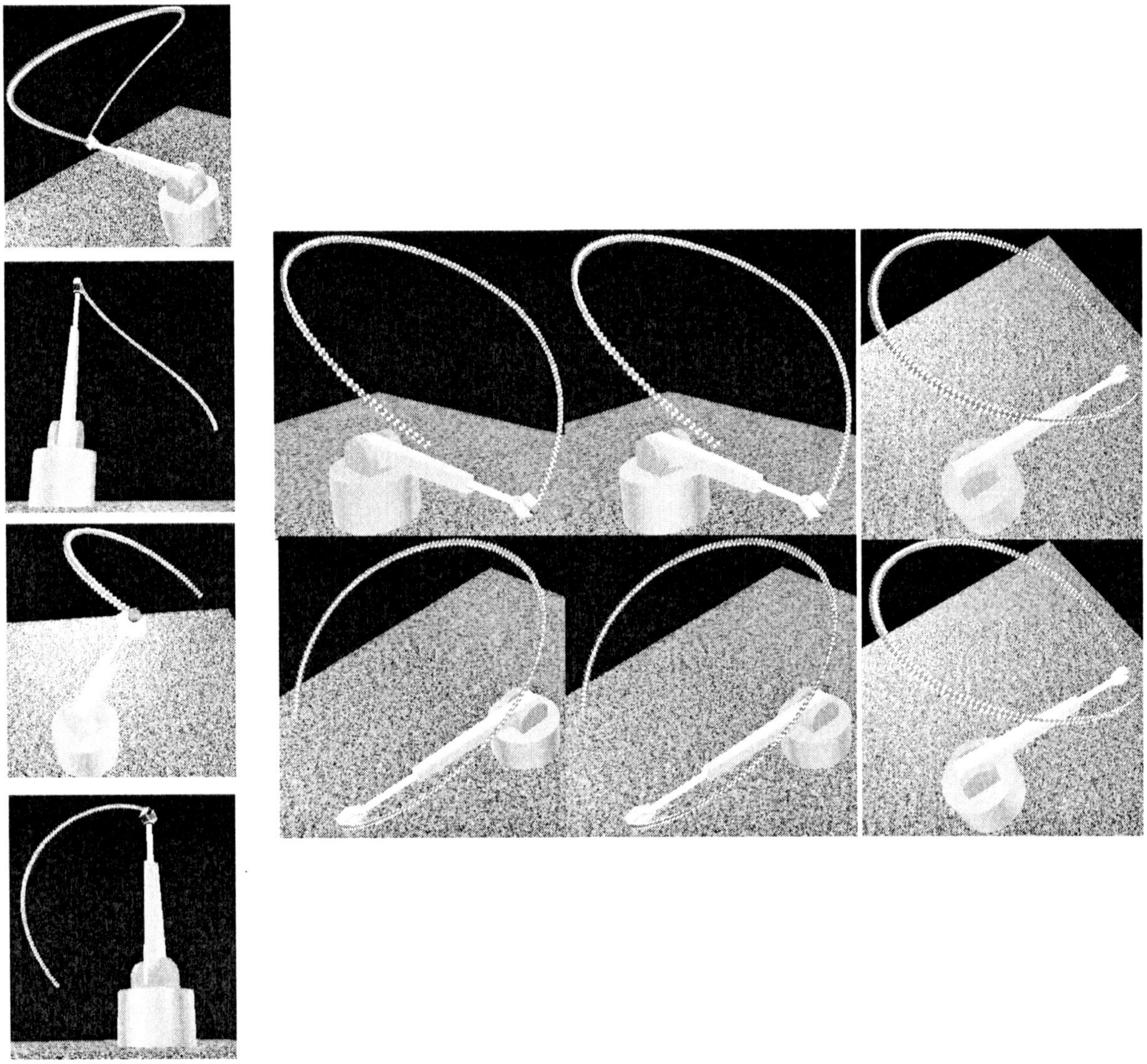

Appendix O

NEEDS COURSEWARE QUALITY REVIEW FORM: DRAFT

Criteria for Peer-Review of Engineering Courseware on the NEEDS Database

PART A. Author Supplied Information
Title of Courseware:____
Author Name: ___________________________________
Engineering subject areas addressed by this courseware.
Format of courseware: Case Study
Tutorial
Technical Reference
Laboratory/Experiment Support
Demonstration
Educational Game
Practice problems/Examination
Other ______________________
Prerequisite knowledge for intended student audience:
Context for Courseware's Instructional Use: Stand-alone initial instruction
Stand-alone supplementary instruction
Instructor-guided classroom instruction
Collaborative learning environment
Independent practice
Other __________________
Educational Goals of Courseware:
Recommended Pedagogical Use of Courseware
References to documentation regarding courseware:
Author Comments:

Part B: Courseware Gestalt
Title of Courseware: ____
Author Name: ____
Reviewer Name: ____

The purpose of this review is to determine if the above-referenced engineering courseware is of sufficient quality that it could easily be incorporated by an instructor (other than the author) into a course and enhance the students' learning experience. Please answer the questions below on a separate sheet of paper. Feel free to add any other comments you may have regarding the courseware.

Was the engineering content error-free?

Is the author's statement of target audience and educational goals consistent with the courseware's content?

Was the courseware visually appealing?

Did the courseware perform without technical errors? Was it complete and fully functional?

Are the media elements (graphics, images, sound, video) of average or better quality?

Do you think an instructor, other than the author of the courseware, would consider utilizing the courseware in his/her classroom?

Are you aware of any copyright infringements that may have occurred within this courseware?

Do you recommend that the courseware accepted as an Endorsed piece of courseware on the NEEDS Database?
YES NO (if no, please justify on a separate sheet.)

Is this courseware of such exceptional quality that it should be considered for the Premier Designation (which entails a more rigorous evaluation)? YES NO

Appendix P

Analysis of TAPS Packages
Questionnaire

Please go through the questions in this document and try to answer them as carefully as possible

Most of the questions relate
to the computer packages you have just used.

All information will be treated with the strictest confidence

On the following two sections are a list of questions about yourself and your familiarity with computers. Please respond either by ticking the appropriate box or by writing the relevant answer on the dotted lines provided.

1. Please indicate your name here..
2. Your major...

A. EXPERIENCE WITH COMPUTING

3. (a) Do you have a computer at HOME?

   Yes ☐ No ☐

(b) If YES, how much do you use it?

Daily ☐ Weekly ☐ Monthly ☐ Rarely ☐ Never ☐

4. (a) Do you have access to a computer at University/Hostel?

Yes ☐ No ☐

(b) If YES, how much do you use it?

Daily ☐ Weekly ☐ Monthly ☐ Rarely ☐ Never ☐

5. On the whole, can you operate it without help of a user manual or other person?

Yes ☐ No ☐ Not applicable ☐

6. How much do you enjoy using a computer?

Not at all	☐	Quite a lot	☐
Not much	☐	Very much	☐
Unsure	☐	Not applicable	☐

7. How often did you use a computer at school?

Daily ☐ Weekly ☐ Monthly ☐ Rarely ☐ Never ☐

8. (a) Have you received any computer related educational training/course?

Yes ☐ No ☐

(b) Which type(s) of educational training/course did you receive?

Education: Certificate in Information Technology ☐

Software applications:

- Database ☐
- Spreadsheet ☐
- Word processing ☐
- Internet ☐
- Multimedia ☐
- Graphics ☐
- AutoCAD ☐

If other, please specify......................................

9. Can you program in any computer languages?

Yes ☐ No ☐

10. (a) Have you used any computer aided learning package before?

Yes ☐ No ☐

(b) If YES, was it in engineering subjects?

Yes ☐ No ☐ Not applicable ☐

Please specify the topics covered...

1	2	3	4	5
Strongly Agree	***Agree***	***Uncertain***	***Disagree***	***Strongly Disagree***

B. EVALUATION OF THE TAPS PACKAGES

Given below are a number of statements about the TAPS packages, please indicate your response for each statement, ***using the scale above***. For example, if you feel that you ***strongly agree*** about the statement, then tick ✓ in the box next to the statement. There are no right and wrong answers, please give the response closest to your own feelings.

COMPUTING KNOWLEDGE		1	2	3	4	5
11	The packages do not require any knowledge of computing.	☐	☐	☐	☐	☐
12	I could not learn much because I spend time getting to know how to use the package.	☐	☐	☐	☐	☐
13	After using the TAPS packages I became more confident in learning from a computer.	☐	☐	☐	☐	☐
14	It was difficult enough to tackle the selected engineering problems without having to struggle with the computer.	☐	☐	☐	☐	☐
15	The system can be used sufficiently without any manuals.	☐	☐	☐	☐	☐
16	Normally I find these types of problems to be very simple, but I could not see how to solve the problems by using a computer.	☐	☐	☐	☐	☐
COMPARISON WITH OTHER MEDIA		1	2	3	4	5
17	I would prefer to learn from a human tutor than from these packages.	☐	☐	☐	☐	☐
18	It would have been a better improvement to have a tutor close by, so that I could ask questions.	☐	☐	☐	☐	☐
19	I liked problem solving in the subject matter with these packages, because it's like being one to one basis with the tutor.	☐	☐	☐	☐	☐
20	I was left with a lot of unanswered questions after using the packages.	☐	☐	☐	☐	☐
21	This form of problem solving and learning was really clear, because unlike an instructor, the program does not miss out a lot of steps.	☐	☐	☐	☐	☐
22	It is easier to get involved solving Engineering Mechanics problems using these packages than in a classroom tutorial.	☐	☐	☐	☐	☐
23	I prefer a textbook because I can flip it forward and backwards.	☐	☐	☐	☐	☐

Appendix P

1	2	3	4	5
Strongly Agree	***Agree***	***Uncertain***	***Disagree***	***Strongly Disagree***

	COMPARISON WITH OTHER MEDIA	1	2	3	4	5
24	It's much easier to understand the problem solving steps by using these packages than reading an Engineering Mechanics textbook because these packages are interactive and uses multimedia (2-D/3-D animations) in it's contents.	☐	☐	☐	☐	☐
25	It was good to be able to discover the relationships between variables and engineering concepts, which I could not see in the textbooks.	☐	☐	☐	☐	☐
26	I would rather learn Engineering Mechanics by working problems using pencil and paper.	☐	☐	☐	☐	☐
27	I like learning and solving Engineering Mechanics problems this way because I can see how it's done, rather than asking someone to explain it.	☐	☐	☐	☐	☐
	INDIVIDUAL FLEXIBILITY	1	2	3	4	5
28	These TAPS packages are only appropriate for advanced students of engineering.	☐	☐	☐	☐	☐
29	The computer usage got harder just as I got better at the problems.	☐	☐	☐	☐	☐
31	I knew how to solve the problems, but the TAPS packages persisted in following its long and laborious steps.	☐	☐	☐	☐	☐
32	The TAPS packages kept adjusting its advice according to my needs and progress.	☐	☐	☐	☐	☐
33	I could not go into the more demanding material without having to go through the easier stuff again.	☐	☐	☐	☐	☐
	ENGINEERING KNOWLEDGE/TEACHING	1	2	3	4	5
34	I did not need any previous knowledge of engineering to use the TAPS packages.	☐	☐	☐	☐	☐
35	I have gained a good introduction to the concept of solving Engineering Mechanics problems.	☐	☐	☐	☐	☐
36	I think the TAPS packages are only useful for students who are already familiar with Engineering Mechanics and want to improve.	☐	☐	☐	☐	☐
37	It is difficult to learn Engineering Mechanics with these TAPS packages.	☐	☐	☐	☐	☐
38	The packages give a very good background on solving the Engineering Mechanics problems.	☐	☐	☐	☐	☐

1	2	3	4	5
Strongly Agree	*Agree*	*Uncertain*	*Disagree*	*Strongly Disagree*

ENGINEERING KNOWLEDGE/TEACHING		1	2	3	4	5
39	Most students would find it hard to solve Engineering Mechanics problems using the TAPS packages.	☐	☐	☐	☐	☐
40	The TAPS packages allow me to fully test my understanding of engineering mechanics.	☐	☐	☐	☐	☐
41	After using the TAPS packages I still do not understand how to solve Engineering Mechanics problems.	☐	☐	☐	☐	☐
42	After using the TAPS packages I can easily use my knowledge to solve Engineering Mechanics problems better.	☐	☐	☐	☐	☐
43	I would find it easy to use these TAPS packages to teach someone else about engineering mechanics.	☐	☐	☐	☐	☐
44	The TAPS packages have helped improve my knowledge by guiding me through the learning process.	☐	☐	☐	☐	☐
45	I would like to learn solving other problems in Engineering Mechanics topics using these packages.	☐	☐	☐	☐	☐
46	The TAPS packages allowed me to learn an accurate problem solving steps of Engineering Mechanics and corrected my mistakes.	☐	☐	☐	☐	☐
47	Even now I do not understand the problems presented in the TAPS packages.	☐	☐	☐	☐	☐
ERROR MESSAGES AND DOCUMENTATION		1	2	3	4	5
48	The computer messages displayed on the screen were good enough to guide me when I got stuck.	☐	☐	☐	☐	☐
49	The error messages were unhelpful.	☐	☐	☐	☐	☐
50	I liked the TAPS packages because it helped by displaying messages about what to do whenever I made any mistakes.	☐	☐	☐	☐	☐
51	The computer messages displayed on the screen were good enough to guide me when I got stuck.	☐	☐	☐	☐	☐
52	The TAPS packages should not allow you to enter any incorrect values.	☐	☐	☐	☐	☐
53	The TAPS packages are so unforgiving, it does not allow me to do anything else, until I provide the correct answer.	☐	☐	☐	☐	☐

1	2	3	4	5
Strongly Agree	*Agree*	*Uncertain*	*Disagree*	*Strongly Disagree*

ERROR MESSAGES AND DOCUMENTATION		1	2	3	4	5
54	There should be an error message if I enter an incorrect negative, positive numbers or symbols (i.e. "/","[").	☐	☐	☐	☐	☐
55	I like the way the TAPS packages gave explanation when I made a mistake.	☐	☐	☐	☐	☐
56	The TAPS packages directly give explanations without giving me another chance.	☐	☐	☐	☐	☐
57	It was a good idea to have the TAPS packages display a small hint whenever I got the answer wrong before giving the correct answer.	☐	☐	☐	☐	☐
ERROR MESSAGES AND DOCUMENTATION		1	2	3	4	5
58	The TAPS packages provide good feedback only when you have entered an answer.	☐	☐	☐	☐	☐
59	The TAPS packages did not give enough praise for the correct answer.	☐	☐	☐	☐	☐
60	The TAPS packages helped me with every step in solving the Engineering Mechanics problems.	☐	☐	☐	☐	☐
61	Through the feedback I immediately knew whether I was right or wrong.	☐	☐	☐	☐	☐
62	The explanations provided by the TAPS packages were hard to understand.	☐	☐	☐	☐	☐
63	The explanations showed clearly why I had got the wrong answer.	☐	☐	☐	☐	☐
64	The TAPS packages gives the same type of feedback at all times regardless of what I have done.	☐	☐	☐	☐	☐
65	The TAPS packages give individual feedback to each student depending upon how far they have progressed to the solution.	☐	☐	☐	☐	☐
66	It would be better if clearer explanation of the aim of the problem solving exercises and the required information to complete the exercises had been given.	☐	☐	☐	☐	☐

1	2	3	4	5
Strongly Agree	*Agree*	*Uncertain*	*Disagree*	*Strongly Disagree*

ROUTE THROUGH TAPS PACKAGES		1	2	3	4	5
67	I never knew what all the items on the various buttons meant.	☐	☐	☐	☐	☐
68	The instructions given on how to proceed through the problem-solving task were unclear.	☐	☐	☐	☐	☐
69	I like learning and problem solving in this approach, because you can check over things and do not repeat the same mistakes again and again.	☐	☐	☐	☐	☐
70	Sometimes I found it hard to keep track about which bits/chunks of tutorial I have completed.	☐	☐	☐	☐	☐
71	The TAPS packages allowed me to solve the problem presented in multiple orders.	☐	☐	☐	☐	☐
72	I would have found it more helpful to be given a suggested route through the TAPS packages.	☐	☐	☐	☐	☐
73	The prompts, asking me to enter text/numeric/symbols, were not clear from the screen.	☐	☐	☐	☐	☐
74	I want to be able to stop the exercise and go back to the problem solving tutorial, and then return where I left off.	☐	☐	☐	☐	☐
75	On screen problem solving tutorial are difficult to follow.	☐	☐	☐	☐	☐
76	It provided little understanding of what you are doing or how to do it.	☐	☐	☐	☐	☐
CONTROL OF LEARNING AND PROBLEM SOLVING		1	2	3	4	5
77	I felt the TAPS packages allowed me to work at my own pace and directions.	☐	☐	☐	☐	☐
78	I found the TAPS packages were too slow because I could do the exercise much more quickly with pencil and paper.	☐	☐	☐	☐	☐
79	It was too quick, I would like more control of how much time I was given to solve the problems.	☐	☐	☐	☐	☐
80	I wanted slower speed in the beginning and quicker when I got some experience, but the TAPS packages kept on at the same speed which was annoying.	☐	☐	☐	☐	☐
81	I learnt a lot about problem solving task in the subject matter by trying different things with the TAPS packages.	☐	☐	☐	☐	☐

1	2	3	4	5
Strongly Agree	*Agree*	*Uncertain*	*Disagree*	*Strongly Disagree*

	RETENTION AND LEARNING	1	2	3	4	5
82	I am sure I will retain most of what I have learnt.	☐	☐	☐	☐	☐
83	I learnt a lot from the TAPS packages, by learning how to solve various Engineering Mechanics problem.	☐	☐	☐	☐	☐
84	If I use the TAPS packages again, I could solve similar problems quicker.	☐	☐	☐	☐	☐
85	I will not be able to solve the problems if they are presented in a different format/order than these packages.	☐	☐	☐	☐	☐
	HELP FACILITY	1	2	3	4	5
86	I would prefer more help from these TAPS packages.	☐	☐	☐	☐	☐
87	The help facility should have given a few hints, rather than the correct answer.	☐	☐	☐	☐	☐
88	The TAPS packages do not offer much help where it is most wanted.	☐	☐	☐	☐	☐
89	The TAPS packages gave a general help every time instead of context sensitive help.	☐	☐	☐	☐	☐
	FUNCTIONS OR KEYS	1	2	3	4	5
90	I found it very difficult to know which buttons/keys corresponded to the command I wanted.	☐	☐	☐	☐	☐
91	When navigating through the packages, I found I was clear which options I was allowed to select.	☐	☐	☐	☐	☐
92	I didn't like some of the buttons/keys, I sometimes got lost because of them.	☐	☐	☐	☐	☐
93	When I was going through the TAPS package(s), pressing the same key would produce a different command.	☐	☐	☐	☐	☐
94	I got confused because same command was available through different keys in different screen.	☐	☐	☐	☐	☐
95	It was easy to exit from a particular route through the packages.	☐	☐	☐	☐	☐
96	I could not skip some parts of the problem solving tutorials.	☐	☐	☐	☐	☐

1	2	3	4	5
Strongly Agree	*Agree*	*Uncertain*	*Disagree*	*Strongly Disagree*

FUNCTIONS OR KEYS		1	2	3	4	5
97	I could not skip some parts of the exercises.	☐	☐	☐	☐	☐
98	It was easy to delete any answer, which, I knew was wrong.	☐	☐	☐	☐	☐
99	I was not allowed to answer a question in the exercises provided more than one time.	☐	☐	☐	☐	☐
USER TOOLS (CALCULATOR / NOTEPAD / GLOSSARY / CHARACTER MAP TABLE)		1	2	3	4	5
100	The calculator facility was very useful.	☐	☐	☐	☐	☐
101	I found it hard to use the calculator to transfer the answer to the package.	☐	☐	☐	☐	☐
102	Calculator should be customized to have more functions, which are needed to solve the problems.	☐	☐	☐	☐	☐
103	The notepad was very useful in that I could type/make important notes and save it for reference.	☐	☐	☐	☐	☐
104	The glossary of commands table was very useful in that I could look up / search for a related word of the problem matter and understand its meaning better.	☐	☐	☐	☐	☐
105	The character map table was very useful as it contain important symbols unavailable on the standard keyboard and allowed me to copy and paste into the textboxes on the screen.	☐	☐	☐	☐	☐
PRESENTATION OF INFORMATION ON SCREEN		1	2	3	4	5
106	The screen layout presented an easy way to communicate with the computer.	☐	☐	☐	☐	☐
107	Sometimes I was confused by the large amount of information on the screen.	☐	☐	☐	☐	☐
108	The way the problem is presented on the screen makes it easy to work out the solution.	☐	☐	☐	☐	☐
109	There was not enough information on the screen for me to solve the exercise questions.	☐	☐	☐	☐	☐
110	The screen was attractive and clear	☐	☐	☐	☐	☐

1	2	3	4	5
|……………	|……………	|……………	|……………	|
Strongly Agree	***Agree***	***Uncertain***	***Disagree***	***Strongly Disagree***

PRESENTATION OF INFORMATION ON SCREEN		1	2	3	4	5
111	The colors on the screen were confusing	☐	☐	☐	☐	☐
112	The TAPS packages looked dull and boring.	☐	☐	☐	☐	☐
113	I like to see the different font types, sizes and colors in the packages.	☐	☐	☐	☐	☐
HOLISTIC SYSTEM		1	2	3	4	5
114	I did not like the TAPS packages on the whole.	☐	☐	☐	☐	☐
115	I really liked using the TAPS packages because it was fun to use.	☐	☐	☐	☐	☐
116	I enjoyed using the packages, but it could have been improved more, so I would be tempted to use it again.	☐	☐	☐	☐	☐
117	I simply used the TAPS packages to read what is on the screen.	☐	☐	☐	☐	☐
118	The learning process went on too long.	☐	☐	☐	☐	☐
119	It is easy to use the packages and does not require much practice to be familiar.	☐	☐	☐	☐	☐
120	I would prefer, if I could be given problem-solving exercises using these sorts of packages.	☐	☐	☐	☐	☐
121	The packages did not manage to convince me that I could learn the problem solving techniques/method better.	☐	☐	☐	☐	☐
122	I liked the way the packages coach in solving the problems.	☐	☐	☐	☐	☐
123	It is a good way to revise something that I have forgotten.	☐	☐	☐	☐	☐
124	The packages are very helpful especially for a student like me.	☐	☐	☐	☐	☐
OVERALL PERCEPTION		1	2	3	4	5
125	Most students may not be interested to use such TAPS packages to assists them in learning the subject matter.	☐	☐	☐	☐	☐
126	I solved some of the exercise questions but was not sure why the answer(s) were incorrect.	☐	☐	☐	☐	☐
127	I would like to have more problem solving examples.	☐	☐	☐	☐	☐

128. Apart from any issues discussed previously, what other features you find contributed to the:

(a) STRENGTH of the TAPS packages……………………………………………

……………………………………………………………………………………..

……………………………………………………………………………………..

(b) WEAKNESS of the TAPS packages..

....

....

(c) SUGGESTIONS for improvement..

....

....

PLEASE CHECK IF YOU HAVE ANSWERED ALL THE QUESTIONS.

THANK YOU VERY MUCH FOR YOUR CO-OPERATION.

About the Author

Manjit Singh Sidhu is currently an Associate Professor and the Head of Graphics and Multimedia Department, College of Computer Science and IT, University Tenaga Nasional, Malaysia. He received his BSc. (Hons) degree in Computer Science from the University of Wolverhampton, UK in 1997, Masters in Information Technology from University Putra Malaysia in 2000 and PhD in Multimedia from the University of Malaya in 2008. His background covers a wide range of computer science and educational experiences, initially teaching undergraduates and supervising postgraduates both Masters and PhDs in graphics and multimedia department. He has written and published several conference proceedings, journals and book chapters both locally and internationally. He has also won a number of research, development and invention awards both locally and internationally. He is a chartered IT professional, UK and a member of the British Computer Society (MBCS), a member of Institute of Electrical and Electronics Engineers (IEEE), Computer and Communications Society, Malaysian National Computer Confederation (MNCC), and Malaysian Scientific Association (MSA). His research focuses on patterns of interactions in multimedia and virtual reality applications, visualization and computer animations.

Index

N

O

P

R

S

T

U

V

W